D0772381

Data Streams
Models and Algorithms

ADVANCES IN DATABASE SYSTEMS

Series Editor

Ahmed K. Elmagarmid

Purdue University
West Lafayette, IN 47907

For a complete listing of books in this series, go to http://www.springer.com

Data Streams
Models and Algorithms

edited by

Charu C. Aggarwal
IBM, T.J. Watson Research Center
Yorktown Heights, NY, USA

 Springer

Charu C. Aggarwal
IBM
Thomas J. Watson Research Center
19 Skyline Drive
Hawthorne NY 10532

Library of Congress Control Number: 2006934111

DATA STREAMS: Models and Algorithms edited by Charu C. Aggarwal

ISBN-10: 0-387-28759-0
ISBN-13: 978-0-387-28759-1
e-ISBN-10: 0-387-47534-6
e-ISBN-13: 978-0-387-47534-9

Cover by Will Ladd, NRL Mapping, Charting and Geodesy Branch
utilizing NRL's GIDB® Portal System that can be utilized at
http://dmap.nrlssc.navy.mil

Printed on acid-free paper.

9 8 7 6 5 4 3 2 1

springer.com

Contents

List of Figures

List of Tables

Preface

In recent years, the progress in hardware technology has made it possible for organizations to store and record large streams of transactional data. Such data sets which continuously and rapidly grow over time are referred to as data streams. In addition, the development of sensor technology has resulted in the possibility of monitoring many events in real time. While data mining has become a fairly well established field now, the data stream problem poses a number of unique challenges which are not easily solved by traditional data mining methods.

The topic of data streams is a very recent one. The first research papers on this topic appeared slightly under a decade ago, and since then this field has grown rapidly. There is a large volume of literature which has been published in this field over the past few years. The work is also of great interest to practitioners in the field who have to mine actionable insights with large volumes of continuously growing data. Because of the large volume of literature in the field, practitioners and researchers may often find it an arduous task of isolating the right literature for a given topic. In addition, from a practitioners point of view, the use of research literature is even more difficult, since much of the relevant material is buried in publications. While handling a real problem, it may often be difficult to know where to look in order to solve the problem.

This book contains contributed chapters from a variety of well known researchers in the data mining field. While the chapters will be written by different researchers, the topics and content will be organized in such a way so as to present the most important models, algorithms, and applications in the data mining field in a structured and concise way. In addition, the book is organized in order to make it more accessible to application driven practitioners. Given the lack of structurally organized information on the topic, the book will provide insights which are not easily accessible otherwise. In addition, the book will be a great help to researchers and graduate students interested in the topic. The popularity and current nature of the topic of data streams is likely to make it an important source of information for researchers interested in the topic. The data mining community has grown rapidly over the past few years, and the topic of data streams is one of the most relevant and current areas of interest to the community. This is because of the rapid advancement of the field of data streams in the past two to three years. While the data stream field clearly falls in the emerging category because of its recency, it is now beginning to reach a maturation and popularity point, where the development of an overview book on the topic becomes both possible and necessary. While this book attempts to provide an overview of the stream mining area, it also tries to discuss current topics of interest so as to be useful to students and researchers. It is hoped that this book will provide a reference to students, researchers and practitioners in both introducing the topic of data streams and understanding the practical and algorithmic aspects of the area.

Chapter 1

AN INTRODUCTION TO DATA STREAMS

Charu C. Aggarwal
IBM T. J. Watson Research Center
Hawthorne, NY 10532
charu@us.ibm.com

Abstract

In recent years, advances in hardware technology have facilitated new ways of collecting data continuously. In many applications such as network monitoring, the volume of such data is so large that it may be impossible to store the data on disk. Furthermore, even when the data can be stored, the volume of the incoming data may be so large that it may be impossible to process any particular record more than once. Therefore, many data mining and database operations such as classification, clustering, frequent pattern mining and indexing become significantly more challenging in this context.

In many cases, the data patterns may evolve continuously, as a result of which it is necessary to design the mining algorithms effectively in order to account for changes in underlying structure of the data stream. This makes the solutions of the underlying problems even more difficult from an algorithmic and computational point of view. This book contains a number of chapters which are carefully chosen in order to discuss the broad research issues in data streams. The purpose of this chapter is to provide an overview of the organization of the stream processing and mining techniques which are covered in this book.

1.   Introduction

In recent years, advances in hardware technology have facilitated the ability to collect data continuously. Simple transactions of everyday life such as using a credit card, a phone or browsing the web lead to automated data storage. Similarly, advances in information technology have lead to large flows of data across IP networks. In many cases, these large volumes of data can be mined for interesting and relevant information in a wide variety of applications. When the volume of the underlying data is very large, it leads to a number of computational and mining challenges:

- With increasing volume of the data, it is no longer possible to process the data efficiently by using multiple passes. Rather, one can process a data item at most once. This leads to constraints on the implementation of the underlying algorithms. Therefore, stream mining algorithms typically need to be designed so that the algorithms work with one pass of the data.

- In most cases, there is an inherent temporal component to the stream mining process. This is because the data may evolve over time. This behavior of data streams is referred to as *temporal locality*. Therefore, a straightforward adaptation of one-pass mining algorithms may not be an effective solution to the task. Stream mining algorithms need to be carefully designed with a clear focus on the evolution of the underlying data.

Another important characteristic of data streams is that they are often mined in a distributed fashion. Furthermore, the individual processors may have limited processing and memory. Examples of such cases include sensor networks, in which it may be desirable to perform in-network processing of data stream with limited processing and memory [8, 19]. This book will also contain a number of chapters devoted to these topics.

This chapter will provide an overview of the different stream mining algorithms covered in this book. We will discuss the challenges associated with each kind of problem, and discuss an overview of the material in the corresponding chapter.

2.   Stream Mining Algorithms

In this section, we will discuss the key stream mining problems and will discuss the challenges associated with each problem. We will also discuss an overview of the material covered in each chapter of this book. The broad topics covered in this book are as follows:

Data Stream Clustering.   Clustering is a widely studied problem in the data mining literature. However, it is more difficult to adapt arbitrary clustering algorithms to data streams because of one-pass constraints on the data set. An interesting adaptation of the k-means algorithm has been discussed in [14] which uses a partitioning based approach on the entire data set. This approach uses an adaptation of a k-means technique in order to create clusters over the entire data stream. In the context of data streams, it may be more desirable to determine clusters in specific user defined horizons rather than on the entire data set. In chapter 2, we discuss the micro-clustering technique [3] which determines clusters over the entire data set. We also discuss a variety of applications of micro-clustering which can perform effective summarization based analysis of the data set. For example, micro-clustering can be extended to the problem of classification on data streams [5]. In many cases, it can also be used for arbitrary data mining applications such as privacy preserving data mining or query estimation.

Data Stream Classification. The problem of classification is perhaps one of the most widely studied in the context of data stream mining. The problem of classification is made more difficult by the evolution of the underlying data stream. Therefore, effective algorithms need to be designed in order to take temporal locality into account. In chapter 3, we discuss a survey of classification algorithms for data streams. A wide variety of data stream classification algorithms are covered in this chapter. Some of these algorithms are designed to be purely one-pass adaptations of conventional classification algorithms [12], whereas others (such as the methods in [5, 16]) are more effective in accounting for the evolution of the underlying data stream. Chapter 3 discusses the different kinds of algorithms and the relative advantages of each.

Frequent Pattern Mining. The problem of frequent pattern mining was first introduced in [6], and was extensively analyzed for the conventional case of disk resident data sets. In the case of data streams, one may wish to find the frequent itemsets either over a sliding window or the entire data stream [15, 17]. In Chapter 4, we discuss an overview of the different frequent pattern mining algorithms, and also provide a detailed discussion of some interesting recent algorithms on the topic.

Change Detection in Data Streams. As discussed earlier, the patterns in a data stream may evolve over time. In many cases, it is desirable to track and analyze the nature of these changes over time. In [1, 11, 18], a number of methods have been discussed for change detection of data streams. In addition, data stream evolution can also affect the behavior of the underlying data mining algorithms since the results can become stale over time. Therefore, in Chapter 5, we have discussed the different methods for change detection data streams. We have also discussed the effect of evolution on data stream mining algorithms.

Stream Cube Analysis of Multi-dimensional Streams. Much of stream data resides at a multi-dimensional space and at rather low level of abstraction, whereas most analysts are interested in relatively high-level dynamic changes in some combination of dimensions. To discover high-level dynamic and evolving characteristics, one may need to perform multi-level, multi-dimensional on-line analytical processing (OLAP) of stream data. Such necessity calls for the investigation of new architectures that may facilitate on-line analytical processing of multi-dimensional stream data [7, 10].

In Chapter 6, an interesting **stream_cube** architecture that effectively performs on-line partial aggregation of multi-dimensional stream data, captures the essential dynamic and evolving characteristics of data streams, and facilitates fast OLAP on stream data. Stream cube architecture facilitates online analytical processing of stream data. It also forms a preliminary structure for online stream mining. The impact of the design and implementation of stream cube in the context of stream mining is also discussed in the chapter.

Loadshedding in Data Streams.   Since data streams are generated by processes which are extraneous to the stream processing application, it is not possible to control the incoming stream rate. As a result, it is necessary for the system to have the ability to quickly adjust to varying incoming stream processing rates. Chapter 7 discusses one particular type of adaptivity: the ability to gracefully degrade performance via "load shedding" (dropping unprocessed tuples to reduce system load) when the demands placed on the system cannot be met in full given available resources. Focusing on aggregation queries, the chapter presents algorithms that determine at what points in a query plan should load shedding be performed and what amount of load should be shed at each point in order to minimize the degree of inaccuracy introduced into query answers.

Sliding Window Computations in Data Streams.   Many of the synopsis structures discussed use the entire data stream in order to construct the corresponding synopsis structure. The sliding-window model of computation is motivated by the assumption that it is more important to use recent data in data stream computation [9]. Therefore, the processing and analysis is only done on a fixed history of the data stream. Chapter 8 formalizes this model of computation and answers questions about how much space and computation time is required to solve certain problems under the sliding-window model.

Synopsis Construction in Data Streams.   The large volume of data streams poses unique space and time constraints on the computation process. Many query processing, database operations, and mining algorithms require efficient execution which can be difficult to achieve with a fast data stream. In many cases, it may be acceptable to generate *approximate solutions* for such problems. In recent years a number of *synopsis structures* have been developed, which can be used in conjunction with a variety of mining and query processing techniques [13]. Some key synopsis methods include those of sampling, wavelets, sketches and histograms. In Chapter 9, a survey of the key synopsis techniques is discussed, and the mining techniques supported by such methods. The chapter discusses the challenges and tradeoffs associated with using different kinds of techniques, and the important research directions for synopsis construction.

Join Processing in Data Streams.   Stream join is a fundamental operation for relating information from different streams. This is especially useful in many applications such as sensor networks in which the streams arriving from different sources may need to be related with one another. In the stream setting, input tuples arrive continuously, and result tuples need to be produced continuously as well. We cannot assume that the input data is already stored or indexed, or that the input rate can be controlled by the query plan. Standard join algorithms that use blocking operations, e.g., sorting, no longer work. Conventional methods for cost estimation and query optimization are also inappropriate, because they assume finite input. Moreover, the long-running nature of stream queries calls for more adaptive processing strategies that can react to changes and fluctuations in data and stream characteristics. The "stateful" nature of stream joins adds another dimension to the challenge. In general, in order to compute the complete result of a stream join, we need to retain all past arrivals as part of the processing state, because a new tuple may join with an arbitrarily old tuple arrived in the past. This problem is exacerbated by unbounded input streams, limited processing resources, and high performance requirements, as it is impossible in the long run to keep all past history in fast memory. Chapter 10 provides an overview of research problems, recent advances, and future research directions in stream join processing.

Indexing Data Streams.   The problem of indexing data streams attempts to create a an indexed representation, so that it is possible to efficiently answer different kinds of queries such as aggregation queries or trend based queries. This is especially important in the data stream case because of the huge volume of the underlying data. Chapter 11 explores the problem of indexing and querying data streams.

Dimensionality Reduction and Forecasting in Data Streams.   Because of the inherent temporal nature of data streams, the problems of dimensionality reduction and forecasting and particularly important. When there are a large number of simultaneous data stream, we can use the correlations between different data streams in order to make effective predictions [20, 21] on the future behavior of the data stream. In Chapter 12, an overview of dimensionality reduction and forecasting methods have been discussed for the problem of data streams. In particular, the well known MUSCLES method [21] has been discussed, and its application to data streams have been explored. In addition, the chapter presents the SPIRIT algorithm, which explores the relationship between dimensionality reduction and forecasting in data streams. In particular, the chapter explores the use of a compact number of hidden variables to comprehensively describe the data stream. This compact representation can also be used for effective forecasting of the data streams.

Distributed Mining of Data Streams. In many instances, streams are generated at multiple distributed computing nodes. Analyzing and monitoring data in such environments requires data mining technology that requires optimization of a variety of criteria such as communication costs across different nodes, as well as computational, memory or storage requirements at each node. A comprehensive survey of the adaptation of different conventional mining algorithms to the distributed case is provided in Chapter 13. In particular, the clustering, classification, outlier detection, frequent pattern mining, and summarization problems are discussed. In Chapter 14, some recent advances in stream mining algorithms are discussed.

Stream Mining in Sensor Networks. With recent advances in hardware technology, it has become possible to track large amounts of data in a distributed fashion with the use of sensor technology. The large amounts of data collected by the sensor nodes makes the problem of monitoring a challenging one from many technological stand points. Sensor nodes have limited local storage, computational power, and battery life, as a result of which it is desirable to minimize the storage, processing and communication from these nodes. The problem is further magnified by the fact that a given network may have millions of sensor nodes and therefore it is very expensive to localize all the data at a given global node for analysis both from a storage and communication point of view. In Chapter 15, we discuss an overview of a number of stream mining issues in the context of sensor networks. This topic is closely related to distributed stream mining, and a number of concepts related to sensor mining have also been discussed in Chapters 13 and 14.

3.   Conclusions and Summary

Data streams are a computational challenge to data mining problems because of the additional algorithmic constraints created by the large volume of data. In addition, the problem of temporal locality leads to a number of unique mining challenges in the data stream case. This chapter provides an overview to the different mining algorithms which are covered in this book. We discussed the different problems and the challenges which are associated with each problem. We also provided an overview of the material in each chapter of the book.

References

[1] Aggarwal C. (2003). A Framework for Diagnosing Changes in Evolving Data Streams. *ACM SIGMOD Conference.*

[2] Aggarwal C (2002). An Intuitive Framework for understanding Changes in Evolving Data Streams. *IEEE ICDE Conference.*

[3] Aggarwal C., Han J., Wang J., Yu P (2003). A Framework for Clustering Evolving Data Streams. *VLDB Conference.*

[4] Aggarwal C., Han J., Wang J., Yu P (2004). A Framework for High Dimensional Projected Clustering of Data Streams. *VLDB Conference.*

[5] Aggarwal C, Han J., Wang J., Yu P. (2004). On-Demand Classification of Data Streams. *ACM KDD Conference.*

[6] Agrawal R., Imielinski T., Swami A. (1993) Mining Association Rules between Sets of items in Large Databases. *ACM SIGMOD Conference.*

[7] Chen Y., Dong G., Han J., Wah B. W., Wang J. (2002) Multi-dimensional regression analysis of time-series data streams. *VLDB Conference.*

[8] Cormode G., Garofalakis M. (2005) Sketching Streams Through the Net: Distributed Approximate Query Tracking. *VLDB Conference.*

[9] Datar M., Gionis A., Indyk P., Motwani R. (2002) Maintaining stream statistics over sliding windows. *SIAM Journal on Computing*, 31(6):1794–1813.

[10] Dong G., Han J., Lam J., Pei J., Wang K. (2001) Mining multi-dimensional constrained gradients in data cubes. *VLDB Conference.*

[11] Dasu T., Krishnan S., Venkatasubramaniam S., Yi K. (2005). An Information-Theoretic Approach to Detecting Changes in Multidimensional data Streams. *Duke University Technical Report CS-2005-06.*

[12] Domingos P. and Hulten G. (2000) Mining High-Speed Data Streams. In *Proceedings of the ACM KDD Conference.*

[13] Garofalakis M., Gehrke J., Rastogi R. (2002) Querying and mining data streams: you only get one look (a tutorial). *SIGMOD Conference.*

[14] Guha S., Mishra N., Motwani R., O'Callaghan L. (2000). Clustering Data Streams. *IEEE FOCS Conference.*

[15] Giannella C., Han J., Pei J., Yan X., and Yu P. (2002) Mining Frequent Patterns in Data Streams at Multiple Time Granularities. *Proceedings of the NSF Workshop on Next Generation Data Mining.*

[16] Hulten G., Spencer L., Domingos P. (2001). Mining Time Changing Data Streams. *ACM KDD Conference.*

[17] Jin R., Agrawal G. (2005) An algorithm for in-core frequent itemset mining on streaming data. *ICDM Conference.*

[18] Kifer D., David S.-B., Gehrke J. (2004). Detecting Change in Data Streams. *VLDB Conference*, 2004.

[19] Kollios G., Byers J., Considine J., Hadjielefttheriou M., Li F. (2005) Robust Aggregation in Sensor Networks. *IEEE Data Engineering Bulletin.*

[20] Sakurai Y., Papadimitriou S., Faloutsos C. (2005). BRAID: Stream mining through group lag correlations. *ACM SIGMOD Conference.*

[21] Yi B.-K., Sidiropoulos N.D., Johnson T., Jagadish, H. V., Faloutsos C., Biliris A. (2000). Online data mining for co-evolving time sequences. *ICDE Conference.*

Chapter 2

ON CLUSTERING MASSIVE DATA STREAMS: A SUMMARIZATION PARADIGM

Charu C. Aggarwal
IBM T. J. Watson Research Center
Hawthorne, NY 10532
charu@us.ibm.com

Jiawei Han
University of Illinois at Urbana-Champaign
Urbana, IL
hanj@cs.uiuc.edu

Jianyong Wang
University of Illinois at Urbana-Champaign
Urbana, IL
jianyong@tsinghua.edu.cn

Philip S. Yu
IBM T. J. Watson Research Center
Hawthorne, NY 10532
psyu@us.ibm.com

Abstract

In recent years, data streams have become ubiquitous because of the large number of applications which generate huge volumes of data in an automated way. Many existing data mining methods cannot be applied directly on data streams because of the fact that the data needs to be mined in one pass. Furthermore, data streams show a considerable amount of temporal locality because of which a direct application of the existing methods may lead to misleading results. In this paper, we develop an efficient and effective approach for mining fast evolving data streams, which integrates the *micro-clustering* technique with the high-level data mining process, and discovers data evolution regularities as well. Our analysis and experiments demonstrate two important data mining problems, namely *stream clustering* and *stream classification*, can be performed effectively using this approach, with high quality mining results. We discuss the use of micro-clustering as a general summarization technology to solve data mining problems on streams. Our discussion illustrates the importance of our approach for a variety of mining problems in the data stream domain.

1.   Introduction

In recent years, advances in hardware technology have allowed us to automatically record transactions and other pieces of information of everyday life at a rapid rate. Such processes generate huge amounts of online data which grow at an unlimited rate. These kinds of online data are referred to as *data streams*. The issues on management and analysis of data streams have been researched extensively in recent years because of its emerging, imminent, and broad applications [11, 14, 17, 23].

Many important problems such as clustering and classification have been widely studied in the data mining community. However, a majority of such methods may not be working effectively on data streams. Data streams pose special challenges to a number of data mining algorithms, not only because of the huge volume of the online data streams, but also because of the fact that the data in the streams may show temporal correlations. Such temporal correlations may help disclose important data evolution characteristics, and they can also be used to develop efficient and effective mining algorithms. Moreover, data streams require *online mining*, in which we wish to mine the data in a continuous fashion. Furthermore, the system needs to have the capability to perform an *offline analysis* as well based on the user interests. This is similar to an online analytical processing (OLAP) framework which uses the paradigm of pre-processing once, querying many times.

Based on the above considerations, we propose a new stream mining framework, which adopts a tilted time window framework, takes micro-clustering as a preprocessing process, and integrates the preprocessing with the incremental, dynamic mining process. Micro-clustering preprocessing effectively compresses the data, preserves the general temporal locality of data, and facilitates both online and offline analysis, as well as the analysis of current data and data evolution regularities.

In this study, we primarily concentrate on the application of this technique to two problems: (1) stream clustering, and (2) stream classification. The heart of the approach is to use an online summarization approach which is efficient and also allows for effective processing of the data streams. We also discuss

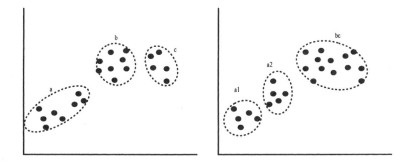

Figure 2.1. Micro-clustering Examples

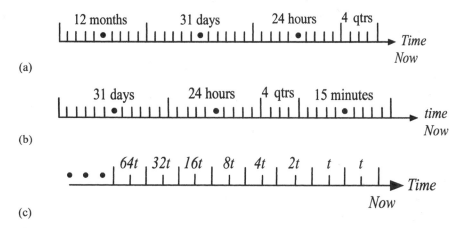

Figure 2.2. Some Simple Time Windows

a number of research directions, in which we show how the approach can be adapted to a variety of other problems.

This paper is organized as follows. In the next section, we will present our micro-clustering based stream mining framework. In section 3, we discuss the stream clustering problem. The classification methods are developed in Section 4. In section 5, we discuss a number of other problems which can be solved with the micro-clustering approach, and other possible research directions. In section 6, we will discuss some empirical results for the clustering and classification problems. In Section 7 we discuss the issues related to our proposed stream mining methodology and compare it with other related work. Section 8 concludes our study.

2. The Micro-clustering Based Stream Mining Framework

In order to apply our technique to a variety of data mining algorithms, we utilize a micro-clustering based stream mining framework. This framework is designed by capturing summary information about the nature of the data stream. This summary information is defined by the following structures:

• **Micro-clusters:** We maintain statistical information about the data locality in terms of micro-clusters. These micro-clusters are defined as a temporal extension of the *cluster feature vector* [24]. The additivity property of the micro-clusters makes them a natural choice for the data stream problem.

• **Pyramidal Time Frame:** The micro-clusters are stored at snapshots in time which follow a pyramidal pattern. This pattern provides an effective trade-off between the storage requirements and the ability to recall summary statistics from different time horizons.

The summary information in the micro-clusters is used by an offline component which is dependent upon a wide variety of user inputs such as the time horizon or the granularity of clustering. In order to define the micro-clusters, we will introduce a few concepts. It is assumed that the data stream consists of a set of multi-dimensional records $\overline{X}_1 \ldots \overline{X}_k \ldots$ arriving at time stamps $T_1 \ldots T_k \ldots$. Each $\overline{X}_i$ is a multi-dimensional record containing d dimensions which are denoted by $\overline{X}_i = (x_i^1 \ldots x_i^d)$.

We will first begin by defining the concept of micro-clusters and pyramidal time frame more precisely.

DEFINITION 2.1 *A micro-cluster for a set of d-dimensional points $X_{i_1} \ldots X_{i_n}$ with time stamps $T_{i_1} \ldots T_{i_n}$ is the $(2 \cdot d + 3)$ tuple $(\overline{CF2^x}, \overline{CF1^x}, CF2^t, CF1^t, n)$, wherein $\overline{CF2^x}$ and $\overline{CF1^x}$ each correspond to a vector of d entries. The definition of each of these entries is as follows:*

• *For each dimension, the sum of the squares of the data values is maintained in $\overline{CF2^x}$. Thus, $\overline{CF2^x}$ contains d values. The p-th entry of $\overline{CF2^x}$ is equal to $\sum_{j=1}^n (x_{i_j}^p)^2$.*

• *For each dimension, the sum of the data values is maintained in $\overline{CF1^x}$. Thus, $\overline{CF1^x}$ contains d values. The p-th entry of $\overline{CF1^x}$ is equal to $\sum_{j=1}^n x_{i_j}^p$.*

• *The sum of the squares of the time stamps $T_{i_1} \ldots T_{i_n}$ is maintained in $CF2^t$.*

• *The sum of the time stamps $T_{i_1} \ldots T_{i_n}$ is maintained in $CF1^t$.*

• *The number of data points is maintained in n.*

We note that the above definition of micro-cluster maintains similar summary information as the cluster feature vector of [24], except for the additional information about time stamps. We will refer to this temporal extension of the cluster feature vector for a set of points C by $\overline{CFT}(C)$. As in [24], this summary information can be expressed in an additive way over the different data points. This makes it a natural choice for use in data stream algorithms.

We note that the maintenance of a large number of micro-clusters is essential in the ability to maintain more detailed information about the micro-clustering process. For example, Figure 2.1 forms 3 clusters, which are denoted by a, b, c. At a later stage, evolution forms 3 different figures a1, a2, bc, with a split into a1 and a2, whereas b and c merged into bc. If we keep micro-clusters (each point represents a micro-cluster), such evolution can be easily captured. However, if we keep only 3 cluster centers a, b, c, it is impossible to derive later a1, a2, bc clusters since the information of more detailed points are already lost.

The data stream clustering algorithm discussed in this paper can generate approximate clusters in any user-specified length of history from the current instant. This is achieved by storing the micro-clusters at particular moments in the stream which are referred to as *snapshots*. At the same time, the current snapshot of micro-clusters is always maintained by the algorithm. The macro-clustering algorithm discussed at a later stage in this paper will use these finer level micro-clusters in order to create higher level clusters which can be more easily understood by the user. Consider for example, the case when the current clock time is t_c and the user wishes to find clusters in the stream based on a history of length h. Then, the macro-clustering algorithm discussed in this paper will use some of the additive properties of the micro-clusters stored at snapshots t_c and $(t_c - h)$ in order to find the higher level clusters in a history or *time horizon* of length h. Of course, since it is not possible to store the snapshots at each and every moment in time, it is important to choose particular instants of time at which it is possible to store the state of the micro-clusters so that clusters in any user specified time horizon $(t_c - h, t_c)$ can be approximated.

We note that some examples of time frames used for the clustering process are the natural time frame (Figure 2.2(a) and (b)), and the logarithmic time frame (Figure 2.2(c)). In the natural time frame the snapshots are stored at regular intervals. We note that the scale of the natural time frame could be based on the application requirements. For example, we could choose days, months or years depending upon the level of granularity required in the analysis. A more flexible approach is to use the logarithmic time frame in which different variations of the time interval can be stored. As illustrated in Figure 2.2(c), we store snapshots at times of $t, 2 \cdot t, 4 \cdot t \ldots$. The danger of this is that we may jump too far between successive levels of granularity. We need an intermediate solution which provides a good balance between storage requirements and the level of approximation which a user specified horizon can be approximated.

In order to achieve this, we will introduce the concept of a pyramidal time frame. In this technique, the snapshots are stored at differing levels of granularity depending upon the recency. Snapshots are classified into different *orders* which can vary from 1 to $\log(T)$, where T is the clock time elapsed since the beginning of the stream. The order of a particular class of snapshots define the level of granularity in time at which the snapshots are maintained. The snapshots of different order are maintained as follows:

- Snapshots of the i-th order occur at time intervals of α^i, where α is an integer and $\alpha \geq 1$. Specifically, each snapshot of the i-th order is taken at a moment in time when the clock value[1] from the beginning of the stream is exactly divisible by α^i.

- At any given moment in time, only the last $\alpha + 1$ snapshots of order i are stored.

We note that the above definition allows for considerable redundancy in storage of snapshots. For example, the clock time of 8 is divisible by 2^0, 2^1, 2^2, and 2^3 (where $\alpha = 2$). Therefore, the state of the micro-clusters at a clock time of 8 simultaneously corresponds to order 0, order 1, order 2 and order 3 snapshots. From an implementation point of view, a snapshot needs to be maintained only once. We make the following observations:

- For a data stream, the maximum order of any snapshot stored at T time units since the beginning of the stream mining process is $\log_\alpha(T)$.

- For a data stream the maximum number of snapshots maintained at T time units since the beginning of the stream mining process is $(\alpha + 1) \cdot \log_\alpha(T)$.

- For any user specified time window of h, at least one stored snapshot can be found within $2 \cdot h$ units of the current time.

While the first two results are quite easy to see, the last one needs to be proven formally.

LEMMA 2.2 *Let h be a user-specified time window, t_c be the current time, and t_s be the time of the last stored snapshot of any order just before the time $t_c - h$. Then $t_c - t_s \leq 2 \cdot h$.*

Proof: Let r be the smallest integer such that $\alpha^r \geq h$. Therefore, we know that $\alpha^{r-1} < h$. Since we know that there are $\alpha + 1$ snapshots of order $(r-1)$, at least one snapshot of order $r-1$ must *always* exist before $t_c - h$. Let t_s be the snapshot of order $r - 1$ which occurs just before $t_c - h$. Then $(t_c - h) - t_s \leq \alpha^{r-1}$. Therefore, we have $t_c - t_s \leq h + \alpha^{r-1} < 2 \cdot h$.

Thus, in this case, it is possible to find a snapshot within a factor of 2 of any user-specified time window. Furthermore, the total number of snapshots which need to be maintained are relatively modest. For example, for a data stream running for 100 years with a clock time granularity of 1 second, the total number of snapshots which need to be maintained are given by $(2 + 1) \cdot \log_2(100 * 365 * 24 * 60 * 60) \approx 95$. This is quite a modest requirement given the fact that a snapshot within a factor of 2 can always be found within any user specified time window.

It is possible to improve the accuracy of time horizon approximation at a modest additional cost. In order to achieve this, we save the $\alpha^l + 1$ snapshots

Order of Snapshots	Clock Times (Last 5 Snapshots)
0	55 54 53 52 51
1	54 52 50 48 46
2	52 48 44 40 36
3	48 40 32 24 16
4	48 32 16
5	32

Table 2.1. An example of snapshots stored for $\alpha = 2$ and $l = 2$

of order r for $l > 1$. In this case, the storage requirement of the technique corresponds to $(\alpha^l + 1) \cdot \log_\alpha(T)$ snapshots. On the other hand, the accuracy of time horizon approximation also increases substantially. In this case, any time horizon can be approximated to a factor of $(1 + 1/\alpha^{l-1})$. We summarize this result as follows:

LEMMA 2.3 *Let h be a user specified time horizon, t_c be the current time, and t_s be the time of the last stored snapshot of any order just before the time $t_c - h$. Then $t_c - t_s \leq (1 + 1/\alpha^{l-1}) \cdot h$.*

Proof: Similar to previous case.

For larger values of l, the time horizon can be approximated as closely as desired. For example, by choosing $l = 10$, it is possible to approximate any time horizon within 0.2%, while a total of only $(2^{10} + 1) \cdot \log_2(100 * 365 * 24 * 60 * 60) \approx 32343$ snapshots are required for 100 years. Since historical snapshots can be stored on disk and only the current snapshot needs to be maintained in main memory, this requirement is quite feasible from a practical point of view. It is also possible to specify the pyramidal time window in accordance with user preferences corresponding to particular moments in time such as beginning of calendar years, months, and days. While the storage requirements and horizon estimation possibilities of such a scheme are different, all the algorithmic descriptions of this paper are directly applicable.

In order to clarify the way in which snapshots are stored, let us consider the case when the stream has been running starting at a clock-time of 1, and a use of $\alpha = 2$ and $l = 2$. Therefore $2^2 + 1 = 5$ snapshots of each order are stored. Then, at a clock time of 55, snapshots at the clock times illustrated in Table 2.1 are stored.

We note that a large number of snapshots are common among different orders. From an implementation point of view, the states of the micro-clusters at times of 16, 24, 32, 36, 40, 44, 46, 48, 50, 51, 52, 53, 54, and 55 are stored. It is easy to see that for more recent clock times, there is less distance between successive snapshots (better granularity). We also note that the storage requirements estimated in this section do not take this redundancy into account. Therefore, the requirements which have been presented so far are actually worst-case requirements.

These redundancies can be eliminated by using a systematic rule described in [6], or by using a more sophisticated geometric time frame. In this technique, snapshots are classified into different *frame numbers* which can vary from 0 to a value no larger than $\log_2(T)$, where T is the maximum length of the stream. The frame number of a particular class of snapshots defines the level of granularity in time at which the snapshots are maintained. Specifically, snapshots of frame number i are stored at clock times which are divisible by 2^i, but not by 2^{i+1}. Therefore, snapshots of frame number 0 are stored only at odd clock times. It is assumed that for each frame number, at most *max_capacity* snapshots are stored.

We note that for a data stream, the maximum frame number of any snapshot stored at T time units since the beginning of the stream mining process is $\log_2(T)$. Since at most *max_capacity* snapshots of any order are stored, this also means that the maximum number of snapshots maintained at T time units since the beginning of the stream mining process is $(max\_capacity) \cdot \log_2(T)$. One interesting characteristic of the geometric time window is that for any user-specified time window of h, at least one stored snapshot can be found within a factor of 2 of the specified horizon. This ensures that sufficient granularity is available for analyzing the behavior of the data stream over different time horizons. We will formalize this result in the lemma below.

LEMMA 2.4 *Let h be a user-specified time window, and t_c be the current time. Let us also assume that max_capacity ≥ 2. Then a snapshot exists at time t_s, such that $h/2 \leq t_c - t_s \leq 2 \cdot h$.*

Proof: Let r be the smallest integer such that $h < 2^{r+1}$. Since r is the smallest such integer, it also means that $h \geq 2^r$. This means that for any interval $(t_c - h, t_c)$ of length h, at least one integer $t' \in (t_c - h, t_c)$ must exist which satisfies the property that $t' \mod 2^{r-1} = 0$ and $t' \mod 2^r \neq 0$. Let t' be the time stamp of the last (most current) such snapshot. This also means the following:

$$h/2 \leq t_c - t' < h \qquad (2.1)$$

Then, if *max_capacity* is at least 2, the second last snapshot of order $(r-1)$ is also stored and has a time-stamp value of $t' - 2^r$. Let us pick the time $t_s = t' - 2^r$. By substituting the value of t_s, we get:

$$t_c - t_s = (t_c - t' + 2^r) \qquad (2.2)$$

Since $(t_c - t') \geq 0$ and $2^r > h/2$, it easily follows from Equation 2.2 that $t_c - t_s > h/2$.

Frame no.	Snapshots (by clock time)
0	69 67 65
1	70 66 62
2	68 60 52
3	56 40 24
4	48 16
5	64 32

Table 2.2.   A geometric time window

Since t' is the position of the latest snapshot of frame $(r-1)$ occurring before the current time t_c, it follows that $(t_c - t') \leq 2^r$. Subsituting this inequality in Equation 2.2, we get $t_c - t_s \leq 2^r + 2^r \leq h + h = 2 \cdot h$. Thus, we have:

$$h/2 \leq t_c - t_s \leq 2 \cdot h \qquad (2.3)$$

The above result ensures that every possible horizon can be closely approximated within a modest level of accuracy. While the geometric time frame shares a number of conceptual similarities with the pyramidal time frame [6], it is actually quite different and also much more efficient. This is because it eliminates the double counting of the snapshots over different frame numbers, as is the case with the pyramidal time frame [6]. In Table 2.2, we present an example of a frame table illustrating snapshots of different frame numbers. The rules for insertion of a snapshot t (at time t) into the snapshot frame table are defined as follows: (1) if $(t \bmod 2^i) = 0$ but $(t \bmod 2^{i+1}) \neq 0$, t is inserted into *frame_number* i (2) each slot has a *max_capacity* (which is 3 in our example). At the insertion of t into *frame_number* i, if the slot already reaches its *max_capacity*, the oldest snapshot in this frame is removed and the new snapshot inserted. For example, at time 70, since $(70 \bmod 2^1) = 0$ but $(70 \bmod 2^2) \neq 0$, 70 is inserted into frame_number 1 which knocks out the oldest snapshot 58 if the slot capacity is 3. Following this rule, when slot capacity is 3, the following snapshots are stored in the geometric time window table: 16, 24, 32, 40, 48, 52, 56, 60, 62, 64, 65, 66, 67, 68, 69, 70, as shown in Table 2.2. From the table, one can see that the closer to the current time, the denser are the snapshots stored.

3.   Clustering Evolving Data Streams: A Micro-clustering Approach

The clustering problem is defined as follows: for a given set of data points, we wish to partition them into one or more groups of similar objects. The similarity of the objects with one another is typically defined with the use of some distance measure or objective function. The clustering problem has been widely researched in the database, data mining and statistics communities [12, 18, 22, 20, 21, 24] because of its use in a wide range of applications. Recently, the clustering problem has also been studied in the context of the data stream environment [17, 23].

A previous algorithm called STREAM [23] assumes that the clusters are to be computed over the entire data stream. While such a task may be useful in many applications, a clustering problem may often be defined only over a portion of a data stream. This is because a data stream should be viewed as an infinite process consisting of data which continuously evolves with time. As a result, the underlying clusters may also change considerably with time. The nature of the clusters may vary with both the moment at which they are computed as well as the time horizon over which they are measured. For example, a data analyst may wish to examine clusters occurring in the last month, last year, or last decade. Such clusters may be considerably different. Therefore, we assume that one of the inputs to the clustering algorithm is a time horizon over which the clusters are found. Next, we will discuss CluStream, the online algorithm used for clustering data streams.

3.1   Micro-clustering Challenges

We note that since stream data naturally imposes a one-pass constraint on the design of the algorithms, it becomes more difficult to provide such a flexibility in computing clusters over different kinds of time horizons using conventional algorithms. For example, a direct extension of the stream based k-means algorithm in [23] to such a case would require a simultaneous maintenance of the intermediate results of clustering algorithms over all possible time horizons. Such a computational burden increases with progression of the data stream and can rapidly become a bottleneck for online implementation. Furthermore, in many cases, an analyst may wish to determine the clusters at a previous moment in time, and compare them to the current clusters. This requires even greater book-keeping and can rapidly become unwieldy for fast data streams.

Since a data stream cannot be revisited over the course of the computation, the clustering algorithm needs to maintain a substantial amount of information so that important details are not lost. For example, the algorithm in [23] is implemented as a continuous version of k-means algorithm which continues to maintain a number of cluster centers which change or merge as necessary throughout the execution of the algorithm. Such an approach is especially risky when the characteristics of the stream change over time. This is because the amount of information maintained by a k-means type approach is too approximate in granularity, and once two cluster centers are joined, there is no way to informatively split the clusters when required by the changes in the stream at a later stage.

Therefore a natural design to stream clustering would be separate out the process into an online micro-clustering component and an offline macro-clustering component. The online micro-clustering component requires a very efficient process for storage of appropriate summary statistics in a fast data stream. The offline component uses these summary statistics in conjunction with other user input in order to provide the user with a quick understanding of the clusters whenever required. Since the offline component requires only the summary statistics as input, it turns out to be very efficient in practice. This leads to several challenges:

- What is the nature of the summary information which can be stored efficiently in a continuous data stream? The summary statistics should provide sufficient temporal and spatial information for a horizon specific offline clustering process, while being prone to an efficient (online) update process.

- At what moments in time should the summary information be stored away on disk? How can an effective trade-off be achieved between the storage requirements of such a periodic process and the ability to cluster for a specific time horizon to within a desired level of approximation?

- How can the periodic summary statistics be used to provide clustering and evolution insights over user-specified time horizons?

3.2 Online Micro-cluster Maintenance: The CluStream Algorithm

The micro-clustering phase is the online statistical data collection portion of the algorithm. This process is not dependent on any user input such as the time horizon or the required granularity of the clustering process. The aim is to maintain statistics at a sufficiently high level of (temporal and spatial) granularity so that it can be effectively used by the offline components such as horizon-specific macro-clustering as well as evolution analysis. The basic concept of the micro-cluster maintenance algorithm derives ideas from the k-means and nearest neighbor algorithms. The algorithm works in an iterative fashion, by always maintaining a current set of micro-clusters. It is assumed that a total of q micro-clusters are stored at any moment by the algorithm. We will denote these micro-clusters by $\mathcal{M}_1 \ldots \mathcal{M}_q$. Associated with each micro-cluster i, we create a unique id whenever it is first created. If two micro-clusters are merged (as will become evident from the details of our maintenance algorithm), a *list* of ids is created in order to identify the constituent micro-clusters. The value of q is determined by the amount of main memory available in order to store the micro-clusters. Therefore, typical values of q are significantly larger than the natural number of clusters in the data but are also significantly smaller than the number of data points arriving in a long period of time for a massive data stream. These micro-clusters represent the current snapshot of clusters which change over the course of the stream as new points arrive. Their status is stored away on disk whenever the clock time is divisible by α^i for any integer i. At the same time any micro-clusters of order r which were stored at a time in the past more remote than α^{l+r} units are deleted by the algorithm.

We first need to create the initial q micro-clusters. This is done using an offline process at the very beginning of the data stream computation process. At the very beginning of the data stream, we store the first $InitNumber$ points on disk and use a standard k-means clustering algorithm in order to create the q initial micro-clusters. The value of $InitNumber$ is chosen to be as large as permitted by the computational complexity of a k-means algorithm creating q clusters.

Once these initial micro-clusters have been established, the online process of updating the micro-clusters is initiated. Whenever a new data point $\overline{X_{i_k}}$ arrives, the micro-clusters are updated in order to reflect the changes. Each data point either needs to be absorbed by a micro-cluster, or it needs to be put in a cluster of its own. The first preference is to absorb the data point into a currently existing micro-cluster. We first find the distance of each data point to the micro-cluster centroids $\mathcal{M}_1 \ldots \mathcal{M}_q$. Let us denote this distance value of the data point $\overline{X_{i_k}}$ to the centroid of the micro-cluster $\mathcal{M}_j$ by $dist(\mathcal{M}_j, \overline{X_{i_k}})$. Since the centroid of the micro-cluster is available in the cluster feature vector, this value can be computed relatively easily.

We find the closest cluster $\mathcal{M}_p$ to the data point $\overline{X_{i_k}}$. We note that in many cases, the point $\overline{X_{i_k}}$ does not naturally belong to the cluster $\mathcal{M}_p$. These cases are as follows:

- The data point $\overline{X_{i_k}}$ corresponds to an outlier.
- The data point $\overline{X_{i_k}}$ corresponds to the beginning of a new cluster because of evolution of the data stream.

While the two cases above cannot be distinguished until more data points arrive, the data point $\overline{X_{i_k}}$ needs to be assigned a (new) micro-cluster of its own with a unique id. How do we decide whether a completely new cluster should be created? In order to make this decision, we use the cluster feature vector of $\mathcal{M}_p$ to decide if this data point falls within the *maximum boundary* of the micro-cluster $\mathcal{M}_p$. If so, then the data point $\overline{X_{i_k}}$ is added to the micro-cluster $\mathcal{M}_p$ using the CF additivity property. The maximum boundary of the micro-cluster $\mathcal{M}_p$ is defined as a factor of t of the RMS deviation of the data points in $\mathcal{M}_p$ from the centroid. We define this as the *maximal boundary factor*. We note that the RMS deviation can only be defined for a cluster with more than 1 point. For a cluster with only 1 previous point, the maximum boundary is defined in a heuristic way. Specifically, we choose it to be r times that of the next closest cluster.

If the data point does not lie within the maximum boundary of the nearest micro-cluster, then a new micro-cluster must be created containing the data point $\mathcal{X}_{i_k}$. This newly created micro-cluster is assigned a new id which can identify it uniquely at any future stage of the data steam process. However, in order to create this new micro-cluster, the number of other clusters must be reduced by one in order to create memory space. This can be achieved by either deleting an old cluster or joining two of the old clusters. Our maintenance algorithm first determines if it is safe to delete any of the current micro-clusters as outliers. If not, then a merge of two micro-clusters is initiated.

The first step is to identify if any of the old micro-clusters are possibly outliers which can be safely deleted by the algorithm. While it might be tempting to simply pick the micro-cluster with the fewest number of points as the micro-cluster to be deleted, this may often lead to misleading results. In many cases, a given micro-cluster might correspond to a point of considerable cluster presence in the past history of the stream, but may no longer be an active cluster in the recent stream activity. Such a micro-cluster can be considered an outlier from the current point of view. An ideal goal would be to estimate the average timestamp of the last m arrivals in each micro-cluster [2], and delete the micro-cluster with the least recent timestamp. While the above estimation can be achieved by simply storing the last m points in each micro-cluster, this increases the memory requirements of a micro-cluster by a factor of m. Such a requirement reduces the number of micro-clusters that can be stored by the available memory and therefore reduces the effectiveness of the algorithm.

We will find a way to approximate the average timestamp of the last m data points of the cluster $\mathcal{M}$. This will be achieved by using the data about the timestamps stored in the micro-cluster $\mathcal{M}$. We note that the timestamp data allows us to calculate the mean and standard deviation[3] of the arrival times of points in a given micro-cluster $\mathcal{M}$. Let these values be denoted by $\mu\mathcal{M}$ and $\sigma\mathcal{M}$ respectively. Then, we find the time of arrival of the $m/(2 \cdot n)$-th percentile of the points in $\mathcal{M}$ assuming that the timestamps are normally distributed. This timestamp is used as the approximate value of the recency. We shall call this value as the *relevance stamp* of cluster $\mathcal{M}$. When the least relevance stamp of any micro-cluster is below a user-defined threshold δ, it can be eliminated and a new micro-cluster can be created with a unique id corresponding to the newly arrived data point $\overline{X_{i_k}}$.

In some cases, none of the micro-clusters can be readily eliminated. This happens when all relevance stamps are sufficiently recent and lie above the user-defined threshold δ. In such a case, two of the micro-clusters need to be merged. We merge the two micro-clusters which are closest to one another. The new micro-cluster no longer corresponds to one id. Instead, an $idlist$ is created which is a union of the the ids in the individual micro-clusters. Thus, any micro-cluster which is result of one or more merging operations can be identified in terms of the individual micro-clusters merged into it.

While the above process of updating is executed at the arrival of each data point, an additional process is executed at each clock time which is divisible by α^i for any integer i. At each such time, we store away the current set of micro-clusters (possibly on disk) together with their id list, and indexed by their time of storage. We also delete the least recent snapshot of order i, if $\alpha^l + 1$ snapshots of such order had already been stored on disk, and if the clock time for this snapshot is not divisible by α^{i+1}. (In the latter case, the snapshot continues to be a viable snapshot of order $(i + 1)$.) These micro-clusters can then be used to form higher level clusters or an evolution analysis of the data stream.

3.3   High Dimensional Projected Stream Clustering

The method can also be extended to the case of high dimensional projected stream clustering . The algorithms is referred to as HPSTREAM. The high-dimensional case presents a special challenge to clustering algorithms even in the traditional domain of static data sets. This is because of the sparsity of the data in the high-dimensional case. In high-dimensional space, all pairs of points tend to be almost equidistant from one another. As a result, it is often unrealistic to define distance-based clusters in a meaningful way. Some recent work on high-dimensional data uses techniques for *projected clustering* which can determine clusters for a specific subset of dimensions [1, 4]. In these methods, the definitions of the clusters are such that each cluster is specific to a particular group of dimensions. This alleviates the sparsity problem in high-dimensional space to some extent. Even though a cluster may not be meaningfully defined on all the dimensions because of the sparsity of the data, some subset of the dimensions can always be found on which particular subsets of points form high quality and meaningful clusters. Of course, these subsets of dimensions may vary over the different clusters. Such clusters are referred to as *projected clusters* [1].

In [8], we have discussed methods for high dimensional projected clustering of data streams. The basic idea is to use an (incremental) algorithm in which we associate a set of dimensions with each cluster. The set of dimensions is represented as a d-dimensional bit vector $\mathcal{B}(\mathcal{C}_i)$ for each cluster structure in $\mathcal{FCS}$. This bit vector contains a 1 bit for each dimension which is included in cluster $\mathcal{C}_i$. In addition, the maximum number of clusters k and the average cluster dimensionality l is used as an input parameter. The average cluster dimensionality l represents the average number of dimensions used in the cluster projection. An iterative approach is used in which the dimensions are used to update the clusters and vice-versa. The structure in $\mathcal{FCS}$ uses a decay-based mechanism in order to adjust for evolution in the underlying data stream. Details are discussed in [8].

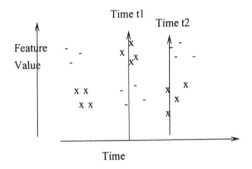

Figure 2.3. Varying Horizons for the classification process

4. Classification of Data Streams: A Micro-clustering Approach

One important data mining problem which has been studied in the context of data streams is that of stream classification [15]. The main thrust on data stream mining in the context of classification has been that of one-pass mining [14, 19]. In general, the use of one-pass mining does not recognize the changes which have occurred in the model since the beginning of the stream construction process [5]. While the work in [19] works on time changing data streams, the focus is on providing effective methods for incremental updating of the classification model. We note that the accuracy of such a model cannot be greater than the best sliding window model on a data stream. For example, in the case illustrated in Figure 2.3, we have illustrated two classes (labeled by 'x' and '-') whose distribution changes over time. Correspondingly, the best horizon at times t_1 and t_2 will also be different. As our empirical results will show, the true behavior of the data stream is captured in a temporal model which is sensitive to the level of evolution of the data stream.

The classification process may require simultaneous model construction and testing in an environment which constantly evolves over time. We assume that the testing process is performed concurrently with the training process. This is often the case in many practical applications, in which only a portion of the data is labeled, whereas the remaining is not. Therefore, such data can be separated out into the (labeled) training stream, and the (unlabeled) testing stream. The main difference in the construction of the micro-clusters is that the micro-clusters are associated with a class label; therefore an incoming data point in the training stream can only be added to a micro-cluster belonging to the same class. Therefore, we construct micro-clusters in almost the same way as the unsupervised algorithm, with an additional class-label restriction.

From the testing perspective, the important point to be noted is that the most effective classification model does not stay constant over time, but varies with progression of the data stream. If a static classification model were used for an evolving test stream, the accuracy of the underlying classification process is likely to drop suddenly when there is a sudden burst of records belonging to a particular class. In such a case, a classification model which is constructed using a smaller history of data is likely to provide better accuracy. In other cases, a longer history of training provides greater robustness.

In the classification process of an evolving data stream, either the short term or long term behavior of the stream may be more important, and it often cannot be known a-priori as to which one is more important. How do we decide the window or horizon of the training data to use so as to obtain the best classification accuracy? While techniques such as decision trees are useful for one-pass mining of data streams [14, 19], these cannot be easily used in the context of an *on-demand classifier* in an evolving environment. This is because such a classifier requires rapid variation in the horizon selection process due to data stream evolution. Furthermore, it is too expensive to keep track of the entire history of the data in its original fine granularity. Therefore, the on-demand classification process still requires the appropriate machinery for efficient statistical data collection in order to perform the classification process.

4.1   On-Demand Stream Classification

We use the micro-clusters to perform an *On Demand Stream Classification Process*. In order to perform effective classification of the stream, it is important to find the correct time-horizon which should be used for classification. How do we find the most effective horizon for classification at a given moment in time? In order to do so, a small portion of the training stream is not used for the creation of the micro-clusters. This portion of the training stream is referred to as the horizon fitting stream segment. The number of points in the stream used for horizon fitting is denoted by k_{fit}. The remaining portion of the training stream is used for the creation and maintenance of the class-specific micro-clusters as discussed in the previous section.

Since the micro-clusters are based on the entire history of the stream, they cannot directly be used to test the effectiveness of the classification process over different time horizons. This is essential, since we would like to find the time horizon which provides the greatest accuracy during the classification process. We will denote the set of micro-clusters at time t_c and horizon h by $\mathcal{N}(t_c, h)$. This set of micro-clusters is determined by subtracting out the micro-clusters at time $t_c - h$ from the micro-clusters at time t_c. The subtraction operation is naturally defined for the micro-clustering approach. The essential idea is to match the micro-clusters at time t_c to the micro-clusters at time $t_c - h$, and subtract out the corresponding statistics. The additive property of micro-clusters ensures that the resulting clusters correspond to the horizon $(t_c - h, t_c)$. More details can be found in [6].

Once the micro-clusters for a particular time horizon have been determined, they are utilized to determine the classification accuracy of that particular horizon. This process is executed periodically in order to adjust for the changes which have occurred in the stream in recent time periods. For this purpose, we use the horizon fitting stream segment. The last k_{fit} points which have arrived in the horizon fitting stream segment are utilized in order to test the classification accuracy of that particular horizon. The value of k_{fit} is chosen while taking into consideration the computational complexity of the horizon accuracy estimation. In addition, the value of k_{fit} should be small enough so that the points in it reflect the immediate locality of t_c. Typically, the value of k_{fit} should be chosen in such a way that the least recent point should be no larger than a pre-specified number of time units from the current time t_c. Let us denote this set of points by Q_{fit}. Note that since Q_{fit} is a part of the training stream, the class labels are known a-priori.

In order to test the classification accuracy of the process, each point $\overline{X} \in Q_{fit}$ is used in the following nearest neighbor classification procedure:

- We find the closest micro-cluster in $\mathcal{N}(t_c, h)$ to $\overline{X}$.
- We determine the class label of this micro-cluster and compare it to the true class label of $\overline{X}$. The accuracy over all the points in Q_{fit} is then determined. This provides the accuracy over that particular time horizon.

The accuracy of all the time horizons which are tracked by the geometric time frame are determined. The p time horizons which provide the greatest dynamic classification accuracy (using the last k_{fit} points) are selected for the classification of the stream. Let us denote the corresponding horizon values by $\mathcal{H} = \{h_1 \ldots h_p\}$. We note that since k_{fit} represents only a small locality of the points within the current time period t_c, it would seem at first sight that the system would always pick the smallest possible horizons in order to maximize the accuracy of classification. However, this is often not the case for evolving data streams. Consider for example, a data stream in which the records for a given class arrive for a period, and then subsequently start arriving again after a time interval in which the records for another class have arrived. In such a case, the horizon which includes previous occurrences of the same class is likely to provide higher accuracy than shorter horizons. Thus, such a system dynamically adapts to the most effective horizon for classification of data streams. In addition, for a stable stream the system is also likely to pick larger horizons because of the greater accuracy resulting from use of larger data sizes.

The classification of the test stream is a separate process which is executed continuously throughout the algorithm. For each given test instance $\overline{X_t}$, the above described nearest neighbor classification process is applied using each $h_i \in \mathcal{H}$. It is often possible that in the case of a rapidly evolving data stream, different horizons may report result in the determination of different class labels. The majority class among these p class labels is reported as the relevant class. More details on the technique may be found in [7].

5. Other Applications of Micro-clustering and Research Directions

While this paper discusses two applications of micro-clustering, we note that a number of other problems can be handled with the micro-clustering approach. This is because the process of micro-clustering creates a summary of the data which can be leveraged in a variety of ways for other problems in data mining. Some examples of such problems are as follows:

- **Privacy Preserving Data Mining:** In the problem of privacy preserving data mining, we create condensed representations [3] of the data which show k-anonymity. These condensed representations are like micro-clusters, except that each cluster has a minimum cardinality threshold on the number of data points in it. Thus, each cluster contains at least k data-points, and we ensure that the each record in the data cannot be distinguished from at least k other records. For this purpose, we only maintain the summary statistics for the data points in the clusters as opposed to the individual data points themselves. In addition to the first and second order moments we also maintain the covariance matrix for the data in each cluster. We note that the covariance matrix provides a complete overview of the distribution of in the data. This covariance matrix can be used in order to generate the pseudo-points which match the distribution behavior of the data in each micro-cluster. For relatively small micro-clusters, it is possible to match the probabilistic distribution in the data fairly closely. The pseudo-points can be used as a surrogate for the actual data points in the clusters in order to generate the relevant data mining results. Since the pseudo-points match the original distribution quite closely, they can be used for the purpose of a variety of data mining algorithms. In [3], we have illustrated the use of the privacy-preserving technique in the context of the classification problem. Our results show that the classification accuracy is not significantly reduced because of the use of pseudo-points instead of the individual data points.

- **Query Estimation:** Since micro-clusters encode summary information about the data, they can also be used for query estimation . A typical example of such a technique is that of estimating the selectivity of queries.

In such cases, the summary statistics of micro-clusters can be used in order to estimate the number of data points which lie within a certain interval such as a range query. Such an approach can be very efficient in a variety of applications since voluminous data streams are difficult to use if they need to be utilized for query estimation. However, the micro-clustering approach can condense the data into summary statistics, so that it is possible to efficiently use it for various kinds of queries. We note that the technique is quite flexible as long as it can be used for different kinds of queries. An example of such a technique is illustrated in [9], in which we use the micro-clustering technique (with some modifications on the tracked statistics) for futuristic query processing in data streams.

- **Statistical Forecasting:** Since micro-clusters contain temporal and condensed information, they can be used for methods such as statistical forecasting of streams . While it can be computationally intensive to use standard forecasting methods with large volumes of data points, the micro-clustering approach provides a methodology in which the condensed data can be used as a surrogate for the original data points. For example, for a standard regression problem, it is possible to use the centroids of different micro-clusters over the various temporal time frames in order to estimate the values of the data points. These values can then be used for making aggregate statistical observations about the future. We note that this is a useful approach in many applications since it is often not possible to effectively make forecasts about the future using the large volume of the data in the stream. In [9], it has been shown how to use the technique for querying and analysis of future behavior of data streams.

In addition, we believe that the micro-clustering approach is powerful enough to accomodate a wide variety of problems which require information about the summary distribution of the data. In general, since many new data mining problems require summary information about the data, it is conceivable that the micro-clustering approach can be used as a methodology to store condensed statistics for general data mining and exploration applications.

6.   Performance Study and Experimental Results

All of our experiments are conducted on a PC with Intel Pentium III processor and 512 MB memory, which runs Windows XP professional operating system. For testing the accuracy and efficiency of the CluStream algorithm, we compare CluStream with the STREAM algorithm [17, 23], the best algorithm reported so far for clustering data streams. CluStream is implemented according to the description in this paper, and the STREAM K-means is done strictly according to [23], which shows better accuracy than BIRCH [24]. To make the comparison fair, both CluStream and STREAM K-means use the same amount of memory.

Specifically, they use the same stream incoming speed, the same amount of memory to store intermediate clusters (called Micro-clusters in CluStream), and the same amount of memory to store the final clusters (called Macro-clusters in CluStream).

Because the synthetic datasets can be generated by controlling the number of data points, the dimensionality, and the number of clusters, with different distribution or evolution characteristics, they are used to evaluate the scalability in our experiments. However, since synthetic datasets are usually rather different from real ones, we will mainly use real datasets to test accuracy, cluster evolution, and outlier detection.

Real datasets. First, we need to find some real datasets that evolve significantly over time in order to test the effectiveness of CluStream. A good candidate for such testing is the KDD-CUP'99 Network Intrusion Detection stream data set which has been used earlier [23] to evaluate STREAM accuracy with respect to BIRCH. This data set corresponds to the important problem of automatic and real-time detection of cyber attacks. This is also a challenging problem for dynamic stream clustering in its own right. The offline clustering algorithms cannot detect such intrusions in real time. Even the recently proposed stream clustering algorithms such as BIRCH and STREAM cannot be very effective because the clusters reported by these algorithms are all generated from the entire history of data stream, whereas the current cases may have evolved significantly.

The Network Intrusion Detection dataset consists of a series of TCP connection records from two weeks of LAN network traffic managed by MIT Lincoln Labs. Each n record can either correspond to a normal connection, or an intrusion or attack. The attacks fall into four main categories: DOS (i.e., denial-of-service), R2L (i.e., unauthorized access from a remote machine), U2R (i.e., unauthorized access to local superuser privileges), and PROBING (i.e., surveillance and other probing). As a result, the data contains a total of five clusters including the class for *"normal connections"*. The attack-types are further classified into one of 24 types, such as buffer-overflow, guess-passwd, neptune, portsweep, rootkit, smurf, warezclient, spy, and so on. It is evident that each specific attack type can be treated as a sub-cluster. Most of the connections in this dataset are *normal*, but occasionally there could be a burst of attacks at certain times. Also, each connection record in this dataset contains 42 attributes, such as duration of the connection, the number of data bytes transmitted from source to destination (and vice versa), percentile of connections that have "SYN" errors, the number of "root" accesses, etc. As in [23], all 34 continuous attributes will be used for clustering and one outlier point has been removed.

Second, besides testing on the rapidly evolving network intrusion data stream, we also test our method over relatively stable streams. Since previously reported stream clustering algorithms work on the entire history of stream data, we believe that they should perform effectively for some data sets with stable distribution over time. An example of such a data set is the KDD-CUP'98 Charitable Donation data set. We will show that even for such datasets, the CluStream can consistently beat the STREAM algorithm.

The KDD-CUP'98 Charitable Donation data set has also been used in evaluating several one-scan clustering algorithms, such as [16]. This data set contains 95412 records of information about people who have made charitable donations in response to direct mailing requests, and clustering can be used to group donors showing similar donation behavior. As in [16], we will only use 56 fields which can be extracted from the total 481 fields of each record. This data set is converted into a data stream by taking the data input order as the order of streaming and assuming that they flow-in with a uniform speed.

Synthetic datasets. To test the scalability of CluStream, we generate some synthetic datasets by varying base size from 100K to 1000K points, the number of clusters from 4 to 64, and the dimensionality in the range of 10 to 100. Because we know the true cluster distribution a priori, we can compare the clusters found with the true clusters. The data points of each synthetic dataset will follow a series of Gaussian distributions, and to reflect the evolution of the stream data over time, we change the mean and variance of the current Gaussian distribution every 10K points in the synthetic data generation.

The quality of clustering on the real data sets was measured using the sum of square distance (SSQ), defined as follows. Assume that there are a total of N points in the past horizon at current time T_c. For each point p_i, we find the centroid C_{p_i} of its closest macro-cluster, and compute $d(p_i, C_{p_i})$, the distance between p_i and C_{p_i}. Then the SSQ at time T_c with horizon H (denoted as $SSQ(T_c, H)$) is equal to the sum of $d^2(p_i, C_{p_i})$ for all the N points within the previous horizon H. Unless otherwise mentioned, the algorithm parameters were set at $\alpha = 2$, $l = 10$, $InitNumber = 2000$, and $t = 2$.

We compare the clustering quality of CluStream with that of STREAM for different horizons at different times using the Network Intrusion dataset and the Charitable donation data set. The results are illustrated in Figures 2.4 and 2.5. We run each algorithm 5 times and compute their average SSQs. The results show that CluStream is almost always better than STREAM. All experiments for these datasets have shown that CluStream has substantially higher quality than STREAM. However the Network Intrusion data set showed significantly better results than the charitable donation data set because of the fact the network intrusion data set was a highly evolving data set. For such cases, the evolution sensitive CluStream algorithm was much more effective than the STREAM algorithm.

We also tested the accuracy of the *On-Demand Stream Classifier*. The first test was performed on the Network Intrusion Data Set. The first experiment

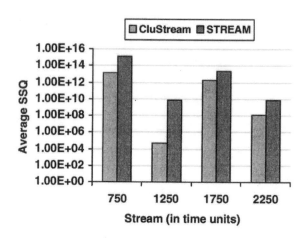

Figure 2.4.   Quality comparison (Network Intrusion dataset, horizon=256, stream_speed=200)

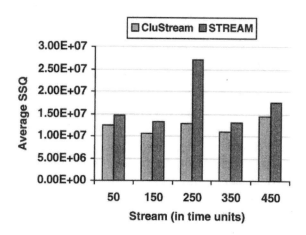

Figure 2.5.   Quality comparison (Charitable Donation dataset, horizon=4, stream_speed=200)

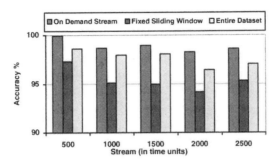

Figure 2.6. Accuracy comparison (Network Intrusion dataset, stream_speed=80, buffer_size=1600, k_{fit}=80, $init\_number$=400)

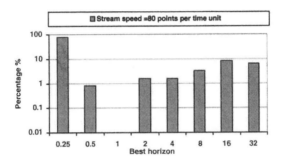

Figure 2.7. Distribution of the (smallest) best horizon (Network Intrusion dataset, Time units=2500, buffer_size=1600, k_{fit}=80, $init\_number$=400)

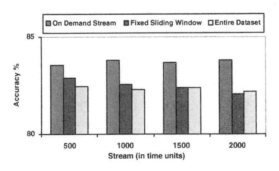

Figure 2.8. Accuracy comparison (Synthetic dataset B300kC5D20, stream_speed=100, buffer_size=500, k_{fit}=25, $init\_number$=400)

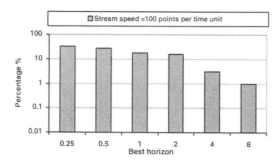

Figure 2.9.　Distribution of the (smallest) best horizon (Synthetic dataset B300kC5D20, Time units=2000, buffer_size=500, k_{fit}=25, $init\_number$=400)

was conducted with a stream speed at 80 connections per time unit (i.e., there are 40 training stream points and 40 test stream points per time unit). We set the buffer_size at 1600 points, which means upon receiving 1600 points (including both training and test stream points) we'll use a small set of the training data points (In this case k_{fit} =80) to choose the best horizon. We compared the accuracy of the *On-Demand-Stream classifier* with two simple one-pass stream classifiers over the entire data stream and the selected sliding window (i.e., sliding window $H = 8$). Figure 2.6 shows the accuracy comparison among the three algorithms. We can see the *On-Demand-Stream classifier* consistently beats the two simple one-pass classifiers. For example, at time unit 2000, the *On-Demand-Stream classifier*'s accuracy is about 4% higher than the classifier with fixed sliding window, and is about 2% higher than the classifier with the entire dataset. Because the class distribution of this dataset evolves significantly over time, either the entire dataset or a fixed sliding window may not always capture the underlying stream evolution nature. As a result, they always have a worse accuracy than the *On-Demand-Stream classifier* which always dynamically chooses the best horizon for classifying.

Figure 2.7 shows the distribution of the best horizons (They are the smallest ones if there exist several best horizons at the same time). Although about 78.4% of the (smallest) best horizons have a value 1/4, there do exist about 21.6% best horizons ranging from 1/2 to 32 (e.g., about 6.4% of the best horizons have a value 32). This also illustrates that there is no fixed sliding window that can achieve the best accuracy and the reason why the *On-Demand-Stream classifier* can outperform the simple one-pass classifiers over either the entire dataset or a fixed sliding window.

We have also generated one synthetic dataset B300kC5D20 to test the classification accuracy of these algorithms. This dataset contains 5 class labels and 300K data points with 20 dimensions. We first set the stream speed at 100 points

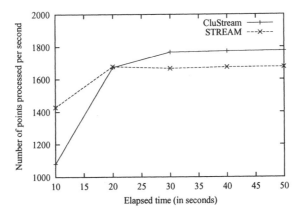

Figure 2.10.   Stream Proc. Rate (Charit. Donation data, stream_speed=2000)

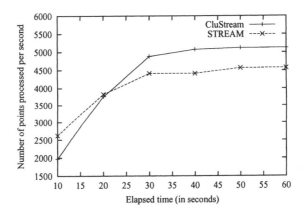

Figure 2.11.   Stream Proc. Rate (Ntwk. Intrusion data, stream_speed=2000)

per time unit. Figure 2.8 shows the accuracy comparison among the three algortihms: The *On-Demand-Stream classifier* always has much better accuracy than the other two classifiers. Figure 2.9 shows the distribution of the (smallest) best horizons which can explain very well why the *On-Demand-Stream classifier* has better accuracy.

We also tested the efficiency of the micro-cluster maintenance algorithm with respect to STREAM on the real data sets. We note that this maintenance process needs to be performed both for the clustering and classificiation algorithms with minor differences. Therefore, we present the results for the case of clustering. By setting the number of micro-clusters to 10 times the number of natural clusters, Figures 2.10 and 2.11 show the stream processing rate (i.e.,

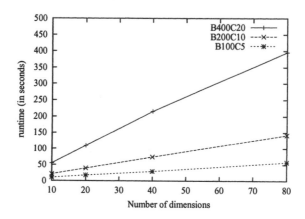

Figure 2.12.   Scalability with Data Dimensionality (stream_speed=2000)

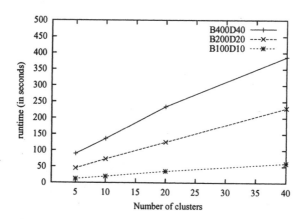

Figure 2.13.   Scalability with Number of Clusters (stream_speed=2000)

the number of points processed per second) as opposed to the running time for two real data sets. Since CluStream requires some time to compute the initial set of micro-clusters, its precessing rate is lower than STREAM at the very beginning. However, once steady state is reached, CluStream becomes faster than STREAM in spite of the fact that it needs to store the snapshots to disk periodically. This is because STREAM takes a few iterations to make k-means clustering converge, whereas CluStream just needs to judge whether a set of points will be absorbed by the existing micro-clusters and insert into them appropriately.

The key to the success of micro-cluster maintenance is high scalability. This is because this process is exposed to a potentially large volume of incoming data and needs to be implemented in an efficient and online fashion. The most time-consuming and frequent operation during micro-cluster maintenance is that of finding the closest micro-cluster for each newly arrived data point. It is clear that the complexity of this operation increases linearly with the number of micro-clusters. It is also evident that the number of micro-clusters maintained should be sufficiently larger than the number of input clusters in the data in order to obtain a high quality clustering. While the number of input clusters cannot be known a priori, it is instructive to examine the scalability behavior when the number of micro-clusters was fixed at a constant large factor of the number of input clusters. Therefore, for all the experiments in this section, we will fix the number of micro-clusters to 10 times the number of input clusters. We will present the scalability behavior of the CluStream algorithm with data dimensionality, and the number of natural clusters.

The first series of data sets were generated by varying the dimensionality from 10 to 80, while fixing the number of points and input clusters. The first data set series B100C5 indicates that it contains 100K points and 5 clusters. The same notational convention is used for the second data set series B200C10 and the third one B400C20. Figure 2.12 shows the experimental results, from which one can see that CluStream has linear scalability with data dimensionality. For example, for dataset series B400C20, when the dimensionality increases from 10 to 80, the running time increases less than 8 times from 55 seconds to 396 seconds.

Another three series of datasets were generated to test the scalability against the number of clusters by varying the number of input clusters from 5 to 40, while fixing the stream size and dimensionality. For example, the first data set series B100D10 indicates it contains 100K points and 10 dimensions. The same convention is used for the other data sets. Figure 2.13 demonstrates that CluStream has linear scalability with the number of input clusters.

7.　　Discussion

In this paper, we have discussed effective and efficient methods for clustering and classification of data streams. The techniques discussed in this paper utilize a micro-clustering approach in conjunction with a pyramidal time window. The technique can be used to cluster different kinds of data streams, as well as create a classifier for the data. The methods have clear advantages over recent techniques which try to cluster the whole stream at one time rather than viewing the stream as a changing process over time. The CluStream model provides a wide variety of functionality in characterizing data stream clusters over different time horizons in an evolving environment.

This is achieved through a careful division of labor between the online statistical data collection component and an offline analytical component. Thus, the process provides considerable flexibility to an analyst in a real-time and changing environment. In order to achieve these goals, we needed to the design the statistical storage process of the online component very carefully. The use of a *pyramidal time window* assures that the essential statistics of *evolving* data streams can be captured without sacrificing the underlying *space- and time-efficiency* of the stream clustering process.

The essential idea behind the CluStream model is to perform effective data summarization so that the underlying summary data can be used for a host of tasks such as clustering and classification. Therefore, the technique provides a framework upon which many other data mining tasks can be built.

Notes

1. Without loss of generality, we can assume that one unit of clock time is the smallest level of granularity. Thus, the 0-th order snapshots measure the time intervals at the smallest level of granularity.

2. If the micro-cluster contains fewer than $2 \cdot m$ points, then we simply find the average timestamp of all points in the cluster.

3. The mean is equal to $CF1^t/n$. The standard deviation is equal to $\sqrt{CF2^t/n - (CF1^t/n)^2}$.

References

[1] Aggarwal C., Procopiuc C., Wolf J., Yu P., Park J.-S. (1999). Fast algorithms for projected clustering. *ACM SIGMOD Conference.*

[2] Aggarwal C., Yu P. (2000). Finding Generalized Projected Clusters in High Dimensional Spaces, *ACM SIGMOD Conference.*

[3] Aggarwal C., Yu P.. (2004). A Condensation Approach to Privacy Preserving Data Mining. *EDBT Conference.*

[4] Agrawal R., Gehrke J., Gunopulos D., Raghavan P (1998). Automatic Subspace Clustering of High Dimensional Data for Data Mining Applications. *ACM SIGMOD Conference.*

[5] Aggarwal C (2003). A Framework for Diagnosing Changes in Evolving Data Streams. ACM SIGMOD Conference.

[6] Aggarwal C., Han J., Wang J., Yu P (2003). A Framework for Clustering Evolving Data Streams. *VLDB Conference.*

[7] Aggarwal C, Han J., Wang J., Yu P. (2004). On-Demand Classification of Evolving Data Streams. *ACM KDD Conference.*

[8] Aggarwal C., Han J., Wang J., Yu P. (2004). A Framework for Projected Clustering of High Dimensional Data Streams. *VLDB Conference.*

[9] Aggarwal C. (2006) on Futuristic Query Processing in Data Streams. *EDBT Conference.*

[10] Ankerst M., Breunig M., Kriegel H.-P., Sander J. (1999). OPTICS: Ordering Points To Identify the Clustering Structure. *ACM SIGMOD Conference.*

[11] Babcock B., Babu S., Datar M., Motwani R., Widom J. (2002). Models and Issues in Data Stream Systems, *ACM PODS Conference.*

[12] Bradley P., Fayyad U., Reina C. (1998) Scaling Clustering Algorithms to Large Databases. *SIGKDD Conference.*

[13] Cortes C., Fisher K., Pregibon D., Rogers A., Smith F. (2000). Hancock: A Language for Extracting Signatures from Data Streams. ACM SIGKDD Conference.

[14] Domingos P., Hulten G. (2000). Mining High-Speed Data Streams. *ACM SIGKDD Conference.*

[15] Duda R., Hart P (1973). *Pattern Classification and Scene Analysis*, Wiley, New York.

[16] Farnstrom F,, Lewis J., Elkan C. (2000). Scalability for Clustering Algorithms Revisited. *SIGKDD Explorations*, 2(1):pp. 51–57.

[17] Guha S., Mishra N., Motwani R., O'Callaghan L. (2000). Clustering Data Streams. *IEEE FOCS Conference.*

[18] Guha S., Rastogi R., Shim K. (1998). CURE: An Efficient Clustering Algorithm for Large Databases. *ACM SIGMOD Conference.*

[19] Hulten G., Spencer L., Domingos P. (2001). Mining Time Changing Data Streams. *ACM KDD Conference.*

[20] Jain A., Dubes R. (1998). Algorithms for Clustering Data, *Prentice Hall,* New Jersey.

[21] Kaufman L., Rousseuw P. (1990). Finding Groups in Data- An Introduction to Cluster Analysis. *Wiley Series in Probability and Math. Sciences.*

[22] Ng R., Han J (1994). Efficient and Effective Clustering Methods for Spatial Data Mining. *Very Large Data Bases Conference.*

[23] O'Callaghan L., Mishra N., Meyerson A., Guha S., Motwani R (2002). Streaming-Data Algorithms For High-Quality Clustering. *ICDE Conference*.

[24] Zhang T., Ramakrishnan R., and Livny M (1996). BIRCH: An Efficient Data Clustering Method for Very Large Databases. *ACM SIGMOD Conference*.

Chapter 3

A SURVEY OF CLASSIFICATION METHODS IN DATA STREAMS

Mohamed Medhat Gaber, Arkady Zaslavsky and Shonali Krishnaswamy
Caulfield School of Information Technology
Monash University,
900 Dandenong Rd, Caulfield East,
Melbourne VIC3145, Australia

{ Mohamed.Medhat.Gaber, Arkady.Zaslavsky, Shonali.Krishnaswamy } @infotech.monash.edu.au

Abstract   With the advance in both hardware and software technologies, automated data generation and storage has become faster than ever. Such data is referred to as data streams. Streaming data is ubiquitous today and it is often a challenging task to store, analyze and visualize such rapid large volumes of data. Most conventional data mining techniques have to be adapted to run in a streaming environment, because of the underlying resource constraints in terms of memory and running time. Furthermore, the data stream may often show concept drift, because of which adaptation of conventional algorithms becomes more challenging. One such important conventional data mining problem is that of classification. In the classification problem, we attempt to model the class variable on the basis of one or more feature variables. While this problem has been extensively studied from a conventional mining perspective, it is a much more challenging problem in the data stream domain. In this chapter, we will re-visit the problem of classification from the data stream perspective. The techniques for this problem need to be thoroughly re-designed to address the issue of resource constraints and concept drift. This chapter reviews the state-of-the-art techniques in the literature along with their corresponding advantages and disadvantages.

1.   Introduction

Classification problems [19, 20] have been studied thoroughly as a major category of the data analysis tasks in machine learning, statistical inference [18] and data mining. Classification methods represent the set of supervised learning techniques where a set of dependent variables needs to be predicted based on another set of input attributes. There are two main distinctive approaches under the supervised learning category: classification and regression. Classification is mainly concerned with categorical attributes as dependent variables; however regression is concerned with numerical attributes as its output. The classification process is divided into two phases: model building and model testing. In model building, a learning algorithm runs over a data set to induce a model that could be used in estimating an output. The quality of this estimation is assessed in the model testing phase. The model building process is referred to as training as well. Classification techniques [19, 20] have attracted the attention of researchers due to the significance of their applications. A variety of methods such as decision trees, rule based methods, and neural networks are used for the classification problem. Many of these techniques have been designed to build classification models from static data sets where several passes over the stored data is possible. This is not possible in the case of data streams, in which it is necessary to process the entire data set in one pass. Furthermore, the classification problem needs to be re-designed in the context of *concept drift*, a unique problem in the case of data streams.

The applications of data stream classification can vary from critical astronomical and geophysical applications [6] to real-time decision support in business and industrial applications [24, 25]. There are several potential scenarios for such applications. For example, classification and analysis of biosensor measurements around a city for security reasons is an important emerging application. The analysis of simulation results and on-board sensor readings in scientific applications has its potential in changing the mission plan or the experimental settings in real time. Web log and clickstream analysis is an important application in the electronic commerce domain. The classification of data streams generated from the marketplace such as stock market streaming information is another appealing application. Decision trees created from stock market data in distributed streaming environment have been used in MobiMine [24, 25].

The process of adapting classification models to many of the above applications is often non-trivial. The most important challenge with regard to classification is that of concept drifting of evolving data streams. The process of *concept drift* results from the natural tendency of the underlying data to evolve over time. The classifier is most likely to be outdated after a time window due to the continuous change of the streaming information on a temporal basis. We discuss this issue along with a number of other challenges for the classification problem. Solution approaches used in addressing these issues are summarized in order to emphasize their advantages, and drawbacks. This summarization also provides some insight for other stream mining techniques due to the shared research issues across different applications. A thorough discussion of classification techniques in data streams is given as a guide to researchers as well as practitioners. The techniques are presented in an easy way with illustrative figures depicting each algorithm in a diagrammatic way.

The chapter is organized as follows. Research issues with regard to the stream classification problems are discussed in section 2. Section 3 represents the approaches proposed as solutions to address the previous research issues. Section 4 provides a survey of the classification techniques in stream mining literature. Techniques surveyed include the Ensemble-based Classification [30], Very Fast Decision Trees (VFDT) [9] with its extensions [22], [23], On-Demand Classification [3], On-Line Information Network (OLIN) [26], Lightweight Classification (LWClass) [14], Scalable Classification Algorithm by Learning decisiOn Patterns (SCALLOP) [12] and Adaptive Nearest Neighbor Classification for Data-streams (ANNCAD) [27]. This selection of techniques is based on the soundness of the techniques and how well the techniques addresses important research challenges. Finally, the chapter is concluded with a summary in section 5.

2.   Research Issues

In this section, we will address the primary research issues encountered in the context of stream mining. While many of these issues are shared across different stream mining applications, we discuss these issues with a special emphasis on the problem of classification [4, 9, 10, 13, 14, 16, 17, 21, 28].

- **High Speed Nature of Data Streams:** The inherent characteristic of data streams is its high speed. The algorithm should be able to adapt to the high speed nature of streaming information. The rate of building a classification model should be higher than the data rate. Furthermore, it is not possible to scan the data more than once. This is referred to as the *one-pass* constraint.

- **Unbounded Memory Requirements:** Classification techniques require data to be resident in memory for building the model. The huge amounts of data streams generated rapidly dictate the need for unbounded memory. This challenge has been addressed using load shedding, sampling, aggregation, and creating data synopsis. The memory issue is an important motivation behind many of the developed techniques in the area.

- **Concept Drifting:** Concept drifts change the classifier results over time. This is because of the change in the underlying data patterns. It is also referred to as data stream *evolution* [1]. This results in the model becoming stale and less relevant over time. The capture of such changes would help in updating the classifier model effectively. The use of an outdated model could lead to a very low classification accuracy.

- **Tradeoff between Accuracy and Efficiency:** The main tradeoff in data stream mining algorithms is between the accuracy of the output with regard to the application and the time and space complexity. In many cases, approximation algorithms can guarantee error bounds, while maintaining a high level of efficiency.

- **Challenges in Distributed Applications:** A significant number of data stream applications run in mobile environments with limited bandwidth such as sensor networks and handheld devices. Thus knowledge structure representation is an important issue. After extracting models and patterns locally from data stream generators or receivers, it is important to transfer the data mining output to the user. The user could be a mobile user or a stationary one getting the results from mobile nodes. This is often a challenge because of the bandwidth limits in transferring data. Kargupta et al. [24] have addressed this problem by using Fourier transformations to efficiently represent decision trees for the purpose of transmission over limited bandwidth links.

- **Visualization of data stream mining results:** Visualization of traditional data mining results on a desktop has been a research issue for more than a decade. Visualization of mining results in small screens of a Personal Digital Assistant (PDA) for example is a real challenge and an open research problem. Given a scenario for a businessman on a move and the data are being streamed and analyzed on his PDA. The results of this analysis should be efficiently visualized in a way that allows him to take a quick decision. The pioneering work on representation of decision trees in a mobile device has been suggested by Kargupta et al [24].

- **Modelling change of mining results over time:** In some cases, the user is not interested in mining data stream results, but how these results are changing over a temporal basis. The classification changes could help in understanding the change in data streams over time.

- **Interactive Mining environment to satisfy user results:** Mining data streams is a highly application oriented field. For example, the user should be able to change the classification parameters to serve the special needs of the user under the current context. The fast nature of data streams often makes it more difficult to incorporate user-interaction.

- **The integration of data stream management systems and data stream mining approaches:** The integration among storage, querying, mining and reasoning of the incoming stream would realize robust streaming systems that could be used in different applications [5, 7]. Along this line, current database management systems have achieved this goal over static stored data sets. However, this goal has not been fully realized for the case of data streams. An important future research issue is to integrate the stream mining algorithms with known stream management systems in order to design complete systems for stream processing.

- **Hardware and other Technological Issues:** The technological issue of mining data streams is an important one. How do we represent the data in such an environment in a compressed way? Which platforms are best suited such special real-time applications? Hardware issues are of special concerns. Small devices generating data streams are not designed for complex computations. Currently emulators are used for such tasks and it is a real burden for data stream mining applications which run in resource-constrained environments. Novel hardware solutions are required to address this issue.

- **Real time accuracy evaluation and formalization:** In many cases, resource constrained methods work with a trade-off between accuracy and efficiency of the designed method. Therefore, we need a feedback of the current achieved accuracy with relation to the available resources. This is needed to adjust the algorithm parameters according to the available resources. This formalization would also help in making decisions about the reliability of the output.

Among the above-mentioned issues, the first three are of special significance. Thus, we will use them as the basis for comparing different stream classification techniques in this chapter. We also note that many of these issues are shared among all mining techniques in streaming environment. The following section concisely summarizes the approaches that are used as solutions addressing the above issues.

3. Solution Approaches

Many of the afore-mentioned issues can be solved using well-established statistical and computational approaches. While, specific methods for stream classification will be discussed later, it is useful to understand the broad characteristics of different methods which are used to adapt conventional classification techniques to the case of data streams. We can categorize these solutions as data-based and task-based ones. In data-based solutions, the idea is to examine only a subset of the whole data set or to transform the data vertically or horizontally to an approximate smaller size data representation. Such an approach allows us to utilize many known data mining techniques to the case of data streams. On the other hand, in task based solutions, some standard algorithmic modification techniques can be used to achieve time and space efficient solutions [13]. Table 3.1 shows the data-based techniques, while Table 3.2 shows

Technique	Definition	Pros	Cons
Sampling	Choosing a data subset for analysis	Error Bounds Guaranteed	Poor for anomaly detection
Load Shedding	Ignoring a chunk of data	Efficient for queries	Very poor for anomaly detection
Sketching	Random projection on feature set	Extremely Efficient	May ignore Relevant features
Synopsis Structure	Quick Transformation	Analysis Task Independent	Not sufficient for very fast stream
Aggregation	Compiling summary statistics	Analysis Task Independent	May ignore Relevant features

Table 3.1.   Data Based Techniques

Technique	Definition	Pros	Cons
Approximation Algorithms	Algorithms with Error Bounds	Efficient	Resource adaptivity with data rates not always possible
Sliding Window	Analyzing most recent streams	General	Ignores part of stream
Algorithm Output Granularity	Highly Resource aware technique with memory and fluctuating data rates	General	Cost overhead of resource aware component

Table 3.2.   Task Based Techniques

the task-based techniques. Each table provides a definition, advantages and disadvantages of each technique.

While the methods in Tables 3.1 and 3.2 provide an overview of the broad methods which can be used to adapt conventional methods to classification, it is more useful to study specific techniques which are expressly designed for the purpose of classification. In the next section, we will provide a review of these methods.

4.   Classification Techniques

This section reviews the state-of-the-art of data stream classification techniques. We have provided an overview of some of the key methods, how well they address the research problems discussed earlier.

4.1 Ensemble Based Classification

Wang et al. [30] have proposed a generic framework for mining concept drifting data streams. The framework is based on the observation that many data stream mining algorithms do not address the issue of concept drift in the evolving data. The idea is based on using an ensemble of classification models such as decision trees using C4.5, RIPPER, naïve Bayesian and others to vote for the classification output to increase the accuracy of the predicted output.

This framework was developed to address three research challenges in data stream classification:

1. **Concept Drift:** The accuracy of the output of many classifiers is very sensitive to concept drifts in the evolving streams. At the same time, one does not want to remove excessive parts of the stream, when there is no concept drift. Therefore, a method needs to be designed to decide which part of the stream to be used for the classification process.

2. **Efficiency:** The process of building classifiers is a complex computational task and the update of the model due to concept drifts is a complicated process. This is especially relevant in the case of high speed data streams.

3. **Robustness:** Ensemble based classification has traditionally been used in order to improve robustness. The key idea is to avoid the problem of overfitting of individual classifiers. However, it is often a challenging task to use the ensemble effectively because of the high speed nature of the data streams.

An important motivation behind the framework is to deal with the expiration of old data streams. The idea of using the most recent data streams to build and use the developed classifiers may not be valid for most applications. Although the old streams can affect the accuracy of the classification model in a negative way, it is still important to keep track of this data in the current model. The work in [30] shows that it is possible to use weighted ensemble classifiers in order to achieve this goal.

The work in [30] uses weighted classifier ensembles according to the current accuracy of each classifier used in the ensemble. The weight of any classifier is calculated and contributed to predict the final output. The weight of each classifier may vary as the data stream evolves, and a given classifier may become more or less important on a particular sequential chunk of the data. The framework has outperformed single classifiers experimentally. This is partly because of the greater robustness of the ensemble, and partly because of more effective tracking of the change in the underlying structure of the data. More interesting variations of similar concepts may be found in [11]. Figure 3.1 depicts the proposed framework.

4.2   Very Fast Decision Trees (VFDT)

Domingos and Hulten [9, 22] have developed a decision tree approach which is referred to as *Very Fast Decision Trees (VFDT)*. It is a decision tree learning system based on Hoeffding trees. It splits the tree using the current best attribute taking into consideration that the number of examples used satisfies the Hoeffding bound. Such a technique has the property that its output is (asymptotically) nearly identical to that of a conventional learner. VFDT is an extended version of such a method which can address the research issues of data streams. These research issues are:

- **Ties of attributes:** Such ties occur when two or more attributes have close values of the splitting criteria such as information gain or gini index. We note that at such a moment of the decision tree growth phase, one must make a decision between two or more attributes based on only the set of records received so far. While it is undesirable to delay such split decisions indefinitely, we would like to do so at a point when the errors are acceptable.

- **Bounded memory:** The tree can grow till the algorithm runs out of memory. This results in a number of issues related to effective maintenance of the tree.

- **Efficiency and Accuracy:** This is an inherent characteristic of all data stream algorithms.

The extension of Hoeffding trees in VFDT has been done using the following techniques.

- The key question during the construction of the decision tree is the choice of attributes to be used for splits. Approximate ties on attributes are broken using a user-specified threshold of acceptable error measure for the output. By using this approach, a crisp criterion can be determined on when a split (based on the inherently incomplete information from the current data stream) provides acceptable error. In particular, the Hoeffding inequality provides the necessary bound on the correctness of the choice of split variable. It can be shown for any small value of δ, that a particular choice of the split variable is the correct choice (same as conventional learner) with probability at least $1 - \delta$, if a sufficient number of stream records have been processed. This "sufficient number" increases at the relatively modest rate of $\log(1/\delta)$. The bound on the accuracy of each split can then be extrapolated to the behavior of the entire decision tree. We note that the stream decision tree will provide the same result as the conventional decision tree, if for every node along the path for given test instance, the same choice of split is used. This can be used to show that the behavior of the stream decision tree for a particular test instance differs from the conventional decision tree with probability at most $1 - \delta/p$, where p is the probability that a record is assigned to a leaf at each level.

- Bounded memory has been addressed by de-activating the least promising leaves and ignoring the poor attributes. The calculation of these poor attributes is done through the difference between the splitting criteria of the highest and lowest attributes. If the difference is greater than a pre-specified value, the attribute with the lowest splitting measure will be freed from memory.

- The VFDT system is inherently I/O bound; in other words, the time for processing the example is lower than the time required to read it from disk. This is because of the Hoeffding tree-based approach with a crisp criterion for tree growth and splits. Such an approach can make clear decisions at various points of the tree construction algorithm without having to re-scan the data. Furthermore, the computation of the splitting criteria is done in a batch processing mode rather than online processing. This significantly saves the time of recalculating the criteria for all the attributes with each incoming record of the stream. The accuracy of the output can be further improved using multiple scans in the case of low data rates.

All the above improvements have been tested using special synthetic data sets. The experiments have proved efficiency of these improvements. Figure 3.2 depicts the VFDT learning system. The VFDT has been extended to address the problem of concept drift in evolving data streams. The new framework has been termed as CVFDT [22]. It runs VFDT over fixed sliding windows in order to have the most updated classifier. The change occurs when the splitting criteria changes significantly among the input attributes.

Jin and Agrawal [23] have extended the VFDT algorithm to efficiently process numerical attributes and reduce the sample size calculated using the Hoeffding bound. The former objective has been addressed using their Numerical Interval Pruning (NIP) technique. The pruning is done by first creating a histogram for each interval of numbers. The least promising intervals to be branched are pruned to reduce the memory space. The experimental results show an average of 39% of space reduction by using NIP. The reduction of sample size is done by using properties of information gain functions. The derived method using multivariate delta method has a guarantee of a reduction of sample size over the Hoeffding inequality with the same accuracy. The experiments show a reduction of 37% of the sample size by using the proposed method.

4.3   On Demand Classification

Aggarwal et al. have adopted the idea of micro-clusters introduced in CluStream [2] in On-Demand classification in [3]. The on-demand classification method divides the classification approach into two components. One component continuously stores summarized statistics about the data streams and the second one continuously uses the summary statistics to perform the classification. The summary statistics are represented in the form of class-label specific micro-clusters. This means that each micro-cluster is associated with a specific class label which defines the class label of the points in it. We note that both components of the approach can be used in online fashion, and therefore the approach is referred to as an *on-demand classification method*. This is because the set of test instances could arrive in the form of a data stream and can be classified efficiently on demand. At the same time, the summary statistics (and therefore training model) can be efficiently updated whenever new data arrives. The great flexibility of such an approach can be very useful in a variety of applications.

At any given moment in time, the current set of micro-clusters can be used to perform the classification. The main motivation behind the technique is that the classification model should be defined over a time horizon which depends on the nature of the concept drift and data evolution. When there is smaller concept drift, we need a larger time horizon in order to ensure robustness. In the event of greater concept drift, we require smaller time horizons. One key property of micro-clusters (referred to as the *subtractive property*) ensures that it is possible to compute horizon-specific statistics. As a result it is possible to perform the classification over a wide variety of time horizons. A hold out training stream is used to decide the size of the horizon on which the classification is performed. By using a well-chosen horizon it is possible to achieve a high level of classification accuracy. Figure 3.3 depicts the classification on demand framework.

4.4   Online Information Network (OLIN)

Last [26] has proposed an online classification system which can adapt to concept drift. The system re-builds the classification model with the most recent examples. By using the error-rate as a guide to concept drift, the frequency of model building and the window size is adjusted over time.

The system uses info-fuzzy techniques for building a tree-like classification model. It uses information theory to calculate the window size. The main idea behind the system is to change the sliding window of the model reconstruction according to the classification error rate. If the model is stable, the window size increases. Thus the frequency of model building decreases. The info-fuzzy technique for building a tree-like classification model is referred to as the Info-

A_1	A_2	...	A_n	Class	Weight
$Value(A_1)$	$Value(A_2)$	...	$Value(A_n)$	Class Category	X = # items Contributing
...	...	...	...	...	...

Table 3.3. Typical LWClass Training Results

Fuzzy Network (IFN). The tree is different than conventional decision trees in that each level of the tree represents only one attribute except the root node layer. The nodes represent different values of the attribute. The process of inducing the class label is similar to the one of conventional decision trees. The process of constructing this tree has been termed as Information Network (IN). The IN technique uses a similar procedure of building conventional decision trees by determining if the split of an attribute would decrease the entropy or not. The measure used is mutual conditional information that assesses the dependency between the current input attribute under examination and the output attribute. At each iteration, the algorithm chooses the attribute with the maximum mutual information and adds a layer with each node represents a different value of this attribute. The iterations stop once there is no increase in the mutual information measure for any of the remaining attributes that have not been considered in the tree. OLIN system repeatedly uses the IN algorithm for building a new classification model. The system uses the information theory to calculate the window size (refers to number of examples). It uses a less conservative measure than Hoeffding bound used in VFDT [9, 22] reviewed earlier in this chapter. This measure is derived from the mutual conditional information in the IN algorithm by applying the likelihood ratio test to assess the statistical significance of the mutual information. Subsequently, we change the window size of the model reconstruction according to the classification error rate. The error rate is calculated by measuring the difference between the error rate during the training at one hand and the error rate during the model validation at the other hand. A significance increase in the error rate indicates a high probability of a concept drift. The window size changes according to the value of this increase. Figure 3.4 shows a simple flow chart of the OLIN system.

4.5   LWClass Algorithm

Gaber et al [14] have proposed Lightweight Classification techniques termed as LWClass. LWClass is based on Algorithm Output Granularity. The algorithm output granularity (AOG) introduces the first resource-aware data analysis approach that can cope with fluctuating data rates according to the available memory and the processing speed. The AOG performs the local data analysis on resource constrained devices that generate or receive streams of information. AOG has three stages of mining, adaptation and knowledge integration as shown in Figure 3.5 [14].

LWClass starts with determining the number of instances that could be resident in memory according to the available space. Once a classified data record arrives, the algorithm searches for the nearest instance already stored in the main memory. This is done using a pre-specified distance threshold. This threshold represents the similarity measure acceptable by the algorithm to consider two or more data records as an entry into a matrix. This matrix is a summarized version of the original data set. If the algorithm finds a nearest neighbor, it checks the class label. If the class label is the same, it increases the weight for this instance by one, otherwise it decrements the weight by one. If the weight is decremented down to zero, this entry will be released from the memory conserving the limited memory on streaming applications. The algorithm output granularity is controlled by the distance threshold value and is changing over time to cope with the high speed of the incoming data elements. The algorithm procedure could be described as follows:

1. Each record in the data stream contains attribute values for $a_1, a_2, ..., a_n$ attributes and the class category.

2. According to the data rate and the available memory, the algorithm output granularity is applied as follows:

   2.1 Measure the distance between the new record and the stored ones.

   2.2 If the distance is less than a threshold, store the average of these two records and increase the weight for this average as an entry by 1. (The threshold value determines the algorithm accuracy and is chosen according to the available memory and data rate that determines the algorithm rate). This is in case that both items have the same class category. If they have different class categories, the weight is decreased by 1 and released from memory if the weight reaches zero.

   2.3 After a time threshold for the training, we come up with a matrix represented in Table 3.3.

3. Using Table 3.3, the unlabeled data records could be classified as follows. According to the available time for the classification process, we choose nearest K-table entries and these entries are variable according to the time needed by the process.

4. Find the majority class category taking into account the calculated weights from the K entries. This will be the output for this classification task.

4.6   ANNCAD Algorithm

Law et al [27] have proposed an incremental classification algorithm termed as Adaptive Nearest Neighbor Classification for Data-streams (ANNCAD). The algorithm uses Haar Wavelets Transformation for multi-resolution data representation. A grid-based representation at each level is used.

The process of classification starts with attempting to classify the data record according to the majority nearest neighbors at finer levels. If the finer levels are unable to differentiate between the classes with a pre-specified threshold, the coarser levels are used in a hierarchical way. To address the concept drift problem of the evolving data streams, an exponential fade factor is used to decrease the weight of old data in the classification process. Ensemble classifiers are used to overcome the errors of initial quantization of data. Figure 3.6 depicts the ANNCAD framework.

Experimental results over real data sets have proved the achieved accuracy over the VFDT and CVFDT discussed earlier in this section. The drawback of this technique represented in inability of dealing with sudden concept drifts as the exponential fade factor takes a while to have its effect felt. In fact, the choice of the exponential fade factor is an inherent flexibility which could lead to over-estimation or under-estimation of the rate of concept drift. Both errors would result in a reduction in accuracy.

4.7   SCALLOP Algorithm

Ferrer-Troyano et al. [12] have proposed a scalable classification algorithm for numerical data streams. This is one of the few rule-based classifiers for data streams. It is inherently difficult to construct rule based classifiers for data streams, because of the difficulty in maintaining the underlying rule statistics. The algorithm has been termed as Scalable Classification Algorithm by Learning decisiOn Patterns (SCALLOP).

The algorithm starts by reading a number of user-specified labeled records. A number of rules are created for each class from these records. Subsequently, the key issue is to effectively maintain the rule set after arrival of each new record. On the arrival of a new record, there are three cases:

a) **Positive covering:** This is the case of a new record that strengthens a current discovered rule.

b) **Possible expansion:** This is the case of a new record that is associated with at least one rule, but is not covered by any currently discovered rule.

c) **Negative covering:** This is the case of a new record that weakens a currently discovered rule.

For each of the above cases, a different procedure is used as follows:

a) **Positive covering:** The positive support and confidence of the existing rule is re-calculated.

b) **Possible expansion:** In this case, the rule is extended if it satisfies two conditions:

   – It is bounded within a user-specified growth bounds to avoid a possible wrong expansion of the rule.

   – There is no intersection between the expanded rule and any already discovered rule associated with the same class label.

c) **Negative covering:** In this case, the negative support and confidence is re-calculated. If the confidence is less than a minimum user-specified threshold, a new rule is added.

After reading a pre-defined number of records, the process of rule refining is performed. Rules in the same class and within a user-defined acceptable distance measure are merged. At the same time, care is taken to ensure that these rules do not intersect with rules associated with other class labels. The resulting hypercube of the merged rules should also be within certain growth bounds. The algorithm also has a refinement stage. This stage releases the uninteresting rules from the current model. In particular, the rules that have less than the minimum positive support are released. Furthermore, the rules that are not covered by at least one of the records of the last user-defined number of received records are released. Figure 3.7 shows an illustration of the basic process.

Finally a voting-based classification technique is used to classify the unlabeled records. If there is a rule covers the current record, the label associated with that rule is used as the classifier output. Otherwise, a voting over the current rules within the growth bounds is used to infer the class label.

5.   Summary

Stream classification techniques have several important applications in business, industry and science. This chapter reviews the research problems in data stream classification. Several approaches in the literature have been summarized with their advantages and drawbacks. While the selection of the techniques is based on the performance and quality of addressing the research challenges, there are a number of other methods [11, 8, 15, 22, 31] which we have not discussed in greater detail in this chapter. Many of these techniques are developed along similar lines as one or more techniques presented in this chapter.

The major research challenges in data stream classification are represented in concept drifting, resource adaptivity, high data rates, and the unbounded memory requirements. While many methods have been proposed to address some of

Method	Concept Drift	High Speed	Memory Req.
Ensemble-based Classification	X		
VFDT		X	X
On-Demand Classification	X	X	X
Online Information Network	X		
LWClass		X	X
ANNCAD	X		
SCALLOP	X		

Table 3.4. Summary of Reviewed Techniques

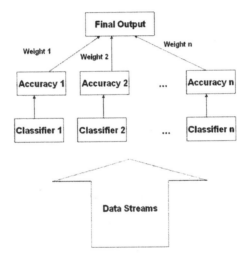

Figure 3.1. The ensemble based classification method

these issues, they are often unable to address these issues simultaneously. Table 3.4 summarizes the previously reviewed techniques in terms of addressing the above challenges. The area of data stream classification is still in its infancy. A number of open challenges still remain in stream classification algorithms; particular in respect to concept drift and resource adaptive classification.

References

[1] Aggarwal C. (2003) A Framework for Diagnosing Changes in Evolving Data Streams. *Proceedings of the ACM SIGMOD Conference.*

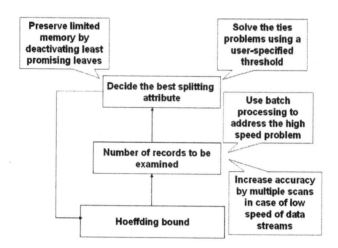

Figure 3.2.   VFDT Learning Systems

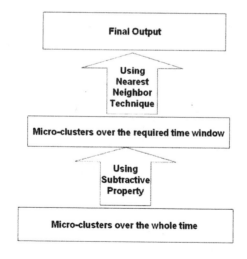

Figure 3.3.   On Demand Classification

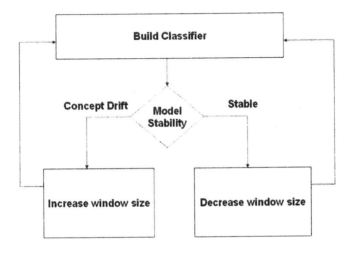

Figure 3.4.   Online Information Network System

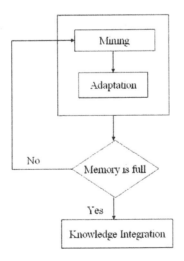

Figure 3.5.   Algorithm Output Granularity

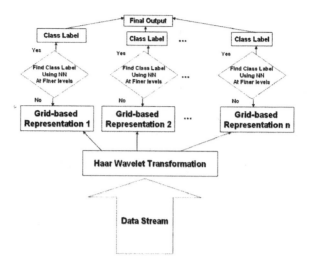

Figure 3.6.   ANNCAD Framework

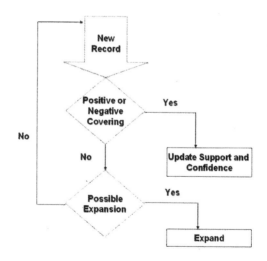

Figure 3.7.   SCALLOP Process

[2] Aggarwal C., Han J., Wang J., Yu P. S., (2003) A Framework for Clustering Evolving Data Streams, *Proc. 2003 Int. Conf. on Very Large Data Bases (VLDB '03), Berlin, Germany, Sept. 2003.*

[3] Aggarwal C., Han J., Wang J., Yu P. S., (2004) On Demand Classification of Data Streams, *Proc. 2004 Int. Conf. on Knowledge Discovery and Data Mining (KDD '04), Seattle, WA.*

[4] Babcock B., Babu S., Datar M., Motwani R., and Widom J. (2002) Models and issues in data stream systems. In *Proceedings of PODS.*

[5] Babcock B., Datar M., and Motwani R. (2003) Load Shedding Techniques for Data Stream Systems (short paper) In *Proc. of the 2003 Workshop on Management and Processing of Data Streams (MPDS 2003).*

[6] Burl M., Fowlkes C., Roden J., Stechert A., and Mukhtar S. (1999), Diamond Eye: A distributed architecture for image data mining, in *SPIE DMKD, Orlando.*

[7] Cai Y. D., Clutter D., Pape G., Han J., Welge M., Auvil L. (2004) MAIDS: Mining Alarming Incidents from Data Streams. *Proceedings of the 23rd ACM SIGMOD (International Conference on Management of Data).*

[8] Ding Q., Ding Q, and Perrizo W., (2002) Decision Tree Classification of Spatial Data Streams Using Peano Count Trees, *Proceedings of the ACM 124 Symposium on Applied Computing,* Madrid, Spain, pp. 413–417.

[9] Domingos P. and Hulten G. (2000) Mining High-Speed Data Streams. In *Proceedings of the Association for Computing Machinery Sixth International Conference on Knowledge Discovery and Data Mining.*

[10] Dong G., Han J., Lakshmanan L. V. S., Pei J., Wang H. and Yu P. S. (2003) Online mining of changes from data streams: Research problems and preliminary results, In *Proceedings of the 2003 ACM SIGMOD Workshop on Management and Processing of Data Streams.*

[11] Fan W. (2004) Systematic data selection to mine concept-drifting data streams. *ACM KDD Conference,* pp. 128-137.

[12] Ferrer-Troyano F. J., Aguilar-Ruiz J. S. and Riquelme J. C. (2004) Discovering Decision Rules from Numerical Data Streams, *ACM Symposium on Applied Computing,* pp. 649-653.

[13] Gaber, M, M., Zaslavsky, A., and Krishnaswamy, S. (2005) Mining Data Streams: A Review. *ACM SIGMOD Record,* Vol. 34, No. 1, June 2005, ISSN: 0163-5808.

[14] Gaber, M, M., Krishnaswamy, S., and Zaslavsky, A., (2005). On-board Mining of Data Streams in Sensor Networks, Accepted as a chapter in the forthcoming book *Advanced Methods of Knowledge Discovery from Complex Data,* (Eds.) Sanghamitra Badhyopadhyay, Ujjwal Maulik, Lawrence Holder and Diane Cook, Springer Verlag, to appear.

[15] Gama J., Rocha R. and Medas P. (2003), Accurate Decision Trees for Mining High-Speed Data Streams, *Proceedings of the Ninth International Conference on Knowledge Discovery and Data Mining.*

[16] Garofalakis M., Gehrke J., Rastogi R. (2002) Querying and mining data streams: you only get one look a tutorial. SIGMOD Conference, 635.

[17] Golab L. and Ozsu T. M. (2003) Issues in Data Stream Management. In *SIGMOD Record*, Volume 32, Number 2, pp. 5–14.

[18] Hand D. J. (1999) Statistics and Data Mining: Intersecting Disciplines *ACM SIGKDD Explorations*, 1, 1, pp. 16-19.

[19] Hand D.J., Mannila H., and Smyth P. (2001) *Principles of data mining*, MIT Press.

[20] Hastie T., Tibshirani R., Friedman J. (2001) *The elements of statistical learning: data mining, inference, and prediction*, New York: Springer.

[21] Henzinger M., Raghavan P. and Rajagopalan S. (1998), Computing on data streams , Technical Note 1998-011, Digital Systems Research Center, Palo Alto, CA.

[22] Hulten G., Spencer L., and Domingos P. (2001) Mining Time-Changing Data Streams. *ACM SIGKDD Conference.*

[23] Jin R. and Agrawal G. (2003), Efficient Decision Tree Construction on Streaming Data, in *Proceedings of ACM SIGKDD Conference.*

[24] Kargupta, H., Park, B., Pittie, S., Liu, L., Kushraj, D. and Sarkar, K. (2002). MobiMine: Monitoring the Stock Market from a PDA. *ACM SIGKDD Explorations*, Volume 3, Issue 2. Pages 37–46. ACM Press.

[25] Kargupta H., Bhargava R., Liu K., Powers M., Blair S., Bushra S., Dull J., Sarkar K., Klein M., Vasa M., and Handy D. (2004) VEDAS: A Mobile and Distributed Data Stream Mining System for Real-Time Vehicle Monitoring. *Proceedings of SIAM International Conference on Data Mining.*

[26] Last M. (2002) Online Classification of Nonstationary Data Streams,*Intelligent Data Analysis*, Vol. 6, No. 2, pp. 129-147.

[27] Law Y., Zaniolo C. (2005) An Adaptive Nearest Neighbor Classification Algorithm for Data Streams, *Proceedings of the 9th European Conference on the Principals and Practice of Knowledge Discovery in Databases*, Springer Verlag, Porto, Portugal.

[28] Muthukrishnan S. (2003) Data streams: algorithms and applications. *Proceedings of the fourteenth annual ACM-SIAM symposium on discrete algorithms.*

[29] Park B. and Kargupta H. (2002) Distributed Data Mining: Algorithms, Systems, and Applications. To be published in the Data Mining Handbook. Editor: Nong Ye.

[30] Wang H., Fan W., Yu P. and Han J. (2003) Mining Concept-Drifting Data Streams using Ensemble Classifiers, in the *9th ACM International Conference on Knowledge Discovery and Data Mining (SIGKDD), Washington DC, USA.*

[31] Wang K., Zhou S., Fu A., Yu J. (2003) Mining changes of classification by correspondence tracing. *SIAM International Conference on Data Mining.*

Chapter 4

FREQUENT PATTERN MINING IN DATA STREAMS

Ruoming Jin
Computer Science Department
Kent State University
jin@cs.kent.edu

Gagan Agrawal
Department of Computer Science and Engineering
The Ohio State University
agrawal@cse.ohio-state.edu

Abstract   Frequent pattern mining is a core data mining operation and has been extensively studied over the last decade. Recently, mining frequent patterns over data streams have attracted a lot of research interests. Compared with other streaming queries, frequent pattern mining poses great challenges due to high memory and computational costs, and accuracy requirement of the mining results.

In this chapter, we overview the state-of-art techniques to mine frequent patterns over data streams. We also introduce a new approach for this problem, which makes two major contributions. First, this one pass algorithm for frequent itemset mining has deterministic bounds on the accuracy, and does not require any out-of-core summary structure. Second, because the one pass algorithm does not produce any false negatives, it can be easily extended to a two pass accurate algorithm. The two pass algorithm is very memory efficient.

1.   Introduction

Frequent pattern mining focuses on discovering frequently occurring patterns from different types of datasets, including unstructured ones, such as transaction and text datasets, semi-structured ones, such as XML datasets, and structured ones, such as graph datasets. The patterns can be itemsets, sequences, subtrees, or subgraphs, etc., depending on the mining tasks and targeting datasets. Frequent patterns can not only effectively summarize the underlying datasets, providing key sights into the data, but also serve as the basic tool for many other data mining tasks, including association rule mining, classification, clustering, and change detection among others [21, 37, 20, 24].

Many efficient frequent pattern algorithms have been developed in the last decade [1, 17, 18, 35, 26, 33, 36]. These algorithms typically require datasets to be stored in persistent storage and involve two or more passes over the dataset. Recently, there has been much interest in data arriving in the form of continuous and infinite data streams. In a streaming environment, a mining algorithm must take only a single pass over the data [4]. Such algorithms can only guarantee an approximate result.

Compared with other stream processing tasks, the unique challenges in discovering frequent patterns are in three-fold. First, frequent pattern mining needs to search a space with an exponential number of patterns. The cardinality of the answering set itself which contains all frequent patterns can be very large too. In particular, it can cost much more space to generate an approximate answering set for frequent patterns in a streaming environment. Therefore, the mining algorithm needs to be very memory-efficient. Second, frequent pattern mining relies on the down-closure property to prune infrequent patterns and generate the frequent ones. This process (even without the streaming constraint) is very compute-intensive. Consequently, keeping up the pace with high- speed data streams can be very hard for a frequent pattern-mining task. Given these challenges, a more important issue is the quality of the approximate mining results. The more accurate results usually require more memory and computations. What should be the acceptable mining results to a data miner? To deal with this problem, a mining algorithm needs to provide users the flexibility to control the accuracy of the final mining results.

In the last several years, several new mining algorithms have been proposed to find frequent patterns over data streams. In the next chapter, we will overview these new algorithms.

2. Overview

2.1 Frequent Pattern Mining: Problem Definition

Let the dataset D be a collection of objects, i.e. $D = \{o_1, o_2, \cdots, o_{|D|}\}$. Let P be the set of all possible (interesting) patterns occurring in D, g be the counting function $g : P \times O \rightarrow N$, where O is the set of objects, and N is the set of nonnegative integers. Given parameters $p \in P$, and $o \in O$, $g(p, o)$ returns the number of times p occurs in o. The support of a pattern $p \in P$ in the dataset D is defined as

$$supp(p) = \sum_{j=0}^{j=|D|} I(g(p, o_j))$$

where, I is an *indicator* function: if $g(p, o_j) > 0$, $I(g(p, o_j)) = 1$; otherwise, $I(g(p, o_j)) = 0$. Given a support level θ, the frequent patterns of P in D is the set of patterns in P which have support greater than or equal to the θ.

The first and arguably the most important frequent pattern-mining task is *frequent itemsets mining*, proposed by Rakesh Agrawal *et.al.* in 1993 [2]. In this setting, the objects in the dataset D are transactions or sets of items. Let *Item* be the set of all possible items in the dataset D. Then the dataset D can be represented as $D = \{I_1, \cdots, I_{|D|}\}$, where $I_j \subseteq Item, \forall j, 1 \leq j \leq |D|$. The set of all possible patterns P is the power-set of *Item*. Note that the set of all possible objects O is the same as P in this setting. The counting function g is defined upon on the set containing ($\subseteq$) relationship. In other words, if the itemset p is contained in I_j ($p \subseteq I_j$), the function $g(p, I_j)$ returns 1; otherwise, it returns 0. For instance, given a dataset $D=\{\{A,B,D,E\}, \{B,C,E\},\{A,B,E\}, \{A,B,C\}, \{A,C\}, \{B,C\}\}$, and a support level $\theta = 50\%$, the frequent patterns are $\{A\}, \{B\}, \{C\}$, and $\{B, C\}$.

The majority of work in mining frequent patterns over data streams focuses on frequent itemsets mining. Many techniques developed in this setting can be served as a basis for mining other more complicated pattern mining tasks, such as graph mining [21]. To simplify the discussion, this chapter will focus on mining frequent itemsets over data streams.

2.2   Data Streams

In a data stream, transactions arrive continuously and the volume of transactions can be potentially infinite. Formally, a data stream D can be defined as a sequence of transactions, $D = (t_1, t_2, \cdots, t_i, \cdots)$, where t_i is the i-th arrived transaction. To process and mine data streams, different window models are often used. A window is a subsequence between i-th and j-th arrived transactions, denoted as $W[i, j] = (t_i, t_{i+1}, \cdots, t_j), i \leq j$. A user can ask different types of frequent pattern-mining questions over different type of window models.

Landmark window: In this model, we are interested in, from a starting timepoint i to the current timepoint t, what are the frequent itemsets. In other words, we are trying to find the frequent itemsets over the window $W[i, t]$. A special case of the landmark window is when $i = 1$. In this case, we are interested in the frequent itemsets over the entire data stream. Clearly, the difficulty in solving the special case is essentially the same as the more general cases, and all of them require an efficient single-pass mining algorithm. For simplicity, we will focus on the case where the *Entire Data Stream* is the target.

Note that in this model, we treat each time-point after the starting point equally important. However, in many cases, we are more interested in the *recent* time-points. The following two models focus on such cases:

Sliding window: Given the length of the sliding window w and current time-point t, we are interested in the frequent pattern time in the window $W[t - w + 1, t]$. As time changes, the window will keep its size and move along with the current time point. In this model, we are not interested in the data which arrived before the timepoint $t - w + 1$.

Damped window model: This model assigns more weights to the recently arrived transactions. A simple way to do that is to define a *decay rate* [7], and use this rate to update the previously arrived transactions (by multiplication) as a new transaction arrives. Correspondingly, the count of an itemset is also defined based on the weight of each transaction.

In the next subsection, we will overview the algorithms for mining frequent itemsets on these three different window models over data streams. In addition, we would like to point out that besides the three windows we introduced above, Jiawei Han *et. al.* proposed another model called *tilted-time window* model. In this model, we are interested in frequent itemsets over a set of windows. Each window corresponds to different time granularity based on their recency. For example, we are interested in every minute for the last hour, every five minutes for the previous hour, every ten minutes for the hour before that. Moreover, the transactions inside each window are also weighted. Such model can allow us to pose more complicated queries over data stream. Giannella *et. al.* have developed a variant of FP-tree, called FP-stream, for dynamically updating frequent patterns on streaming data and answering the approximate frequent itemsets for even arbitrary time intervals [15].

2.3   Mining Algorithms

	All	Closed
Entire Data Stream	Lossy Counting [28] FPDM [34]	
Sliding Window		Moment [11]
Damped Window	estDec [7]	

Closed: Closed frequent itemsets

Table 4.1.   Algorithms for Frequent Itemsets Mining over Data Streams

Table 4.1 lists the algorithms which have been proposed for mining frequent itemsets in the last several years. Note that *All* suggests to find all of the frequent itemsets given support level θ. Closed frequent itemsets are itemsets that are frequent but have higher frequency than all of their supersets. (If an itemset p

is a subset of itemset q, q is called the superset of p.) In the following, we will briefly introduce these algorithms and their basic ideas.

Mining Algorithms for Entire Data Stream Manku and Motwani proposed the first one-pass algorithm, Lossy Counting, to find all frequent itemsets over a data stream [28]. Their algorithm is *false-positive oriented* in the sense that it does not allow false negatives, and has a provable bound on false positives. It uses a user-defined error parameter ϵ to control the quality of the answering set for a given support level θ. More precisely, its answering set is guaranteed to have all itemsets whose frequency exceeds θ, and contains no itemsets whose true frequency is less than $\theta - \epsilon$. In other words, the itemsets whose frequency are between $\theta - \epsilon$ and θ possibly appear in the answering set, and are the false positives.

Recently, Yu and his colleagues proposed FPDM, which is a false-negative oriented approach, for mining frequent itemsets over data streams [34]. Their algorithm does not allow false positive, and has a high-probability to find itemsets which are truly frequent. In particular, they use a user-defined parameter δ to control the probability to find the frequent itemsets at support level θ. Specifically, the answering set does not include any itemsets whose frequency is less than θ, and include any itemsets whose frequency exceeds θ with probability of at least $1 - \delta$. It utilizes the Chernoff bound to achieve such quality control for the answering set.

Both algorithms logically partitioned the data stream into equally sized segments, and find the potentially frequent itemsets for each segment. They aggregate these locally frequent itemsets and further prune the infrequent ones. However, the number of transactions in each segment as well as the method to define potentially frequent itemsets is different for these two methods. In Lossy Counting, the number of transactions in a segment is $k \times \lceil 1/\epsilon \rceil$, and an itemset which occurs more than k times in a segment is potentially frequent. In FDPM, the number of transactions in a segment is $k \times n_0$, where n_0 is the required number of observations in order to achieve Chernoff bound with the user-defined parameter δ. In this case, an itemset whose frequency exceeds $\theta - \epsilon$, where ϵ is computed by Chernoff bound in terms of δ and the number of observations ($k \times n_0$). Note that k is a parameter (batch size) to control the size of each segment.

To theoretically estimate the space requirement for both algorithms, we consider each transaction including only a single item, and the number of transactions in the entire data stream is $|D|$. Lossy Counting will take $O(1/\epsilon log(\epsilon |D|))$ to find frequent items (1-itemsets), and FPDM-1 (the simple version of FPDM on finding frequent items) will need $O((2 + 2ln(2/\delta))/\theta)$.

Note that different approaches have different advantages and disadvantages. For instance, for the false positive approach, if a second pass is allowed, we can easily eliminate false positives. For the false negative approach, we can have a small answering set which have almost all the frequent itemsets, but might miss some of them (with very small probability controlled by δ).

Sliding Window Chi *et al.* have studied the problem on mining closed frequent itemsets over a sliding window of a data stream [12]. In particular, they assume the width of sliding window is not very large, therefore, the transactions of each sliding window could be held in the main memory. Clear, such assumption is very close to the problem setting of the *incremental* association rule mining [10]. But their focus is on how to maintain the closed frequent itemsets in an efficient way.

To deal with this problem, they proposed a new mining algorithm, called MOMENT. It utilizes the heuristic that in most cases, the sets of frequent itemsets are relatively stable for the consecutive sliding windows in the data stream. Specifically, such stability can be expressed as the fact that the boundary between the frequent itemsets and infrequent itemsets, and the boundary between closed frequent itemsets and the rest of itemsets move very slowly. Therefore, instead of generating all closed frequent itemsets for each window, they focus on monitoring such boundaries. As the key of this algorithm, an in-memory data structure, the *closed enumeration tree* (CET), is developed to efficiently monitor closed frequent itemsets as well as itemsets that form the boundary between the closed frequent itemsets and the rest of the itemsets. An efficient mechanism has been proposed to update the CET as the sliding window moves so that the boundary maintains for each sliding window.

Damped Window Model Chang and Lee studied the problem to find recently frequent itemsets over data streams using the damped window model. Specifically, in their model, the weight for an existing transaction in the data stream reduces by a decay factor, d, as a new transaction arrives. For example, the initial weight of a newly arrived transaction has weight 1, and after another transaction arrives, it will be reduced as $d = (1 \times d)$.

To keep tracking down the frequent itemsets in such a setting, they propose a new algorithm, *estDec*, which process the transaction one by one. It maintains a lattice for recording the potentially frequent itemsets and their counts, and updates the lattice for each new transaction correspondingly. Note that theoretically, the count of each itemset in the lattice will change as a new transaction arrives. But by recording an additional information for each itemset p, the time-point of the most recent transaction contains p, the algorithm only needs to update the counts for the itemsets which are the subsets of newly arrived transaction. It will reduce their counts using the constant factor d, and then increases them by one. Further, it inserts the subsets of the current transaction which are potentially frequent into the lattice. It uses a method similar to Carma [19] to estimate the frequency of these newly inserted itemsets.

Discussion Among the above three different problem settings, we can see that the first one, finding the frequent itemsets over the entire data stream, is the most challenging and fundamentally important one. It can often serve as the basis to solve the latter two. For example, the current sliding window model studied in MOMENT is very similar to the incremental data mining. A more difficult problem is the case when the data in a sliding window cannot be held in the main memory. Clearly, in such case, we need single-pass algorithms for even for a single sliding window. The main difference between the damped window model and entire data stream is that the counts of itemsets need to be adjusted for each new arrival transactions in the damped window model even the itemsets do not appear in these transactions.

In the next Section, we will introduce a new mining algorithm *StreamMining* we proposed recently to find frequent itemsets over the entire data stream. It is a false-positive approach (similar to Lossy Counting). It has provable (user-defined) deterministic bounds on accuracy and very memory efficient. In Section 4, we will review research works closely related with the field on frequent pattern mining over data streams. Finally, we will conclude this chapter and discuss directions for future work (Section 5).

3. New Algorithm

This section describes our new algorithm for mining frequent itemsets in a stream. Initially, we discuss a new approach for finding frequent items from Karp *et al.* [25]. We then discuss the challenges in extending this idea to frequent itemset mining, and finally outline our ideas for addressing these issues.

3.1 KPS's algorithm

Our work is derived from the recent work by Karp, Papadimitriou and Shenker on finding frequent elements (or 1-itemset) [25]. Formally, given a sequence of length N and a threshold θ $(0 < \theta < 1)$, the goal of their work is to determine the elements that occur with frequency greater than $N\theta$.

A trivial algorithm for this will involve counting the frequency of all distinct elements, and checking if any of them has the desired frequency. If there are n distinct elements, this will require $O(n)$ memory.

Their approach requires only $O(1/\theta)$ memory. Their approach can be viewed as a generalization of the following simple algorithm for finding the *majority element* in a sequence. A majority element is an element that appears more than half the time in an entire sequence. We find two distinct elements and eliminate them from the sequence. We repeat this process until only one distinct element remains in the sequence. If a majority element exists in the sequence, it will be left after this elimination. At the same time, any element remaining in the sequence is not necessarily the majority element. We can take another pass over the original sequence and check if the frequency of the remaining element is greater than $N/2$.

FindingFrequentItems(Sequence $\mathcal{S}$, θ)
  global *Set* $\mathcal{P}$; // *Set of Potentially*
  $\mathcal{P} \leftarrow \emptyset$;   // *Frequent Items*
  foreach $(s \in \mathcal{S})$ // *each item in* $\mathcal{S}$
    if $s \in \mathcal{P}$
        $s.count + +$;
    else
        $\mathcal{P} \leftarrow \{s\} \cup \mathcal{P}$;
        $s.count = 1$;
        if $|\mathcal{P}| \geq \lceil 1/\theta \rceil$
            foreach $(p \in \mathcal{P})$
                $p.count - -$;
                if $p.count = 0$
                    $\mathcal{P} \leftarrow \mathcal{P} - \{p\}$;
  $Output(\mathcal{P})$;

Figure 4.1.   Karp *et al.* Algorithm to Find Frequent Items

The idea can be generalized to an arbitrary θ. We can proceed as follows. We pick any $1/\theta$ distinct elements in the sequence and eliminate them together. This can be repeated until no more than $1/\theta$ distinct elements remain in the sequence. It can be claimed that any element appearing more than $N\theta$ times will be left in the sequence. The reason is that the elimination can only be performed at most $N/(1/\theta) = N\theta$ times. During each such elimination, any distinct element is removed at most once. Hence, for each distinct element, the total number of eliminations during the entire process is at most $N\theta$. Any element appearing more than $N\theta$ times will remain in the sequence. Note, however, the elements left in the sequence do not necessarily appear with frequency greater than $N\theta$. Thus, this approach will provide a superset of the elements which occur more than $N\theta$ times.

Such processing can be performed to take only a single pass on the sequence, as we show in Figure 4.1. $\mathcal{P}$ is the set of potentially frequent items. We maintain a *count* for each item in the set $\mathcal{P}$. This set is initially empty. As we process a new item from a sequence, we check if it is in the set $\mathcal{P}$. If yes, its count is incremented, otherwise, it is inserted with a count of 1. When the size of the set $\mathcal{P}$ becomes larger than $\lceil 1/\theta \rceil$, we decrement the count of each item in $\mathcal{P}$, and eliminate any item whose count has now become 0. This processing is equivalent to the eliminations we described earlier. Note that this algorithm requires only $\Omega(1/\theta)$ space. It computes a superset of frequent items. To find the precise set of frequent items, another pass can be taken on the sequence, and the frequency of all remaining elements can be counted.

3.2   Issues In Frequent Itemset Mining

In this paper, we build a frequent itemset mining algorithm using the above basic idea. There are three main challenges when we apply this idea to mining frequent itemsets, which we summarize below.

1 *Dealing with Transaction Sequences:* The algorithm from Karp *et al.* assumes that a sequence is comprised of elements, i.e., each transaction in the sequence only contains one-items. In frequent itemset mining, each transaction has a number of items, and the length of every transaction can also be different.

2 *Dealing with k-itemsets:* Karp *et al.*'s algorithm only finds the frequent items, or 1-itemsets. In a frequent itemset mining algorithm, we need to find all k-itemsets, $k \geq 1$, in a single pass.

   Note that their algorithm can be directly extended to find i-itemsets in the case where each transaction has a fixed length, l. This can be done by eliminating a group of $(1/\theta) \times \binom{l}{i}$ different i-itemsets together. This, however, requires $\Omega((1/\theta) \times \binom{l}{i})$ space, which becomes extremely high when l and i are large. Furthermore, in our problem, we have to find all i-itemsets, $i \geq 1$, in a single pass.

3 *Providing an Accuracy Bound:* Karp *et al.*'s algorithm can provably find a superset of the frequent items. However, no accuracy bound is provided for the item(set)s in the superset, which we call the potential frequent item(set)s. For example, even if an item appears just a single time, it can still possibly appear in the superset reported by the algorithm. In frequent itemset mining, we will like to improve above result, and provide a bound on the frequency of the itemsets that are reported by the algorithm.

3.3   Key Ideas

We now outline how we can address the three challenges we listed above. **Dealing with k-itemsets in a Stream of Transactions:** Compared with the problem of finding frequent items, the challenges in finding frequent itemsets from a transaction sequence mainly arise due to the large number of potential frequent itemsets. This also results in high memory costs. As we stated previously, a direct application of the idea from Karp *et al.* will require $\Omega((1/\theta) \times \binom{l}{i})$ space to find potential frequent i-itemsets, where l is the length of each transaction. This approach is prohibitively expensive when l and i are large, but can be feasible when i is small, such as 2 or 3.

Recall that most of the existing work on frequent itemset mining uses the *Apriori* property [1], i.e., an *i*-itemset can be frequent only if all subsets of this itemset are frequent. One of the drawbacks of this approach has been the large number of 2-itemsets, especially when the number of distinct items is large, and θ is small.

Our idea is to use a *hybrid* approach to mine frequent itemsets from a transaction stream. We use the idea from Karp *et al.* to determine the potential frequent 2-itemsets. Then, we use the set of potential frequent 2-itemsets and the Apriori property to generate the potential i-itemsets, for $i > 2$. This approach finds a set of potential frequent itemsets, which is guaranteed to contain all the *true* frequent itemsets, in a single pass of the stream.

Also, if a second pass of the data stream is allowed, we can eliminate all the *false* frequent itemsets from our result set. The second pass is very easy to implement, and in the rest of our discussion, we will only focus on the first pass of our algorithm.

Bounding False Positives: In order to have a accuracy bound, we propose the following criteria for the reported potential frequent itemsets after the first pass. Besides reporting all items or itemsets that occur with frequency more than $N\theta$, we want to report only the items or itemsets which appear with frequency at least $N\theta(1 - \epsilon)$, where $0 < \epsilon \leq 1$. This criteria is similar to the one proposed by Manku and Motwani [28].

We can achieve this goal by modifying the algorithm as shown in Figure 4.2. In the *first* step, we invoke the algorithm from Karp *et al.* with the frequency level $\theta\epsilon$. This will report a superset of items occurring with frequency more than $N\theta\epsilon$. We also record the number of eliminations, c, that occur in this step. Clearly, c is bounded by $N\theta\epsilon$. In the *second* step, we remove all items whose reported frequency is less than $N\theta - c \geq N\theta(1 - \epsilon)$.

We have two claims about the above process: 1) it reports all items that occur with frequency more than $N\theta$, and 2) it only reports items which appear with frequency more than $N\theta(1 - \epsilon)$. The reason for this is as follows. Consider any element that appears with frequency $N\theta$. After the first step, it will be reported in the superset with a frequency greater than $c, c \leq N\theta\epsilon$. Therefore, it will remain in the set after the second step also. Similarly, consider any item that appears with frequency less than $N\theta(1 - \epsilon)$. If this item is present in the superset reported after the first step, it will be removed during the second step since $N\theta - c \geq N\theta(1 - \epsilon)$. This idea can be used for frequent itemset mining also.

In the next Section, we introduce our algorithm for mining frequent itemsets from streaming data based on the above two ideas.

3.4   Algorithm Overview

We now introduce our new algorithm in three steps. In Subsection 3.5, we describe an algorithm for mining frequent itemsets from a data stream, which assumes that each transaction has the same length. In Subsection 3.6, we extend

FindingFrequentItemsBounded(Sequence $\mathcal{S}$, θ, ϵ)
 global *Set $\mathcal{P}$*;
 $\mathcal{P} \leftarrow \emptyset$;
 $c \leftarrow 0;$ *// Number of Elimination*
 foreach $(s \in \mathcal{S})$
   if $s \in \mathcal{P}$
     $s.count + +$;
   else
     $\mathcal{P} \leftarrow \{s\} \cup \mathcal{P}$;
     $s.count = 1$;
     if $|\mathcal{P}| \geq \lceil 1/(\theta\epsilon) \rceil$
       $c + +;$ *// Count Eliminations*
       foreach $(p \in \mathcal{P})$
         $p.count - -$;
         if $p.count = 0$
           $\mathcal{P} \leftarrow \mathcal{P} - \{p\}$;
 foreach $(p \in \mathcal{P})$
   if $p.count \leq (N\theta - c)$
     $\mathcal{P} \leftarrow \mathcal{P} - \{p\}$;
 $Output(\mathcal{P})$;

Figure 4.2. Improving Algorithm with An Accuracy Bound

this algorithm to provide an accuracy bound on the potential frequent itemsets computed after one pass. In Subsection 3.7, we further extend the algorithm to deal with transactions of variable length.

Before detailing each algorithm, we first introduce some terminology. We are mining a stream of transactions $\mathcal{D}$. Each transaction t in this stream comprises a set of *items*, and has the length $|t|$. Let the number of transactions in $\mathcal{D}$ be $|\mathcal{D}|$. Each algorithm takes the support level θ as one parameter. An itemset in $\mathcal{D}$ to be considered frequent should occur more than $\theta|\mathcal{D}|$ times.

To store and manipulate the candidate frequent itemsets during any stage of every algorithm, a lattice $\mathcal{L}$ is maintained.

$$\mathcal{L} = \mathcal{L}_1 \cup \mathcal{L}_2 \cup \ldots \cup \mathcal{L}_k$$

where, k is largest frequent itemset, and $\mathcal{L}_i, 1 \leq i \leq k$ comprises the potential frequent i-itemsets. Note that in mining frequent itemsets, the size of the set $\mathcal{L}_1$, which is bound by the number of distinct items in the dataset, is typically not very large. Therefore, in order to simplify our discussion, we will not consider $\mathcal{L}_1$ in the following algorithms, and assume we can find the exact frequent 1-itemsets in the stream $\mathcal{D}$. Also, we will directly extend the idea from Karp *et al.* to find the potential frequent 2-itemsets.

As we stated in the previous section, we deal with all k-itemsets, $k > 2$, using the Apriori property. To facilitate this, we keep a buffer $\mathcal{T}$ in each algorithm to store the recently received transactions. The buffer will be accessed several times to find the potential frequent k-itemsets, $k > 2$.

3.5 Mining Frequent Itemsets from Fixed Length Transactions

The algorithm we present here mines frequent itemsets from a stream, under the assumption that each transaction has the same length $|t|$. The algorithm has two interleaved phases. The *first* phase deals with 2-itemsets, and the *second* phase deals with k-itemsets, $k > 2$. The main algorithm and the associated subroutines are shown in Figures 4.3 and 4.4, respectively. Note that the two subroutines, *Update* and *ReducFreq*, are used by all the algorithms discussed in this section.

The first phase extends the Karp *et al.*'s algorithm to deal with 2-itemsets. As we stated previously, the algorithm maintains a buffer $\mathcal{T}$ which stores the recently received transactions. Initially, the buffer is empty. When a new transaction t arrives, we put it in $\mathcal{T}$. Next, we call the *Update* routine to increment counts in $\mathcal{L}_2$. This routine simply updates the count of 2-itemsets that are already in $\mathcal{L}_2$. Other 2-itemsets that are in the transaction t are inserted in the sets $\mathcal{L}_2$.

StreamMining-Fixed(Stream $\mathcal{D}$, θ)
  global *Lattice* $\mathcal{L}$;
  local *Buffer* $\mathcal{T}$;
  local *Transaction* t;
  $\mathcal{L} \leftarrow \emptyset$; $\mathcal{T} \leftarrow \emptyset$;
  $f \leftarrow |t| * (|t| - 1)/2$;
  foreach $(t \in \mathcal{D})$
      $\mathcal{T} \leftarrow \mathcal{T} \cup \{t\}$;
      $Update(t, \mathcal{L}, 2)$;
      if $|\mathcal{L}_2| \geq \lceil 1/\theta \rceil \cdot f$
          $ReducFreq(\mathcal{L}, 2)$;
          $\{* \ Deal \ with \ k - itemsets, \ k > 2 *\}$
          $i \leftarrow 2$;
          while $\mathcal{L}_i \neq \emptyset$
              $i + +$;
              foreach $(t \in \mathcal{T})$
                  $Update(t, \mathcal{L}, i)$;
              $ReducFreq(\mathcal{L}, i)$;
          $\mathcal{T} \leftarrow \emptyset$;
  $Output(\mathcal{L})$;

Figure 4.3. StreamMining-Fixed: Algorithm Assuming Fixed Length Transactions

Update(Transaction t, Lattice $\mathcal{L}$, i)
  for *all* i *subsets* s *of* t
      if $s \in \mathcal{L}_i$
          $s.count + +$;
      else if $i \leq 2$
          $\mathcal{L}_i.insert(s)$;
      else if *all* $i - 1$ *subsets of* $s \in L_{i-1}$
          $\mathcal{L}_i.insert(s)$;

ReducFreq(Lattice $\mathcal{L}$, i)
  foreach i *itemsets* $s \in \mathcal{L}_i$
      $s.count - -$;
      if $s.count = 0$
          $\mathcal{L}_i.delete(s)$;

Figure 4.4. Subroutines Description

When the size of $\mathcal{L}_2$ is beyond the *threshold*, $\lceil 1/\theta \rceil f$, where f is the number of 2-itemsets per transaction, we call the procedure *ReducFreq* to reduce the count of each 2-itemsets in $\mathcal{L}_2$, and the itemsets whose count becomes zero are deleted. Invoking *ReducFreq* on $\mathcal{L}_2$ triggers the *second* phase.

The second phase of the algorithm deals with all k-itemsets, $k > 2$. This process is carried out level-wise, i.e, it proceeds from 3-itemsets to the largest potential frequent itemsets. For each transaction in the buffer $\mathcal{T}$, we enumerate all i-subsets. For any i-subset that is already in $\mathcal{L}$, the process will be the same as for a 2-itemset, i.e, we will simply increment the count. However, an i-subset that is not in $\mathcal{L}$ will be inserted in $\mathcal{L}$ only if all of its $i - 1$ subsets are in $\mathcal{L}$ as well. Thus, we use the *Apriori* property.

After updating i-itemsets in $\mathcal{L}$, we will invoke the *ReducFreq* routine. Thus, the itemsets whose count is only 1 will be deleted from the lattice. This procedure will continue until there are no frequent k-itemsets in $\mathcal{L}$. At the end of this, we clear the buffer, and start processing new transactions in the stream. This will restart the first phase of our algorithm to deal with 2-itemsets.

We next discuss the correctness and the memory costs of our algorithm. Let $\mathcal{L}_i^\theta$ be the set of frequent i-itemsets with support level θ in $\mathcal{D}$, and $\mathcal{L}_i$ be the set of potential frequent i-itemsets provided by this algorithm.

THEOREM 1 *In using the algorithm StreamMining-Fixed on a set of transactions with a fixed length, for any $k \geq 2$, $\mathcal{L}_k^\theta \subseteq \mathcal{L}_k$.*

LEMMA 4.1 *In using the algorithm StreamMining-Fixed on a set of transactions with a fixed length, the size of $\mathcal{L}_2$ is bounded by $(\lceil 1/\theta \rceil + 1)\binom{|t|}{2}$.*

The proofs for the Theorem 1 and the Lemma 4.1 are available in a technical report [23]. Theorem 1 implies that any frequent k-itemset is guaranteed to be in the output of our algorithm. Lemma 4.1 provides an estimate of the memory costs for $\mathcal{L}_2$.

3.6   Providing an Accuracy Bound

We now extend the algorithm from the previous subsection to provide a bound on the accuracy of the reported results. As described in Subsection 3.3, the bound is described by an user-defined parameter, ϵ, where $0 < \epsilon \leq 1$. Based on this parameter, the algorithm ensures that the frequent itemsets reported do occur more than $(1 - \epsilon)\theta|\mathcal{D}|$ times in the dataset.

The basic idea for achieving such a bound on frequent items computation was illustrated in Figure 4.2. We can extend this idea to finding frequent itemsets. Our new algorithm is described in Figure 4.5. Note that we still assume that each transaction has the same length.

This algorithm provides the new bound on accuracy in two steps. In the *first* step, we invoke the algorithm in Figure 4.3 with the frequency level $\theta\epsilon$. This

StreamMining-Bounded(Stream $\mathcal{D}$, θ, ϵ)
  global $Lattice$ $\mathcal{L}$;
  local $Buffer$ $\mathcal{T}$;
  local $Transaction$ t;
  $\mathcal{L} \leftarrow \emptyset$; $\mathcal{T} \leftarrow \emptyset$;
  $f \leftarrow |t| * (|t| - 1)/2$;
  $c \leftarrow 0$; // $Number$ of $ReducFreq$ $Invocations$
  foreach $(t \in \mathcal{D})$
      $\mathcal{T} \leftarrow \mathcal{T} \cup \{t\}$;
      $Update(t, \mathcal{L}, 1)$;
      $Update(t, \mathcal{L}, 2)$;
      if $|\mathcal{L}_2| \geq \lceil 1/\theta\epsilon \rceil \cdot f$
          $ReducFreq(\mathcal{L}, 2)$;
          $c + +$;
          $i \leftarrow 2$;
          while $\mathcal{L}_i \neq \emptyset$
              $i + +$;
              foreach $(t \in \mathcal{T})$
                  $Update(t, \mathcal{L}, i)$;
              $ReducFreq(\mathcal{L}, i)$;
          $\mathcal{T} \leftarrow \emptyset$;
  foreach $s \in \mathcal{L}$
      if $s.count \leq \theta|D| - c$
          $\mathcal{L}_i.delete(s)$;
  $Output(\mathcal{L})$;

Figure 4.5. StreamMining-Bounded: Algorithm with a Bound on Accuracy

will report a superset of itemsets occurring with frequency more than $N\theta\epsilon$. We record the number of invocations of *ReducFreq*, c, in the first step. Clearly, c is bounded by $N\theta\epsilon$. In the *second* step, we remove all items whose reported frequency is less than $N\theta - c \geq N\theta(1-\epsilon)$. This is achieved by the last *foreach* loop.

The new algorithm has the following property: 1) if an itemset has frequency more than θ, it will be reported. 2) if an itemset is reported as a potential frequent itemset, it must have a frequency more than $\theta(1-\epsilon)$. Theorem 2 formally states this property, and its proof is available in a technical report [23].

THEOREM 2 *In using the algorithm StreamMining-Bounded on a set of transactions with a fixed length, for any $k \geq 2$, $\mathcal{L}_k^\theta \subseteq \mathcal{L}_k \subseteq \mathcal{L}_k^{(1-\epsilon)\theta}$.*

Note that the number of invocations of *ReducFreq*, c, is usually much smaller than $N\theta\epsilon$ after processing a data stream. Therefore, an interesting property of this approach is that it produces a very small number of false frequent itemsets, even with relatively large ϵ. The experiments in [22] also support this observation.

The following lemma claims that the memory cost of $\mathcal{L}_2$ is increased by a factor proportional to $1/\epsilon$.

LEMMA 4.2 *In using the algorithm StreamMining-Bounded on a set of transactions with a fixed length, the size of $\mathcal{L}_2$ is bounded by $(\lceil 1/\theta\epsilon \rceil + 1)\binom{|t|}{2})$.*

3.7   Dealing with Variable Length Transactions

In this subsection, we present our final algorithm, which improves upon the algorithm from the previous subsection by dealing with variable length transactions. The algorithm is referred to as *StreamMining* and is illustrated in Figure 4.6.

When each transaction has a different length, the number of 2-itemsets in each transaction also becomes different. Therefore, we cannot simply maintain f, the number of 2-itemsets per transaction, as a constant. Instead, we maintain f as a weighted average of the number of 2-itemsets that each transaction processed so far. This weighted average is computed by giving higher weightage to the recent transactions. The details are shown in the pseudo-code for the routine *TwoItemsetPerTransaction*.

To motivate the need for taking such a weighted average, consider the natural alternative, which will be maintaining f as the average number of 2-itemsets that each transaction seen so far has. This will not work correctly. For example, suppose there are 3 transactions, which have the length 2, 2, and 3, respectively, and θ is 0.5. The first two transactions will have a total of two 2-itemsets, and the third one has 6 2-itemsets. We will preform an elimination when the number of different 2-itemsets is larger than or equal to $(1/\theta) \times f$. When the first two

StreamMining(Stream $\mathcal{D}, \theta, \epsilon$)
  global *Lattice* $\mathcal{L}$;
  local *Buffer* $\mathcal{T}$;
  local *Transaction* t;
  $\mathcal{L} \leftarrow \emptyset; \mathcal{T} \leftarrow \emptyset;$
  $f \leftarrow 0; // \ Average \ 2 - itemset \ per \ transaction$
  $c \leftarrow 0;$
  foreach $(t \in \mathcal{D})$
      $\mathcal{T} \leftarrow \mathcal{T} \cup \{t\};$
      $Update(t, \mathcal{L}, 1);$
      $Update(t, \mathcal{L}, 2);$
      $f \leftarrow TwoItemsetPerTransaction(t);$
      if $|\mathcal{L}_2| \geq \lceil 1/\theta\epsilon \rceil \cdot f$
          $ReducFreq(\mathcal{L}, 2);$
          $c + +;$
          $i \leftarrow 2;$
          while $\mathcal{L}_i \neq \emptyset$
              $i + +;$
              foreach $(t \in \mathcal{T})$
                  $Update(t, \mathcal{L}, i);$
              $ReducFreq(\mathcal{L}, i);$
        $\mathcal{T} \leftarrow \emptyset;$
  foreach $s \in \mathcal{L}$
      if $s.count \leq \theta|D| - c$
          $\mathcal{L}_i.delete(s);$
  $Output(\mathcal{L});$

TwoItemsetPerTransaction(Transaction t)
  global $X; // \ Number \ of \ 2 \ Itemset$
  global $N; // \ Number \ of \ Transactions$
  local f;
  $N + +;$
  $X \leftarrow X + \begin{pmatrix} |t| \\ 2 \end{pmatrix};$
  $f \leftarrow \lceil X/N \rceil;$
  if $|\mathcal{L}_2| \geq \lceil 1/\theta\epsilon \rceil \cdot f$
      $N \leftarrow N - \lceil 1/\theta\epsilon \rceil;$
      $X \leftarrow X - \lceil 1/\theta\epsilon \rceil \cdot f;$
  $return f;$

Figure 4.6. StreamMining: Final Algorithm

transactions arrive, an elimination will happen (assuming that the two 2-itemsets are different). When the third one arrives, the average number of 2-itemsets is less than 3, so another elimination will be performed. Unfortunately, a frequent 2-itemset that appears in both transactions 1 and 3 will be deleted in this way.

In our approach, the number of invocations of *ReducFreq*, c, is less than $|\mathcal{D}|(\theta\epsilon)$, where $|\mathcal{D}|$ is the number of transactions processed so far in the algorithm. Lemma 4.3 formalizes this, and its proof is available in a technical report [23].

LEMMA 4.3 $c < |\mathcal{D}|(\theta\epsilon)$ *is an invariant in the algorithm StreamMining.*

Note that by using the Lemma 4.3, we can deduce that the property of the Theorem 2 still holds for mining a stream of transaction with variable transaction lengths. Formally,

THEOREM 3 *In using the algorithm StreamMining on a stream of transactions with variable lengths, for any $k \geq 2$, $\mathcal{L}_k^\theta \subseteq \mathcal{L}_k \subseteq \mathcal{L}_k^{(1-\epsilon)\theta}$.*

An interesting property of our method is that in the situation where each transaction has the same length, our final algorithm, *StreamMining* will work in the same fashion as the algorithm previously shown in Figure 4.5.

Note, however, that unlike the case with fixed length transactions, the size of $\mathcal{L}_2$ cannot be bound by a closed formula. Also, in all the algorithms discussed in this section, the size of sets $\mathcal{L}_k$, $k > 2$ also cannot be bound in any way. Our algorithms use the Apriori property to reduce their sizes.

Finally, we point out that the new algorithm is very memory efficient. For example, Lossy Counting utilizes an out-of-core (disk-resident) data structure to maintain the potentially frequent itemsets. In comparison, we do not need any such structure. On the T10.I4.N10K dataset used in their paper, we see that with 1 million transactions and a support level of 1%, Lossy Counting requires an out-of-core data-structures on top of even a 44 MB buffer. For datasets ranging from 4 million to 20 million transactions, our algorithm only requires 2.5 MB main memory based summary. In addition, we believe that there are a number of advantages of an algorithm that does not require an out-of-core summary structure. Mining on streaming data may often be performed in mobile, hand-held, or sensor devices, where processors do not have attached disks. It is also well known that additional disk activity increases the power requirements, and battery life is an important issue in mobile, hand-held, or sensor devices. Also, while their algorithm is shown to be currently computation-bound, the disparity between processor speeds and disk speeds continues to grow rapidly. Thus, we can expect a clear advantage from an algorithm that does not require frequent disk accesses.

4. Work on Other Related Problems

In this section, we look at the work on problems that are closely related with the frequent pattern-mining problem defined in Section 2.

Scalable Frequent Pattern Mining Algorithms: A lot of research effort has been dedicated to make frequent pattern mining scalable on very large disk-resident datasets. These techniques usually focus on reducing the passes of the datasets. They typically try to find a superset or an approximate set of frequent itemsets in the first pass, and then find all the frequent itemsets as well as the counts in another one or few passes [29], [31], [19]. However, the first pass algorithms either do not have appropriate guarantees on the accuracy of frequent itemsets [31], or produces a large number of false positives [29, 19]. Therefore, they are not very suitable for the streaming environment.

Mining Frequent Items in Data Streams: Given a potentially infinite sequence of items, this work tries to identify the items which have higher frequencies than a given support level. Clearly, this problem can be viewed a simple version of frequent pattern mining over the entire data stream, and indeed most of the mining algorithms discussed in this chapter are derived from these work. Algorithms in mining frequent items in data streams use different techniques, such as random sketches [8, 13] or sampling [31, 28], and achieve the different space requirement. They also have either false-positive or false-negative properties. Interested user can look at [34] for more detailed comparison.

Finding Top-k Items in Distributed Data Streams: Assuming we have several distributed data streams and each item might carry different weights at each of its arrival, the problem is to find the k items which has the highest global weight. Olston and his colleagues have studied this problem, which they call *top-k monitoring queries* [5, 27]. Clearly, in order to maintain the global top-k items in a distributed streaming environment, frequent communication and synchronization is needed. Therefore, the focus of their research is on reducing the communication cost. They have proposed a method to achieve such goal by constraining each individual data streams with an arithmetic condition. The communication is only necessary when the arithmetic condition is violated.

Finding Heavy Hitters in Data Streams: Cormode *et. al.* studied the problem to efficiently identify *heavy hitters* in data streams [14]. It can be looked as an interesting variation of frequent-items mining problem. In this problem, there is a hierarchy among different items. Given a frequency level ϕ, the count of an item i in the hierarchy include all the items which are descendants of i, and whose counts are less than ϕ. An item whose counts exceeds ϕ is called Hierarchy Heavy Hitter (HHH), and we want to find all HHHs in a data stream. They have presented both deterministic and randomized algorithms to find HHHs.

Frequent Temporal Patterns over Data Streams: Considering a sliding window moves along a data stream, we monitor the counts for an itemset at each time point. Clearly, this sequence of counting information for a given itemset can be formulated as a time series. Inspired by such observation, Teng *et. al.* have developed an algorithm to find frequent patterns in sliding window model and collect sufficient statistics for a regression-based analysis of such time series [30]. They have showed such a framework is applicable in answering itemsets queries with flexible time-intervals, trend identification, and change detection.

Mining Semi-Structured Data Streams: Asai *et. al.* developed an efficient algorithm for mining frequent rooted ordered trees from a semi-structured data stream [3]. In this problem setting, they model a semi-structured dataset as a tree with infinite width but finite height. Traversing the tree with a left-most order generates the data stream. In other words, each item in the data stream is a node in the tree, and its arriving order is decided by the left-most traversing of the tree. They utilize the method in Carma [19] for candidate subtree generation.

5.   Conclusions and Future Directions

In this chapter, we gave an overview of the state-of-art in algorithms for frequent pattern mining over data streams. We also introduced a new approach for frequent itemset mining. We have developed a new one-pass algorithm for streaming environment, which has deterministic bounds on the accuracy. Particularly, it does not require any out-of-core memory structure and is very memory efficient in practice.

Though the existing one-pass mining algorithms have been shown to be very accurate and faster than traditional multi-pass algorithms, the experimental results show that they are still computationally expensive, meaning that if the data arrives too rapidly, the mining algorithms will not able to handle the data. Unfortunately, this can be the case for some high-velocity streams, such as network flow data. Therefore, new techniques are needed to increase the speed of stream mining tasks. We conclude this chapter with a list of future research problems to address this challenge.

mining maximal and other condensed frequent itemsets in data streams: Maximal frequent itemsets (MFI), and other condensed frequent itemsets, such as the $\delta - cover$ proposed in [32], provide good compression of the frequent itemsets. Mining them are very likely to reduce the mining costs in terms of both computation and memory over data streams. However, mining such kinds of compressed pattern set poses new challenges. The existing techniques will logically partition the data stream into segments, and mine potentially frequent itemsets each segment. In many compressed pattern sets, for instance, MFI, if we just mine MFI for each segment, it will be very hard to find the global

MFI. This is because the MFI can be different in each segment, and when we combine them together, we need the counts for the itemsets which are frequent but not maximal. However, estimating the counts for these itemsets can be very difficult. The similar problem occurs for other condensed frequent itemsets mining. Clearly, new techniques are necessary to mine condensed frequent itemsets in data steams.

Online Sampling for Frequent Pattern Mining: The current approaches involve high-computational cost for mining the data streams. One of the main reasons is that all of them try to maintain and deliver the potentially frequent patterns at any time. If the data stream arrives very rapidly, this could be unrealistic. Therefore, one possible approach is to maintain a sample set which best represents the data stream and provide good estimation of the frequent itemsets.

Compared with existing sampling techniques [31, 9, 6] on disk-resident datasets for frequent itemsets mining, sampling data streams brings some new issues. For example, the underlying distribution of the data stream can change from time to time. Therefore, sampling needs to adapt to the data stream. However, it will be quite difficult to monitor such changes if we do not mine the set of frequent itemsets directly. In addition, the space requirement of the sample set can be an issue as well. As pointed by Manku and Motwani [28], methods similar to concise sampling [16] might be helpful to reduce the space and achieve better mining results.

References

[1] R. Agrawal, H. Mannila, R. Srikant, H. Toivonent, and A. Inkeri Verkamo. Fast discovery of association rules. In U. Fayyad and et al, editors, *Advances in Knowledge Discovery and Data Mining*, pages 307–328. AAAI Press, Menlo Park, CA, 1996.

[2] Rakesh Agrawal, Tomasz Imielinski, and Arun Swami. Mining association rules between sets of items in large databases. In *Proceedings of the 1993 ACM SIGMOD Conference*, pages 207–216, May 1993.

[3] Tatsuya Asai, Hiroki Arimura, Kenji Abe, Shinji Kawasoe, and Setsuo Arikawa. Online algorithms for mining semi-structured data stream. In *ICDM*, pages 27–34, 2002.

[4] B. Babcock, S. Babu, M. Datar, R. Motwani, and J. Widom. Models and Issues in Data Stream Systems. In *Proceedings of the 2002 ACM Symposium on Principles of Database Systems (PODS 2002) (Invited Paper)*. ACM Press, June 2002.

[5] B. Babcock, S. Chaudhuri, and G. Das. Dynamic Sampling for Approximate Query Processing. In *Proceedings of the 2003 ACM SIGMOD Conference*. ACM Press, June 2003.

[6] Herve; Bronnimann, Bin Chen, Manoranjan Dash, Peter Haas, and Peter Scheuermann. Efficient data reduction with ease. In *KDD '03: Proceedings of the ninth ACM SIGKDD international conference on Knowledge discovery and data mining*, pages 59–68, 2003.

[7] Joong Hyuk Chang and Won Suk Lee. Finding recent frequent itemsets adaptively over online data streams. In *KDD '03: Proceedings of the ninth ACM SIGKDD international conference on Knowledge discovery and data mining*, 2003.

[8] Moses Charikar, Kevin Chen, and Martin Farach-Colton. Finding frequent items in data streams. In *ICALP '02: Proceedings of the 29th International Colloquium on Automata, Languages and Programming*, 2002.

[9] Bin Chen, Peter Haas, and Peter Scheuermann. A new two-phase sampling based algorithm for discovering association rules. In *KDD '02: Proceedings of the eighth ACM SIGKDD international conference on Knowledge discovery and data mining*, pages 462–468, 2002.

[10] D. Cheung, J. Han, V. NG, and C. Wong. Maintenance of discovered association rules in large databases : an incremental updating technique. In *ICDE*, 1996.

[11] Yun Chi, Haixun Wang, Philip S. Yu, and Richard R. Muntz. Moment: Maintaining closed frequent itemsets over a stream sliding window. In *ICDM*, pages 59–66, 2004.

[12] Yun Chi, Yirong Yang, and Richard R. Muntz. Hybridtreeminer: An efficient algorithm for mining frequent rooted trees and free trees using canonical forms. In *The 16th International Conference on Scientific and Statistical Database Management (SSDBM'04)*, 2004.

[13] G. Cormode, M. Datar, P. Indyk, and S. Muthukrishnan. Comparing Data Streams Using Hamming Norms. In *Proceedings of Conference on Very Large Data Bases (VLDB)*, pages 335–345, 2002.

[14] Graham Cormode, Flip Korn, S. Muthukrishnan, and Divesh Srivastava. Finding hierarchical heavy hitters in data streams. In *VLDB*, pages 464–475, 2003.

[15] C. Giannella, Jiawei Han, Jian Pei, Xifeng Yan, and P. S. Yu. Mining Frequent Patterns in Data Streams at Multiple Time Granularities. In *Proceedings of the NSF Workshop on Next Generation Data Mining*, November 2002.

[16] Phillip B. Gibbons and Yossi Matias. New sampling-based summary statistics for improving approximate query answers. In *ACM SIGMOD*, pages 331–342, 1998.

[17] Bart Goethals and Mohammed J. Zaki. Workshop Report on Workshop on Frequent Itemset Mining Implementations (FIMI). 2003.

[18] J. Han, J. Pei, and Y. Yin. Mining frequent patterns without candidate generation. In *Proceedings of the ACM SIGMOD Conference on Management of Data*, 2000.

[19] C. Hidber. Online Association Rule Mining. In *Proceedings of ACM SIGMOD Conference on Management of Data*, pages 145–156. ACM Press, 1999.

[20] Jun Huan, Wei Wang, Deepak Bandyopadhyay, Jack Snoeyink, Jan Prins, and Alexander Tropsha. Mining protein family-specific residue packing patterns from protein structure graphs. In *Eighth International Conference on Research in Computational Molecular Biology (RECOMB)*, pages 308–315, 2004.

[21] Akihiro Inokuchi, Takashi Washio, and Hiroshi Motoda. An apriori-based algorithm for mining frequent substructures from graph data. In *Principles of Knowledge Discovery and Data Mining (PKDD2000)*, pages 13–23, 2000.

[22] R. Jin and G. Agrawal. An algorithm for in-core frequent itemset mining on streaming data. In *ICDM*, November 2005.

[23] Ruoming Jin and Gagan Agrawal. An algorithm for in-core frequent itemset mining on streaming data. Technical Report OSU-CISRC-2/04-TR14, Ohio State University, 2004.

[24] Ruoming Jin and Gagan Agrawal. A systematic approach for optimizing complex mining tasks on multiple datasets. In *Proceedings of ICDE*, 2005.

[25] Richard M. Karp, Christos H. Papadimitriou, and Scott Shanker. A Simple Algorithm for Finding Frequent Elements in Streams and Bags. Available from http://www.cs.berkeley.edu/ christos/iceberg.ps, 2002.

[26] Michihiro Kuramochi and George Karypis. Frequent subgraph discovery. In *ICDM '01: Proceedings of the 2001 IEEE International Conference on Data Mining*, pages 313–320, 2001.

[27] Amit Manjhi, Vladislav Shkapenyuk, Kedar Dhamdhere, and Christopher Olston. Finding (recently) frequent items in distributed data streams. In *ICDE '05: Proceedings of the 21st International Conference on Data Engineering (ICDE'05)*, pages 767–778, 2005.

[28] G. S. Manku and R. Motwani. Approximate Frequency Counts Over Data Streams. In *Proceedings of Conference on Very Large DataBases (VLDB)*, pages 346 – 357, 2002.

[29] A. Savasere, E. Omiecinski, and S.Navathe. An efficient algorithm for mining association rules in large databases. In *21th VLDB Conf.*, 1995.

[30] Wei-Guang Teng, Ming-Syan Chen, and Philip S. Yu. A regression-based temporal pattern mining scheme for data streams. In *VLDB*, pages 93–104, 2003.

[31] H. Toivonen. Sampling large databases for association rules. In *22nd VLDB Conf.*, 1996.

[32] Dong Xin, Jiawei Han, Xifeng Yan, and Hong Cheng. Mining compressed frequent-pattern sets. In *VLDB*, pages 709–720, 2005.

[33] Xifeng Yan and Jiawei Han. gspan: Graph-based substructure pattern mining. In *ICDM '02: Proceedings of the 2002 IEEE International Conference on Data Mining (ICDM'02)*, page 721, 2002.

[34] Jeffrey Xu Yu, Zhihong Chong, Hongjun Lu, and Aoying Zhou. False positive or false negative: Mining frequent itemsets from high speed transactional data streams. In *Proceedings of the 28th International Conference on Very Large Data Bases (VLDB)*, Toronto, Canada, Aug 2004.

[35] M.J. Zaki, S. Parthasarathy, M. Ogihara, and W.Li. Parallel algorithms for fast discovery of association rules. *Data Mining and Knowledge Discovery: An International Journal*, 1(4):343–373, December 1997.

[36] Mohammed J. Zaki. Efficiently mining frequent trees in a forest. In *KDD '02: Proceedings of the eighth ACM SIGKDD international conference on Knowledge discovery and data mining*, pages 71–80, 2002.

[37] Mohammed J. Zaki and Charu C. Aggarwal. Xrules: an effective structural classifier for xml data. In *KDD '03: Proceedings of the ninth ACM SIGKDD international conference on Knowledge discovery and data mining*, pages 316–325, 2003.

Chapter 5

A SURVEY OF CHANGE DIAGNOSIS ALGORITHMS IN EVOLVING DATA STREAMS

Charu C. Aggarwal

IBM T. J. Watson Research Center
Hawthorne, NY 10532

charu@us.ibm.com

Abstract

An important problem in the field of data stream analysis is change detection and monitoring. In many cases, the data stream can show changes over time which can be used for understanding the nature of several applications. We discuss the concept of velocity density estimation, a technique used to understand, visualize and determine trends in the evolution of fast data streams. We show how to use velocity density estimation in order to create both *temporal velocity profiles* and *spatial velocity profiles* at periodic instants in time. These profiles are then used in order to predict three kinds of data evolution. Methods are proposed to visualize the changing data trends in a single online scan of the data stream, and a computational requirement which is linear in the number of data points. In addition, batch processing techniques are proposed in order to identify combinations of dimensions which show the greatest amount of global evolution. We also discuss the problem of change detection in the context of graph data, and illustrate that it may often be useful to determine communities of evolution in graph environments.

The presence of evolution in data streams may also change the underlying data to the extent that the underlying data mining models may need to be modified to account for the change in data distribution. We discuss a number of methods for micro-clustering which are used to study the effect of evolution on problems such as clustering and classification.

1.   Introduction

In recent years, advances in hardware technology have resulted in automated storage of data from a variety of processes. This results in storage which creates millions of records on a daily basis. Often, the data may show important changes in the trends over time because of changes in the underlying phenomena. This process is referred to as *data evolution*. By understanding the nature of such changes, a user may be able to glean valuable insights into emerging trends in the underlying transactional or spatial activity.

The problem of data evolution is interesting from two perspectives:

- For a given data stream, we would like to find the significant changes which have occurred in the data stream. This includes methods of visualizing the changes in the data and finding the significant regions of data dissolution, coagulation, and shift. The aim of this approach is to provide a *direct understanding* of the underlying changes in the stream. Methods such as those discussed in [3, 11, 15, 18] fall into this category. Such methods may be useful in a number of applications such as network traffic monitoring [21]. In [3], the velocity density estimation method has been proposed which can be used in order to visualize different kinds of trends in the data stream. In [11], the difference between two distributions is characterized using the KL-distance between two distributions. Other methods for trend and change detection in massive data sets may be found in [15]. Methods have also been proposed recently for change detection in graph data streams [2].

- The second class of problems relevant to data evolution is that of updating data mining models when a change has occurred. There is a considerable amount of work in the literature with a focus on incremental maintenance of models in the context of evolving data [10, 12, 24]. However, in the context of fast data streams, it is more important to use the evolution of the data stream in order to measure the nature of the change. Recent work [13, 14] has discussed a general framework for quantifying the changes in evolving data characteristics in the context of several data mining problems and algorithms. The focus of our paper is different from and orthogonal to the work in [13, 14]. Specifically, the work in [13, 14] is focussed on the effects of evolution on data mining models and algorithms. While these results show some interesting results in terms of generalizing existing data mining algorithms, our view is that data streams have special mining requirements which cannot be satisfied by using existing data mining models and algorithms. Rather, it is necessary to tailor the algorithms appropriately to each task. The algorithms discussed in [5, 7] discuss methods for clustering and classification in the presence of evolution of data streams.

This chapter will discuss the issue of data stream change in both these contexts. Specifically, we will discuss the following aspects:

- We discuss methods for quantifying the change at a given point of the data stream. This is done using the concept of velocity density estimation [3].

- We show how to use the velocity density in order to construct visual spatial and temporal profiles of the changes in the underlying data stream. This profiles provide a visual overview to the user about the changes in the underlying data stream.

- We discuss methods for utilizing the velocity density in order to characterize the changes in the underlying data stream. These changes correspond to regions of dissolution, coagulation, and sift in the data stream.

- We show how to use the velocity density to determine the overall level of change in the data stream. This overall level of change is defined in terms of the evolution coefficient of the data stream. The evolution coefficient can be used to find interesting combinations of dimensions with a high level of global evolution. This can be useful in many applications in which we wish to find subsets of dimensions which show a global level of change.

- We discuss how clustering methods can be used to analyze the change in different kinds of data mining applications. We discuss the problem of community evolution in interaction graphs and show how the methods for analyzing interaction graphs can be quite similar to other kinds of multi-dimensional data.

- We discuss the issue of effective application of data mining algorithms such as clustering and classification in the presence of change in data streams. We discuss general desiderata for designing change sensitive data mining algorithms for streams.

A closely related problem is that of mining spatio-temporal or mobile data [19, 20, 22], for which it is useful to have the ability to diagnose aggregate changes in spatial characteristics over time. The results in this paper can be easily generalized to these cases. In such cases, the change trends may also be useful from the perspective of providing physical interpretability to the underlying change patterns.

This chapter is organized as follows. In the next section, we will introduce the velocity density method and show how it can be used to provide different kinds of visual profiles. These visual representations may consist of spatial or temporal velocity profiles. The velocity density method also provides measures which are helpful in measuring evolution in the high dimensional case. In section 3 we will discuss how the process of evolution affects data mining algorithms. We will specifically consider the problems of clustering and classification. We will provide general guidelines as to how evolution can be leveraged in order to improve the quality of the results. In section 4, we discuss the conclusions and summary.

2. The Velocity Density Method

The idea in velocity density is to construct a density based velocity profile of the data. This is analogous to the concept of kernel density estimation in static data sets. In kernel density estimation [23], we provide a continuous estimate of the density of the data at a given point. The value of the density at a given point is estimated as the sum of the smoothed values of kernel functions $K'_h(\cdot)$ associated with each point in the data set. Each kernel function is associated with a kernel width h which determines the level of smoothing created by the function. The kernel estimation $\overline{f}(x)$ based on n data points and kernel function $K'_h(\cdot)$ is defined as follows:

$$\overline{f}(x) = (1/n) \cdot \sum_{i=1}^{n} K'_h(x - X_i) \qquad (5.1)$$

Thus, each discrete point X_i in the data set is replaced by a continuous function $K'_h(\cdot)$ which peaks at X_i and has a variance which is determined by the smoothing parameter h. An example of such a distribution would be a gaussian kernel with width h.

$$K'_h(x - X_i) = (1/\sqrt{2\pi} \cdot h) \cdot e^{-(x-X_i)^2/(2h^2)} \qquad (5.2)$$

The estimation error is defined by the kernel width h which is chosen in a data driven manner. It has been shown [23] that for most smooth functions $K'_h(\cdot)$, when the number of data points goes to infinity, the estimator $\overline{f}(x)$ asymptotically converges to the true density function $f(x)$, provided that the width h is chosen appropriately. For the d-dimensional case, the kernel function is chosen to be the product of d identical kernels $K_i(\cdot)$, each with its own smoothing parameter h_i.

In order to compute the velocity density, we use a temporal window h_t in order to perform the calculations. Intuitively, the temporal window h_t is associated with the time horizon over which the rate of change is measured. Thus, if h_t is chosen to be large, then the velocity density estimation technique provides long term trends, whereas if h_t is chosen to be small then the trends are relatively short term. This provides the user flexibility in analyzing the changes in the data over different kinds of time horizons. In addition, we have

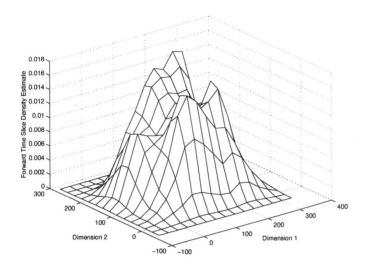

Figure 5.1.   The Forward Time Slice Density Estimate

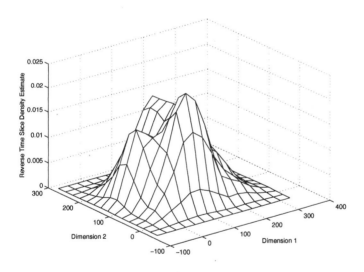

Figure 5.2.   The Reverse Time Slice Density Estimate

a spatial smoothing vector h_s whose function is quite similar to the standard spatial smoothing vector which is used in kernel density estimation.

Let t be the current instant and S be the set of data points which have arrived in the time window $(t - h_t, t)$. We intend to estimate the rate of increase in density at spatial location X and time t by using two sets of estimates: the *forward time slice density estimate* and the *reverse time slice density estimate*. Intuitively, the forward time slice estimate measures the density function for

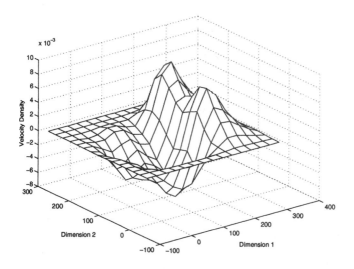

Figure 5.3. The Temporal Velocity Profile

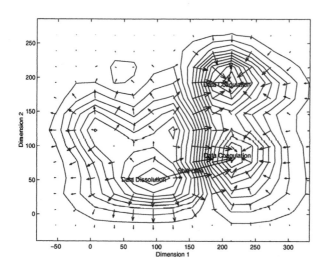

Figure 5.4. The Spatial Velocity Profile

all spatial locations at a given time t based on the set of data points which have arrived in the *past* time window $(t - h_t, t)$. Similarly, the reverse time slice estimate measures the density function at a given time t based on the set of data points which will arrive in the *future* time window $(t, t + h_t)$. Let us assume that the ith data point in S is denoted by (X_i, t_i), where i varies from 1 to $|S|$. Then, the forward time slice estimate $F_{(h_s, h_t)}(X, t)$ of the set S at the spatial location X and time t is given by:

$$F_{(h_s, h_t)}(X, t) = C_f \cdot \sum_{i=1}^{|S|} K_{(h_s, h_t)}(X - X_i, t - t_i) \qquad (5.3)$$

Here $K_{(h_s, h_t)}(\cdot, \cdot)$ is a spatio-temporal kernel smoothing function, h_s is the spatial kernel vector, and h_t is temporal kernel width. The kernel function $K_{(h_s, h_t)}(X - X_i, t - t_i)$ is a smooth distribution which decreases with increasing value of $t - t_i$. The value of C_f is a suitably chosen normalization constant, so that the entire density over the spatial plane is one unit. This is done, because our purpose of calculating the densities at the time slices is to compute the *relative* variations in the density over the different spatial locations. Thus, C_f is chosen such that we have:

$$\int_{\text{All } X} F_{(h_s, h_t)}(X, t)\delta X = 1 \qquad (5.4)$$

The reverse time slice density estimate is also calculated in a somewhat different way to the forward time slice density estimate. We assume that the set of points which have arrived in the time interval $(t, t + h_t)$ is given by U. As before, the value of C_r is chosen as a normalization constant. Correspondingly, we define the value of the reverse time slice density estimate $R_{(h_s, h_t)}(X, t)$ as follows:

$$R_{(h_s, h_t)}(X, t) = C_r \cdot \sum_{i=1}^{|U|} K_{(h_s, h_t)}(X - X_i, t_i - t) \qquad (5.5)$$

Note that in this case, we are using $t_i - t$ in the argument instead of $t - t_i$. Thus, the reverse time-slice density in the interval $(t, t + h_t)$ would be exactly the same as the forward time slice density if we assumed that time was reversed and the data stream arrived in reverse order, starting at $t + h_t$ and ending at t. Examples of the forward and reverse density profiles are illustrated in Figures 5.1 and 5.2 respectively.

For a given spatial location X and time T, let us examine the nature of the functions $F_{(h_s, h_t)}(X, T)$ and $R_{(h_s, h_t)}(X, T - h_t)$. Note that both functions are almost exactly the same, and use the same data points from the interval $(T - h_t, T)$, except that one has been calculated assuming time runs forward, whereas the other has been calculated assuming that the time runs in reverse.

Furthermore, the volumes under each of these curves, when measured over all spatial locations X is equal to one unit because of the normalization. Correspondingly, the density profiles at a given spatial location X would be different between the two depending upon how the relative trends have changed in the interval $(T - h_t, T)$. We define the velocity density $V_{(h_s,h_t)}(X, T)$ at spatial location X and time T as follows:

$$V_{(h_s,h_t)}(X, T) = \frac{F_{(h_s,h_t)}(X, T) - R_{(h_s,h_t)}(X, T - h_t)}{h_t} \qquad (5.6)$$

We note that a positive value of the velocity density corresponds to a increase in the data density of a given point. A negative value of the velocity density corresponds to a reduction in the data density a given point. In general, it has been shown in [3] that when the spatio-temporal kernel function is defined as below, then the velocity density is directly proportional to a rate of change of the data density at a given point.

$$K_{(h_s,h_t)}(X, t) = (1 - t/h_t) \cdot K'_{h_s}(X) \qquad (5.7)$$

This kernel function is only defined for values of t in the range $(0, h_t)$. The gaussian spatial kernel function $K'_{h_s}(\cdot)$ was used because of its well known effectiveness [23]. Specifically, $K'_{h_s}(\cdot)$ is the product of d identical gaussian kernel functions, and $h_s = (h_s^1, \ldots h_s^d)$, where h_s^i is the smoothing parameter for dimension i. Furthermore, for the special case of static snapshots, it is possible to show [3] that he velocity density is proportional to the difference in the spatial kernel densities of the two sets. Thus, the velocity density approach retains its intuitive appeal under a variety of special circumstances.

In general, we utilize a grid partitioning of the data in order to perform the velocity density calculation. We pick a total of β coordinates along each dimension. For a 2-dimensional system, this corresponds to β^2 spatial coordinates. The temporal velocity profile can be calculated by a simple $O(\beta^2)$ additive operations per data point. For each coordinate X_g in the grid, we maintain two sets of counters (corresponding to forward and reverse density counters) which are updated as each point in the data stream is received. When a data point X_i is received at time t_i, then we add the value $K_{(h_s,h_t)}(X_g - X_i, t - t_i)$ to the forward density counter, and the value $K_{(h_s,h_t)}(X_g - X_i, t_i - (t - h_t))$ to the reverse density counter for X_g. At the end of time t, the values computed for each coordinate at the grid need to be normalized. The process of normalization is the same for either the forward or the reverse density profiles. In each case, we sum up the total value in all the β^2 counters, and divide each counter by this total. Thus, for the normalized coordinates the sum of the values over all the β^2 coordinates will be equal to 1. Then the reverse density counters are subtracted from the forward counters in order to compete the computation.

Successive sets of temporal profiles are generated at user-defined time-intervals of of h_t. In order to ensure online computation, the smoothing parameter vector h_s for the time-interval $(T - h_t, T)$ must be available at time $T - h_t$, as soon as the first data point of that interval is scheduled to arrive. Therefore, we need a way of estimating this vector using the data from past intervals. In order to generate the velocity density for the interval $(T - h_t, T)$, the spatial kernel smoothing vector h_s is determined using the Silverman's approximation rule[1] [23] for gaussian kernels on the set of data points which arrived in the interval $(T - 2h_t, T - h_t)$.

2.1 Spatial Velocity Profiles

Even better insight can be obtained by examining the nature of the spatial velocity profiles, which provide an insight into how the data is shifting. For each spatial point, we would like to compute the directions of movements of the data at a given instant. The motivation in developing a spatial velocity profile is to give a user a *spatial* overview of the re-organizations in relative data density at different points. In order to do so, we define an ϵ-perturbation along the ith dimension by $\overline{\epsilon_i} = \epsilon \cdot \overline{e_i}$, where $\overline{e_i}$ is the unit vector along the ith dimension. For a given spatial location X, we first compute the velocity gradient along each of the i dimensions. We denote the velocity gradient along the ith dimension by $\Delta v_i(X, t)$ for spatial location X and time t. This value is computed by subtracting the density at spatial location X from the density at $X + \overline{\epsilon_i}$ (ϵ-perturbation along the ith dimension), and dividing the result by ϵ. The smaller the value of ϵ, the better the approximation. Therefore, we have:

$$\Delta v_i(X, t) = \lim_{\epsilon \Rightarrow 0} \frac{V_{(h_s, h_t)}(X + \overline{\epsilon_i}, t) - V_{(h_s, h_t)}(X, t)}{\epsilon} \qquad (5.8)$$

The value of $\Delta v_i(X, t)$ is negative when the velocity density decreases with increasing value of the ith coordinate of spatial location X. The gradient $\overline{\Delta v(X, t)}$ is given by $(\Delta v_1(X, t) \ldots \Delta v_d(X, t))$. This vector gives the spatial gradient at a given grid point *both* in terms of direction and magnitude. The spatial velocity profile is illustrated by creating a spatial plot which illustrates the directions of the data shifts at different grid points by directed markers which mirror these gradients both in terms of directions and magnitude. An example of a spatial velocity profile is illustrated in Figure 5.4. If desired, the spatial profile can be generated continuously for a fast data stream. This continuous generation of the profile creates spatio-temporal animations which provide a continuous idea of the trend changes in the underlying data. Such animations can also provide real time diagnosis ability for a variety of applications.

An additional useful ability is to be able to concisely diagnose specific trends in given spatial locations. For example, a user may wish to know particular spatial locations in the data at which the data is being reduced, those at which the data is increasing, and those from where the data is shifting to other locations:

DEFINITION 5.1 *A data coagulation for time slice t and user defined threshold min-coag is defined to be a connected region $\mathcal{R}$ in the data space, so that for each point $X \in \mathcal{R}$, we have $V_{(h_s,h_t)}(X,t) > \text{min-coag} > 0$.*

Thus, a data coagulation is a connected region in the data which has velocity density larger than a user-defined noise threshold of *min-coag*. In terms of the temporal velocity profile, these are the connected regions in the data with elevations larger than *min-coag*. Note that there may be multiple such elevated regions in the data, each of which may be disconnected from one another. Each such region is a separate area of data coagulation, since they cannot be connected by a continuous path above the noise threshold. For each such elevated region, we would also have a local peak, which represents the highest density in that locality.

DEFINITION 5.2 *The epicenter of a data coagulation $\mathcal{R}$ at time slice t is defined to be a spatial location X^* such that $X^* \in \mathcal{R}$ and for any $X \in \mathcal{R}$, we have $V_{(h_s,h_t)}(X,t) \leq V_{(h_s,h_t)}(X^*,t)$.*

Similarly regions of data dissolution and corresponding epicenters can be determined.

DEFINITION 5.3 *A data dissolution for time slice t and user defined threshold min-dissol is defined to be a connected region $\mathcal{R}$ in the data space, so that for each point $X \in \mathcal{R}$, we have $V_{(h_s,h_t)}(X,t) < -\text{min-dissol} < 0$.*

We define the epicenter of a data dissolution as follows:

DEFINITION 5.4 *The epicenter of a data dissolution $\mathcal{R}$ at time slice t is defined to be a spatial location X^* such that $X^* \in \mathcal{R}$ and for any $X \in \mathcal{R}$, we have $V_{(h_s,h_t)}(X,t) \geq V_{(h_s,h_t)}(X^*,t)$.*

A region of data dissolution and its epicenter is calculated in an exactly analogous way to the epicenter of a data coagulation. It now remains to discuss how significant shifts in the data can be detected. Many of the epicenters of coagulation and dissolution are connected in a way which results in a funneling of the data from the epicenters of dissolution to the epicenters of coagulation. When this happens, it is clear that the two phenomena of dissolution and coagulation are connected to one another. We refer to such a phenomenon as a *global data shift*. The detection of such shifts can be useful in many problems involving mobile objects. How to find whether a pair of epicenters are connected in this way?

In order to detect such a phenomenon we use the intuition derived from the use of the spatial velocity profiles. Let us consider a directed line drawn from an epicenter to data dissolution to an epicenter of data coagulation. In order for this directed line to be indicative of a global data shift, the spatial velocity profile should be such that the directions of a localized shifts along each of the points in this directed line should be in roughly in the same direction as the line itself. If at any point on this directed line, the direction of the localized shift is in an opposite direction, then it is clear that the these two epicenters are disconnected from one another. In order to facilitate further discussion, we will refer to the line connecting two epicenters as a *potential shift line*.

Recall that the spatial velocity profiles provide an idea of the spatial movements of the data over time. In order to calculate the nature of the data shift, we would need to calculate the projection of the spatial velocity profiles along this potential shift line. In order to do so without scanning the data again, we use the grid points which are closest to this shift line in order to obtain an approximation of the shift velocities at various points along this line. The first step is to find all the elementary rectangles which are intersected by the shift line. Once these rectangles have been found we determine the grid points corresponding to the corners of these rectangles. These are the grid points at which the spatial velocity profiles are examined.

Let the set of n grid points thus discovered be denoted by $Y_1 \ldots Y_n$. Then the corresponding spatial velocities at these grid points at time slice t are $\overline{\Delta v(Y_1, t)} \ldots \overline{\Delta v(Y_n, t)}$. Let $\overline{\mathcal{L}}$ be the unit vector in the direction of the shift line. We assume that this vector is directed from the region of dissolution to the area of coagulation. Then the projections of the spatial velocities in the direction of the shift line are given by $\overline{\mathcal{L}} \cdot \overline{\Delta v(Y_1, t)} \ldots \overline{\mathcal{L}} \cdot \overline{\Delta v(Y_n, t)}$. We shall refer to these values as $p_1 \ldots p_n$ respectively. For a shift line to expose an actual movement of the data, the values of $p_1 \ldots p_n$ must all be substantially positive. In order to quantify this notion, we introduce a user-defined parameter called *min-vel*. A potential shift line is said to be a valid shift when each of values $p_1 \ldots p_n$ is larger than *min-vel*.

Thus, in order to determine the all the possible data shifts, we first find all coagulation and dissolution epicenters for user-defined parameters *min-coag* and *min-dissol* respectively. Then we find all the potential shift lines by connecting each dissolution epicenter to a coagulation epicenter. For each such shift line, we find the grid points which are closest to it using the criteria discussed above. Finally, for each of these grid points, we determine the projection of the corresponding shift velocities along this line and check whether each of them is at least *min-vel*. If so, then this direction is reported as a valid shift line.

2.2   Evolution Computations in High Dimensional Case

In this section, we will discuss how to determine interesting combinations of dimensions with a high level of global evolution. In order to do so, we need to have a measure for the overall level of evolution in a given combination of dimensions. By integrating the value of the velocity density over the entire spatial area, we can obtain the total rate of change over the entire spatial area. In other words, if $E_{(h_s,h_t)}(t)$ be the total evolution in the period $(t - h_t, t)$, then we have:

$$E_{(h_s,h_t)}(t) = h_t \int_{\text{all } X} |V_{(h_s,h_t)}(X,t)| \delta X$$

Intuitively, the evolution coefficient measures the total volume of the evolution in the time horizon $(t - h_t, t)$. It is possible to calculate the evolution coefficients of particular projections of the data by using only the corresponding sets of dimensions in the density calculations. In [3] it has been shown how the computation of the evolution coefficient can be combined with an a-priori like rollup approach in order to find the set of *minimal evolving projections*. In practice, the number of minimal evolving projections is relatively small, and therefore large part of the search space can be pruned. This results in an effective algorithm for finding projections of the data which show a significant amount of evolution. In many applications, the individual attributes may not evolve a lot, but the projections may evolve considerably because of the changes in relationships among the underlying attributes. This can be useful in a number of applications such as target marketing or multi-dimensional trend analysis.

2.3   On the use of clustering for characterizing stream evolution

We note that methods such as clustering can be used to characterize the stream evolution. For this purpose, we utilize the micro-clustering methodology which is discussed[2] in [5]. We note that clustering is a natural choice to study broad changes in trends, since it summarizes the behavior of the data.

In this technique, micro-clusters are utilized in order to determine sudden changes in the data stream. Specifically, new trends in the data show up as new micro-clusters, whereas declining trends correspond to disappearing micro-clusters. In [5], we have illustrated the effectiveness of this kind of technique on an intrusion detection application. In general, the micro-clustering method is useful for change detection in a number of unsupervised applications where training data is not readily available, and anomalies can only be detected as sudden changes in the underlying trends. In the same paper, we have also shown some examples of how the method may be used for intrusion detection.

Such an approach has also been extended to the case of graph and structural data sets. In [2], we use a clustering technique in order to determine community evolution in graph data streams. Such a clustering technique is useful in many cases in which we need to determine changes in interaction over different entities. In such cases, the entities may represent nodes of a graph and the interactions may correspond to edges. A typical example of an interaction may be a phone call between two entities, or the co-authorship of a paper between two entities. In many cases, these trends of interaction may change over time. Such trends include the gradual formation and dissolution of different communities of interaction. In such cases, a user may wish to perform repeated *exploratory querying* of the data for different kinds of user-defined parameters. For example, a user may wish to determine rapidly expanding or contracting communities of interest over different time frames. This is difficult to perform in a fast data stream because of the one-pass constraints on the computations. Some examples of queries which may be performed by a user are as follows:

(1) Find the communities with substantial increase in interaction level in the interval $(t - h, t)$. We refer to such communities as *expanding communities*.

(2) Find the communities with substantial decrease in interaction level in the interval $(t - h, t)$ We refer to such communities as *contracting communities*.

(3) Find the communities with the most stable interaction level in the interval $(t - h, t)$.

In order to resolve such queries, the method in [2] proposes an online analytical processing framework which separates out online data summarization from offline exploratory querying. The process of data summarization stores portions of the graph on disk at specific periods of time. This summarized data is then used in order to resolve different kinds of queries. The result is a method which provides the ability to perform exploratory querying without compromising on the quality of the results. In this context, the clustering of the graph of interactions is a key component. The first step is to create a *differential graph* which represents the significant changes in the data interactions over the user specified horizon. This is done using the summary information stored on the disk. Significant communities of change show up as clusters in this graph. The clustering process is able to find sub-graphs which represent a sudden formation of a cluster of interactions which correspond to the underlying change in the data. It has been shown in [2], that this process can be performed in an efficient and effective way, and can identify both expanding and contracting communities.

3.   On the Effect of Evolution in Data Mining Algorithms

The discussion in this chapter has so far concentrated only on the problem of analyzing and visualizing the change in a data stream directly. In many cases, it is also desirable to analyze the evolution in a more indirect way, when such streams are used in conjunction with data mining algorithms. In this section, we will discuss the effects of evolution on data mining algorithms. The problem of mining incremental data dynamically has often been studied in many data mining scenarios [7, 10, 12, 24]. However, many of these methods are often not designed to work well with data streams since the distribution of the data evolves over time.

Some recent results [13] discuss methods for mining data streams under block evolution. We note that these methods are useful for incrementally updating the model when evolution has taken place. While the method has a number of useful characteristics, it does not attempt to determine the optimal segment of the data to be used for modeling purposes or provide an application-specific method to weight the relative importance of more recent or past data points. In many cases, the user may also desire to have the flexibility to analyze the data mining results over different time horizons. For such cases, it is desirable to use an online analytical processing framework which can store the underlying data in a summarized format over different time horizons. In this respect, it is desirable to store summarized snapshots [5, 7] of the data over different periods of time.

In order to store the data in a summarized format, we need the following two characteristics:

- We need a method for condensing the large number of data points in the stream into condensed summary statistics. In this respect the use of clustering is a natural choice for data condensation.

- We need a method for storing the condensed statistics over different periods of time. This is necessary in order to analyze the characteristics of the data over different time horizons. We note that the storage of the condensed data at each and every time unit can be expensive both in terms of computational resources and storage space. Therefore, a method needs to be used so that a small amount of data storage can retain a high level of accuracy in horizon-recall. This technique is known as the pyramidal or geometric time frame. In this technique, a constant number of snapshots of different orders are stored. The snapshots of the ith order occur at intervals which are divisible by α^i for some $\alpha > 1$. It can be shown that this storage pattern provides constant guarantees on the accuracy of horizon estimation.

Another property of the stored snapshots in [5] is that the corresponding statistics show the additivity property. The additivity property ensures that it is possible to obtain the statistics over a pre-defined time window by subtracting out the statistics of the previous window from those of the current window. Thus, it is possible to examine the evolving behavior of the data over different time horizons.

Once the summarized snapshots are stored in this pattern, they can be leveraged for a variety of data mining algorithms. For example, for the case of the classification problem [7], the underlying data may show significant change trends which result in different optimal time horizons. For this purpose, one can use the statistics over different time horizons. One can use a portion of the training stream to determine the horizon which provides the optimal classification accuracy. This value of the horizon is used in order to perform the final classification. The results in [7] show that there is a significant improvement in accuracy from the use of horizon specific classification.

This technique is useful not just for the classification problem but also for a variety of problem in the evolving scenario. For example, in many cases, one may desire to forecast the future behavior of an evolving data stream. In such cases, the summary statistics can be used to make broad trends about the future behavior of the stream. In general, for the evolving scenario, it is desirable to have the following characteristics for data stream mining algorithms:

- It is desirable to leverage temporal locality in order to improve the mining effectiveness. The concept of temporal locality refers to the fact that the data points in the stream are not randomly distributed. Rather the points at a given period in time are closely correlated, and may show specific levels of evolution in different regions. In many problems such as classification and forecasting, this property can be leveraged in order to improve the quality of the mining process.

- It is desirable to have the flexibility of performing the mining over different time horizons. In many cases, the optimal results can be obtained only after applying the results of the algorithm over a variety of time horizons. An example of this case is illustrated in [7], in which the classification problem is solved by finding the optimal accuracy over different horizons.

- In many problems, it is possible to perform incremental maintenance by using decay-specific algorithms. In such cases, recent points are weighted more heavily than older points during the mining process. The weight of the data points decay according a pre-defined function which is application-specific. This function is typically chosen as an exponential decay function whose decay is defined in terms of the exponential decay parameter. An example of this situation is the high dimensional projected stream clustering algorithm discussed in [6].

- In many cases, synopsis construction algorithms such as sampling may not work very well in the context of an evolving data stream. Traditional reservoir sampling methods [25] may end up summarizing the stale history of the entire data stream. In such cases, it may be desirable to use a biased sampling approach which maintains the temporal stability of the stream sample. The broad idea is to construct a stream sample which maintain the points in proportion to their decay behavior. This is a challenging task for a reservoir construction algorithm, and is not necessarily possible for all decay functions. The method in [8] proposes a new method for reservoir sampling in the case of certain kinds of decay functions.

While the work in [13] proposes methods for monitoring evolving data streams, this framework does not account for the fact that different methodologies may provide the most effective stream analysis in different cases. For some problems, it may be desirable to use a decay based model, and for others it may be desirable to use only a subset of the data for the mining process. In general, the methodology used for a particular algorithm depends upon the details of that particular problem and the data. For example, for some problems such as high dimensional clustering [6], it may be desirable to use a decay-based approach, whereas for other problems such as classification, it may be desirable use the statistics over different time horizons in order to optimize the algorithmic effectiveness. This is because problems such as high dimensional clustering require a large amount of data in order to provide effective results, and historical clusters do provide good insights about the future clusters in the data. Therefore, it makes more sense to use all the data, but with an application specific decay-based approach which provides the new data greater weight than the older data. On the other hand, in problems such as classification, the advantages of using more data is much less relevant to the quality of the result than using the data which is representative of the current trends in the data. The discussion of this section provides clues to the kind of approaches that are useful for re-designing data mining algorithms in the presence of evolution.

4. Conclusions

In this paper, we discussed the issue of change detection in data streams. We discussed different methods for characterizing change in data streams. For thus purpose, we discussed the method of velocity density estimation and its application to different kinds of visual representations of changes in the underlying data. We also discussed the problem of online community evolution in fast data streams. In many of these methods, clustering is a key component since it allows us to summarize the data effectively. We also studied the reverse problem of how data mining models are maintained when the underlying data changes. In this context, we studied the problems of clustering and classification of fast evolving data streams. The key in many of these methods is to use an online analytical processing methodology which preprocesses and summarizes segments of the data stream. These summarized segments can be used for a variety of data mining purposes such as clustering and classification.

Notes

1. According to Silverman's approximation rule, the smoothing parameter for a data set with n points and standard deviation σ is given by $1.06 \cdot \sigma \cdot n^{-1/5}$. For the d-dimensional case, the smoothing parameter

along each dimension is determined independently using the corresponding dimension-specific standard deviation.

2. The methodology is also discussed in an earlier chapter of this book.

References

[1] Aggarwal C., Procopiuc C., Wolf J., Yu P., Park J.-S. (1999). Fast algorithms for projected clustering. *ACM SIGMOD Conference*.

[2] Aggarwal C., Yu P. S (2005). Online Analysis of Community Evolution in Data Streams. *ACM SIAM Conference on Data Mining*.

[3] Aggarwal C (2003). A Framework for Diagnosing Changes in Evolving Data Streams. *ACM SIGMOD Conference*.

[4] Aggarwal C (2002). An Intuitive Framework for understanding Changes in Evolving Data Streams. *IEEE ICDE Conference*.

[5] Aggarwal C., Han J., Wang J., Yu P (2003). A Framework for Clustering Evolving Data Streams. *VLDB Conference*.

[6] Aggarwal C., Han J., Wang J., Yu P (2004). A Framework for High Dimensional Projected Clustering of Data Streams. *VLDB Conference*.

[7] Aggarwal C, Han J., Wang J., Yu P. (2004). On-Demand Classification of Data Streams. *ACM KDD Conference*.

[8] Aggarwal C. (2006). On Biased Reservoir Sampling in the presence of stream evolution. *VLDB Conference*.

[9] Chawathe S., Garcia-Molina H. (1997). Meaningful Change Detection in Structured Data. *ACM SIGMOD Conference Proceedings*.

[10] Cheung D., Han J., Ng V., Wong C. Y. (1996). Maintenance of Discovered Association Rules in Large Databases: An Incremental Updating Technique. *IEEE ICDE Conference Proceedings*.

[11] Dasu T., Krishnan S., Venkatasubramaniam S., Yi K. (2005). An Information-Theoretic Approach to Detecting Changes in Multidimensional data Streams. *Duke University Technical Report CS-2005-06*.

[12] Donjerkovic D., Ioannidis Y. E., Ramakrishnan R. (2000). Dynamic Histograms: Capturing Evolving Data Sets. *IEEE ICDE Conference Proceedings*.

[13] Ganti V., Gehrke J., Ramakrishnan R (2002). Mining Data Streams under Block Evolution. *ACM SIGKDD Explorations*, 3(2), 2002.

[14] Ganti V., Gehrke J., Ramakrishnan R., Loh W.-Y. (1999). A Framework for Measuring Differences in Data Characteristics. *ACM PODS Conference Proceedings*.

[15] Gollapudi S., Sivakumar D. (2004) Framework and Algorithms for Trend Analysis in Massive Temporal Data *ACM CIKM Conference Proceedings*.

[16] Hulten G., Spencer L., Domingos P. (2001). Mining Time Changing Data Streams. *ACM KDD Conference*.

[17] Jain A., Dubes R. (1998). Algorithms for Clustering Data, *Prentice Hall, New Jersey*.

[18] Kifer D., David S.-B., Gehrke J. (2004). Detecting Change in Data Streams. *VLDB Conference*, 2004.

[19] Roddick J. F. et al (2000). Evolution and Change in Data Management: Issues and Directions. *ACM SIGMOD Record*, 29(1): pp. 21–25.

[20] Roddick J. F., Spiliopoulou M (1999). A Bibliography of Temporal, Spatial, and Spatio-Temporal Data Mining Research. *ACM SIGKDD Explorations*, 1(1).

[21] Schweller R., Gupta A., Parsons E., Chen Y. (2004) Reversible Sketches for Efficient and Accurate Change Detection over Network Data Streams. *Internet Measurement Conference Proceedings*.

[22] Sellis T (1999). Research Issues in Spatio-temporal Database Systems. *Symposium on Spatial Databases Proceedings*.

[23] Silverman B. W. (1986). *Density Estimation for Statistics and Data Analysis*. Chapman and Hall.

[24] Thomas S., Bodagala S., Alsabti K., Ranka S. (1997). An Efficient Algorithm for the Incremental Updating of Association Rules in Large Databases. *ACM KDD Conference Proceedings*.

[25] Vitter J. S. (1985) Random Sampling with a Reservoir. *ACM Transactions on Mathematical Software*, Vol. 11(1), pp 37–57.

Chapter 6

MULTI-DIMENSIONAL ANALYSIS OF DATA STREAMS USING STREAM CUBES

Jiawei Han,[1] Y. Dora Cai,[1] Yixin Chen,[2] Guozhu Dong,[3] Jian Pei,[4] Benjamin W. Wah,[1] and Jianyong Wang[5]

[1] *University of Illinois, Urbana, Illinois*
{hanj, ycai, b-wah}@uiuc.edu

[2] *Washington University, St. Louis, Missouri*
chen@cse.wustl.edu

[3] *Wright State University, Dayton, Ohio*
gdong@cs.wright.edu

[4] *Simon Fraser University, British Columbia, Canada*
jpei@cs.sfu.ca

[5] *Tsinghua University, Beijing, China*
jianyong@tsinghua.edu.cn

Abstract   Large volumes of dynamic stream data pose great challenges to its analysis. Besides its dynamic and transient behavior, stream data has another important characteristic: *multi-dimensionality. Much of stream data resides at a multi-dimensional space and at rather low level of abstraction, whereas most analysts are interested in relatively high-level dynamic changes in some combination of dimensions.* To discover high-level dynamic and evolving characteristics, one may need to perform multi-level, multi-dimensional on-line analytical processing (OLAP) of stream data. Such necessity calls for the investigation of new architectures that may facilitate on-line analytical processing of multi-dimensional stream data.

In this chapter, we introduce an interesting **stream_cube** architecture that effectively performs on-line partial aggregation of multi-dimensional stream data, captures the essential dynamic and evolving characteristics of data streams, and facilitates fast OLAP on stream data. Three important techniques are proposed for the design and implementation of stream cubes. First, a *tilted time frame* model is proposed to register time-related data in a multi-resolution model: The more recent data are registered at finer resolution, whereas the more distant data are registered at coarser resolution. This design reduces the overall storage requirements of time-related data and adapts nicely to the data analysis tasks commonly encountered in practice. Second, instead of materializing cuboids at all levels, two *critical* layers: *observation layer* and *minimal interesting layer*, are maintained to support routine as well as flexible analysis with minimal computation cost. Third, an efficient stream data cubing algorithm is developed that computes only the layers (cuboids) along a *popular path* and leaves the other cuboids for on-line, query-driven computation. Based on this design methodology, stream data cube can be constructed and maintained incrementally with reasonable memory space, computation cost, and query response time. This is verified by our substantial performance study.

Stream cube architecture facilitates online analytical processing of stream data. It also forms a preliminary structure for online stream mining. The impact of the design and implementation of stream cube in the context of stream mining is also discussed in the chapter.

Keywords: Data streams, multidimensional analysis, OLAP, data cube, stream cube, tilted time frame, partial materialization.

1.   Introduction

A fundamental difference in the analysis of stream data from that of non-stream one is that the stream data is generated in huge volumes, flowing in-and-out dynamically, and changing rapidly. Due to limited resources available and the usual requirements of fast response, most data streams may not be fully stored and may only be examined in a single pass. These characteristics of stream data have been emphasized and explored in their investigations by many researchers, such as ([6, 8, 17, 18, 16]), and efficient stream data querying, counting, clustering and classification algorithms have been proposed, such as ([2, 3, 22, 17, 18, 16, 25]). However, there is another important characteristic of stream data that has not drawn enough attention: *Most of stream data sets are multidimensional in nature and reside at rather low level of abstraction, whereas an analyst is often more interested in higher levels of abstraction in a small subset of dimension combinations.* Similar to OLAP analysis of static data, multi-level, multi-dimensional on-line analysis should be performed on stream data as well. This can be seen from the following example.

EXAMPLE 6.1 One may observe infinite streams of power usage data in a power supply system. The lowest granularity of such data can be individual household and second. Although there are tremendous varieties at analyzing such data, the most useful online stream data analysis could be the analysis of the fluctuation of power usage at certain dimension combinations and at certain high levels, such as by region and by quarter (of an hour), making timely power supply adjustments and handling unusual situations. ◇

One may easily link such multi-dimensional analysis with the online analytical processing of multi-dimensional nonstream data sets. For analyzing the characteristics of nonstream data, the most influential methodology is to use data warehouse and OLAP technology ([14, 11]). With this technology, data from different sources are integrated, and then aggregated in multi-dimensional space, either completely or partially, generating data cubes. The computed cubes can be stored in the form of relations or multi-dimensional arrays ([1, 31]) to facilitate fast on-line data analysis. In recent years, a large number of data warehouses have been successfully constructed and deployed in applications, and data cube has become an essential component in most data warehouse systems and in some extended relational database systems for multidimensional data analysis and intelligent decision support.

Can we extend the data cube and OLAP technology from the analysis of static, pre-integrated data to that of dynamically changing stream data, including time-series data, scientific and engineering data, and data produced in other dynamic environments, such as power supply, network traffic, stock exchange, telecommunication data flow, Web click streams, weather or environment monitoring? The answer to this question may not be so easy since, as everyone knows, it takes great efforts and substantial storage space to compute and maintain static data cubes. A dynamic stream cube may demand an even greater computing power and storage space. *How can we have sufficient resources to compute and store a dynamic stream cube?*

In this chapter, we examine this issue and propose an interesting architecture, called *stream cube*, for on-line analytical processing of voluminous, infinite, and dynamic stream data, with the following design considerations.

1. For analysis of stream data, it is unrealistic to store and analyze data with an infinitely long and fine scale on time. We propose a *tilted time frame* as the general model of *time dimension*. In the *tilted time frame*, time is registered at different levels of granularity. The most recent time is registered at the finest granularity; the more distant time is registered at coarser granularity; and the level of coarseness depends on the application requirements and on how distant the time point is from the current one. This model is sufficient for most analysis tasks, and at the same time it also ensures that the total amount of data to retain in memory or to be stored on disk is quite limited.

2. With limited memory space in stream data analysis, it is often still too costly to store a precomputed cube, even with the *tilted time frame*. We propose to compute and store only two *critical layers* (which are essentially cuboids) in the cube: (1) an *observation layer*, called *o-layer*,

which is the layer that an analyst would like to check and make decisions for either signaling the exceptions or drilling on the exception cells down to lower layers to find their corresponding lower level exceptions; and (2) the *minimal interesting layer*, called *m-layer*, which is the minimal layer that an analyst would like to examine, since it is often neither cost-effective nor practically interesting to examine the minute detail of stream data. For example, in Example 1, we assume that the *o*-layer is *user-category*, *region*, and *quarter*, while the *m*-layer is *user*, *city-block*, and *minute*.

3 Storing a cube at only two critical layers leaves much room on what and how to compute for the cuboids between the two layers. We propose one method, called **popular-path cubing**, which rolls up the cuboids from the *m*-layer to the *o*-layer, by following the most popular drilling path, materializes only the layers along the path, and leaves other layers to be computed at OLAP query time. An H-tree data structure is used here to facilitate efficient pre- and on-line computation. Our performance study shows that this method achieves a good trade-off between space, computation time, and flexibility, and has both quick aggregation time and query answering time.

The remaining of the paper is organized as follows. In Section 2, we define the basic concepts and introduce the problem. In Section 3, we present an architectural design for on-line analysis of stream data by introducing the concepts of *tilted time frame* and *critical layers*. In Section 4, we present the *popular-path* cubing method, an efficient algorithm for stream data cube computation that supports on-line analytical processing of stream data. Our experiments and performance study of the proposed methods are presented in Section 5. The related work and possible extensions of the model are discussed in Section 6, and our study is concluded in Section 7.

2. Problem Definition

Let $\mathcal{DB}$ be a relational table, called the **base table**, of a given cube. The set of all *attributes* $\mathcal{A}$ in $\mathcal{DB}$ are partitioned into two subsets, the *dimensional attributes* DIM and the *measure attributes* M (so $DIM \cup M = \mathcal{A}$ and $DIM \cap M = \emptyset$). The measure attributes functionally depend on the dimensional attributes in $\mathcal{DB}$ and are defined in the context of data cube using some typical aggregate functions, such as **COUNT, SUM, AVG,** or more sophisticated computational functions, such as standard deviation and regression.

A tuple with schema $\mathcal{A}$ in a multi-dimensional space (i.e., in the context of data cube) is called a **cell**. Given three distinct cells c_1, c_2 and c_3, c_1 is an **ancestor** of c_2, and c_2 a **descendant** of c_1 iff on every dimensional attribute, either c_1 and c_2 share the same value, or c_1's value is a generalized value of

c_2's in the dimension's concept hierarchy. c_2 is a **sibling** of c_3 iff c_2 and c_3 have identical values in all dimensions except one dimension A where $c_2[A]$ and $c_3[A]$ have the same parent in the dimension's domain hierarchy. A cell which has k non-* values is called a k-**d cell**. (We use "*" to indicate "all", i.e., the highest level on any dimension.)

A tuple $c \in \mathcal{D}$ is called a **base cell**. A base cell does not have any descendant. A cell c is an **aggregated cell** iff it is an ancestor of some base cell. For each aggregated cell c, its values on the measure attributes are derived from the complete set of descendant base cells of c. An aggregated cell c is an **iceberg cell** iff its measure value satisfies a specified iceberg condition, such as measure $\geq val_1$. The data cube that consists of all and only the iceberg cells satisfying a specified iceberg condition I is called the **iceberg cube** of a database $\mathcal{DB}$ under condition I.

Notice that in stream data analysis, besides the popularly used SQL aggregate-based measures, such as COUNT, SUM, MAX, MIN, and AVG, *regression* is a useful measure. A stream data cell compression technique LCR (*linearly compressed representation*) is developed in ([12]) to support efficient on-line regression analysis of stream data in data cubes. The study in ([12]) shows that for linear and multiple linear regression analysis, only a small number of *regression measures* rather than the complete stream of data need to be registered. This holds for regression on both the time dimension and the other dimensions. Since it takes a much smaller amount of space and time to handle regression measures in a multi-dimensional space than handling the stream data itself, it is preferable to construct regression(-measured) cubes by computing such regression measures.

A *data stream* is considered as a voluminous, infinite flow of data records, such as power supply streams, Web click streams, and telephone calling streams. The data is collected at the most detailed level in a multi-dimensional space, which may represent time, location, user, and other semantic information. Due to the huge amount of data and the transient behavior of data streams, most of the computations will scan a data stream only once. Moreover, the direct computation of measures at the most detailed level may generate a huge number of results but may not be able to disclose the general characteristics and trends of data streams. Thus data stream analysis will require to consider aggregations and analysis at multi-dimensional and multi-level space.

Our task is to support *efficient, high-level, on-line, multi-dimensional analysis of such data streams in order to find unusual (exceptional) changes of trends, according to users' interest, based on multi-dimensional numerical measures.* This may involve construction of a data cube, if feasible, to facilitate on-line, flexible analysis.

3.   Architecture for On-line Analysis of Data Streams

To facilitate on-line, multi-dimensional analysis of data streams, we propose a stream_cube architecture with the following features: (1) *tilted time frame*, (2) two *critical layers*: a *minimal interesting layer* and an *observation layer*, and (3) *partial computation of data cubes by popular-path cubing*. The stream data cubes so constructed are much smaller than those constructed from the raw stream data but will still be effective for multi-dimensional stream data analysis tasks.

3.1   Tilted time frame

In stream data analysis, people are usually interested in recent changes at a fine scale, but long term changes at a coarse scale. Naturally, one can register time at different levels of granularity. The most recent time is registered at the finest granularity; the more distant time is registered at coarser granularity; and the level of coarseness depends on the application requirements and on how distant the time point is from the current one.

There are many possible ways to design a titled time frame. We adopt three kinds of models: (1) *natural tilted time frame model* (Fig. 6.1), (2) *logarithmic scale tilted time frame model* (Fig. 6.2), and (3) *progressive logarithmic tilted time frame model* (Fig. 6.3).

Figure 6.1.   A tilted time frame with natural time partition

Figure 6.2.   A tilted time frame with logarithmic time partition

A *natural tilted time frame model* is shown in Fig. 6.1, where the time frame is structured in multiple granularity based on natural time scale: the most recent 4 quarters (15 minutes), then the last 24 hours, 31 days, and 12 months (the concrete scale will be determined by applications). Based on this model, one can compute frequent itemsets in the last hour with the precision of quarter of an hour, the last day with the precision of hour, and so on, until the whole year, with the precision of month (we align the time axis with the natural calendar time.

Frame no.	Snapshots (by clock time)
0	69 67 65
1	70 66 62
2	68 60 52
3	56 40 24
4	48 16
5	64 32

Figure 6.3. A tilted time frame with progressive logarithmic time partition

Thus, for each granularity level of the tilt time frame, there might be a partial interval which is less than a full unit at that level.) This model registers only $4 + 24 + 31 + 12 = 71$ units of time for a year instead of $366 \times 24 \times 4 = 35,136$ units, a saving of about 495 times, with an acceptable trade-off of the grain of granularity at a distant time.

The second choice is *logarithmic tilted time model* as shown in Fig. 6.2, where the time frame is structured in multiple granularity according to a logarithmic scale. Suppose the current frame holds the transactions in the current quarter. Then the remaining slots are for the last quarter, the next two quarters, 4 quarters, 8 quarters, 16 quarters, etc., growing at an exponential rate. According to this model, with one year of data and the finest precision at quarter, we will need $\log_2(365 \times 24 \times 4) + 1 = 16.1$ units of time instead of $366 \times 24 \times 4 = 35,136$ units. That is, we will just need 17 time frames to store the compressed information.

The third choice is a *progressive logarithmic tilted time frame*, where snapshots are stored at different levels of granularity depending on the recency. Snapshots are put into different *frame numbers*, varying from 1 to *max_frame*, where $\log_2(T) - max\_capacity \leq max\_frame \leq \log_2(T)$, *max_capacity* is the maximal number of snapshots held in each frame, and T is the clock time elapsed since the beginning of the stream.

Each snapshot is represented by its timestamp. The rules for insertion of a snapshot t (at time t) into the snapshot frame table are defined as follows: (1) if $(t \mod 2^i) = 0$ but $(t \mod 2^{i+1}) \neq 0$, t is inserted into *frame_number* i if $i \leq max\_frame$; otherwise (i.e., $i > max\_frame$), t is inserted into *max_frame*; and (2) each slot has a *max_capacity* (which is 3 in our example of Fig. 6.3). At the insertion of t into *frame_number* i, if the slot already reaches its *max_capacity*, the oldest snapshot in this frame is removed and the new snapshot inserted. For example, at time 70, since $(70 \mod 2^1) = 0$ but $(70 \mod 2^2) \neq 0$, 70 is inserted into frame_number 1 which knocks out the oldest snapshot 58 if the slot capacity is 3. Also, at time 64, since $(64 \mod 2^6) = 0$ but *max_frame* $= 5$, so 64 has to be inserted into frame 5. Following this rule, when slot capacity is 3, the following snapshots are stored in the tilted time frame table: 16, 24, 32, 40, 48, 52, 56, 60, 62, 64, 65, 66, 67, 68, 69, 70, as shown in Fig. 6.3. From the table, one can see that the closer to the current time, the denser are the snapshots stored.

In the logarithmic and progressive logarithmic models discussed above, we have assumed that the base is 2. Similar rules can be applied to any base α, where α is an integer and $\alpha > 1$. The tilted time models shown above are sufficient for usual time-related queries, and at the same time it ensures that the total amount of data to retain in memory and/or to be computed is small.

Both the natural tilted frame model and the progressive logarithmic tilted time frame model provide a natural and systematic way for incremental insertion of data in new frames and gradually fading out the old ones. When fading out the old ones, their measures are properly propagated to their *corresponding* retained timeframe (e.g., from a quarter to its corresponding hour) so that these values are retained in the aggregated form. To simplify our discussion, we will only use the natural titled time frame model in the following discussions. The methods derived from this time frame can be extended either directly or with minor modifications to other time frames.

In our data cube design, we assume that each cell in the base cuboid and in an aggregate cuboid contains a tilted time frame, for storing and propagating measures in the computation. This tilted time frame model is sufficient to handle usual time-related queries and mining, and at the same time it ensures that the total amount of data to retain in memory and/or to be computed is small.

3.2   Critical layers

Even with the *tilted time frame* model, it could still be too costly to dynamically compute and store a full cube since such a cube may have quite a few dimensions, each containing multiple levels with many distinct values. Since stream data analysis has only limited memory space but requires fast response time, a realistic arrangement is to compute and store only some mission-critical cuboids in the cube.

In our design, two critical cuboids are identified due to their conceptual and computational importance in stream data analysis. We call these cuboids layers and suggest to compute and store them dynamically. The first layer, called *m-layer*, is the *minimally interesting layer* that an analyst would like to study. It is necessary to have such a layer since it is often neither cost-effective nor practically interesting to examine the minute detail of stream data. The second layer, called *o-layer*, is the *observation layer* at which an analyst (or an automated system) would like to check and make decisions of either signaling the exceptions, or drilling on the exception cells down to lower layers to find their lower-level exceptional descendants.

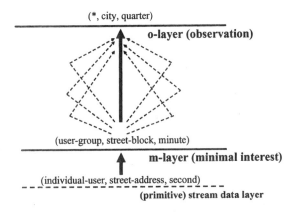

(*, city, quarter)

o-layer (observation)

(user-group, street-block, minute)

m-layer (minimal interest)

(individual-user, street-address, second)

(primitive) stream data layer

Figure 6.4. Two critical layers in the stream cube

Example 3. Assume that "($individual\_user, street\_address, second$)" forms the primitive layer of the input stream data in Ex. 1. With the *natural tilted time frame* shown in Figure 6.1, the two critical layers for power supply analysis are: (1) the m-layer: ($user\_group, street\_block, minute$), and (2) the o-layer: (*, $city, quarter$), as shown in Figure 6.4.

Based on this design, the cuboids lower than the m-layer will not need to be computed since they are beyond the minimal interest of users. Thus the minimal interesting cells that our base cuboid needs to be computed and stored will be the aggregate cells computed with grouping by $user\_group, street\_block$, and $minute$) . This can be done by aggregations (1) on two dimensions, $user$ and $location$, by rolling up from $individual\_user$ to $user\_group$ and from $street\_address$ to $street\_block$, respectively, and (2) on time dimension by rolling up from $second$ to $minute$.

Similarly, the cuboids at the o-layer should be computed dynamically according to the tilted time frame model as well. This is the layer that an analyst takes as an observation deck, watching the changes of the current stream data by examining the slope of changes at this layer to make decisions. The layer can be obtained by rolling up the cube (1) along two dimensions to * (which means *all* user_category) and $city$, respectively, and (2) along time dimension to $quarter$. If something unusual is observed, the analyst can drill down to examine the details and the exceptional cells at low levels. ◇

3.3 Partial materialization of stream cube

Materializing a cube at only two critical layers leaves much room for how to compute the cuboids in between. These cuboids can be precomputed fully, partially, not at all (i.e., leave everything computed on-the-fly). Let us first examine the feasibility of each possible choice in the environment of stream data. Since there may be a large number of cuboids between these two layers and each may contain many cells, it is often too costly in both space and time to fully materialize these cuboids, especially for stream data. On the other hand, materializing nothing forces all the aggregate cells to be computed on-the-fly, which may slow down the response time substantially. Thus, it is clear that partial materialization of a stream cube is a viable choice.

Partial materialization of data cubes has been studied extensively in previous works, such as ([21, 11]). With the concern of both space and on-line computation time, partial computation of dynamic stream cubes poses more challenging issues than its static counterpart: One has to ensure not only the limited pre-computation time and the limited size of a precomputed cube, but also efficient online incremental updating upon the arrival of new stream data, as well as fast online drilling to find interesting aggregates and patterns. Obviously, only careful design may lead to computing a rather small partial stream cube, fast updating such a cube, and fast online drilling. We will examine how to design such a stream cube in the next section.

4.   Stream Data Cube Computation

We first examine whether iceberg cube can be an interesting model for partially materialized stream cube. In data cube computation, *iceberg cube* ([7]) which stores only the aggregate cells that satisfy an iceberg condition has been used popularly as a data cube architecture since it may substantially reduce the size of a data cube when data is sparse. For example, for a sales data cube, one may want to only retain the (cube) cells (i.e., aggregates) containing more than 2 items. This condition is called as an *iceberg condition*, and the cube containing only such cells satisfying the iceberg condition is called an *iceberg cube*. In stream data analysis, people may often be interested in only the substantially important or exceptional cube cells, and such important or exceptional conditions can be formulated as typical *iceberg conditions*. Thus it seems that iceberg cube could be an interesting model for stream cube architecture. Unfortunately, iceberg cube cannot accommodate the incremental update with the constant arrival of new data and thus cannot be used as the architecture of stream cube. We have the following observation.

OBSERVATION (No iceberg cubing for stream data) The iceberg cube model does not fit the stream cube architecture. Nor does the exceptional cube model.

Rationale. With the incremental and gradual arrival of new stream data, as well as the incremental fading of the obsolete data from the time scope of a data cube, it is required that incremental update be performed on such a stream data cube. It is unrealistic to constantly recompute the data cube from scratch upon incremental updates due to the tremendous cost of recomputing the cube on the fly. Unfortunately, such an incremental model does not fit the iceberg cube computation model due to the following observation: Let a cell "$\langle d_i, \ldots, d_k \rangle$: m_{ik}" represent a $k - i + 1$ dimension cell with $d_i, \ldots, d_k$ as its corresponding dimension values and m_{ik} as its measure value. If $SAT(m_{ik}, iceberg\_cond)$ is false, i.e., m_{ik} does not satisfy the iceberg condition, the cell is dropped from the iceberg cube. However, at a later time slot t', the corresponding cube cell may get a new measure m'_{ik} related to t'. However, since m_{ik} has been dropped at a previous instance of time due to its inability to satisfy the iceberg condition, the new measure for this cell cannot be calculated correctly without such information. Thus one cannot use the iceberg architecture to model a stream cube unless recomputing the measure from the based cuboid upon each update. Similar reasoning can be applied to the case of exceptional cell cubes since the exceptional condition can be viewed as a special iceberg condition. ◇

Since iceberg cube cannot be used as a stream cube model, but materializing the full cube is too costly both in computation time and storage space, we propose to compute only a *popular path* of the cube as our partial computation of stream data cube, as described below.

Based on the notions of the minimal interesting layer (the m-layer) and the tilted time frame, stream data can be directly aggregated to this layer according to the tilted time scale. Then the data can be further aggregated following one popular drilling path to reach the observation layer. That is, the *popular path* approach computes and maintains a single popular aggregation path from m-layer to o-layer so that queries directly on those (layers) along the popular path can be answered without further computation, whereas those deviating from the path can be answered with minimal online computation from those reachable from the computed layers. Such cost reduction makes possible the OLAP-styled exploration of cubes in stream data analysis.

To facilitate efficient computation and storage of the popular path of the stream cube, a compact data structure needs to be introduced so that the space taken in the computation of aggregations is minimized. A data structure, called H-tree, a hyper-linked tree structure introduced in ([20]), is revised and adopted here to ensure that a compact structure is maintained in memory for efficient computation of multi-dimensional and multi-level aggregations.

We present these ideas using an example.

Example 4. Suppose the stream data to be analyzed contains 3 dimensions, A, B and C, each with 3 levels of abstraction (excluding the highest level of abstraction "*"), as (A_1, A_2, A_3), (B_1, B_2, B_3), (C_1, C_2, C_3), where the ordering of "$* > A_1 > A_2 > A_3$" forms a high-to-low hierarchy, and so on. The minimal interesting layer (the m-layer) is (A_2, B_2, C_2), and the o-layer is $(A_1, *, C_1)$. From the m-layer (the bottom cuboid) to the o-layer (the top-cuboid to be computed), there are in total $2 \times 3 \times 2 = 12$ cuboids, as shown in Figure 6.5.

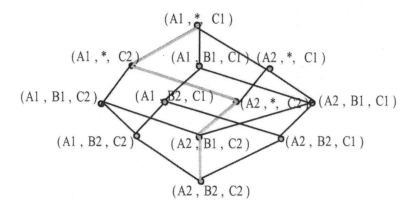

Figure 6.5. Cube structure from the m-layer to the o-layer

Suppose that the popular drilling path is given (which can usually be derived based on domain expert knowledge, query history, and statistical analysis of the sizes of intermediate cuboids). Assume that the given popular path is $\langle (A_1, *, C_1) \rightarrow (A_1, *, C_2) \rightarrow (A_2, *, C_2) \rightarrow (A_2, B_1, C_2) \rightarrow (A_2, B_2, C_2) \rangle$, shown as the dark-line path in Figure 6.5. Then each path of an H-tree from root to leaf is ordered the same as the popular path.

This ordering generates a compact tree because the set of low level nodes that share the same set of high level ancestors will share the same prefix path using the tree structure. Each tuple, which represents the currently in-flow stream data, after being generalized to the m-layer, is inserted into the corresponding path of the H-tree. An example H-tree is shown in Fig. 6.6. In the leaf node of each path, we store relevant measure information of the cells of the m-layer. The measures of the cells at the upper layers are computed using the H-tree and its associated links.

An obvious advantage of the *popular path approach* is that the nonleaf nodes represent the cells of those layers (cuboids) along the popular path. Thus these nonleaf nodes naturally serve as the cells of the cuboids along the path. That is, it serves as a data structure for intermediate computation as well as the storage area for the computed measures of the layers (i.e., cuboids) along the path.

Furthermore, the H-tree structure facilitates the computation of other cuboids or cells in those cuboids. When a query or a drill-down clicking requests to compute cells outside the popular path, one can find the closest lower level computed cells and use such intermediate computation results to compute the measures requested, because the corresponding cells can be found via a linked list of all the corresponding nodes contributing to the cells.   ◇

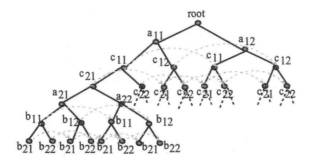

Figure 6.6.   H-tree structure for cube computation

4.1   Algorithms for cube computation

Algorithms related to stream cube in general handle the following three cases: (1) the *initial* computation of (partially materialized) stream cube by popular-path approach, (2) incremental update of stream cube, and (3) online query answering with the popular-path-based stream cube.

First, we present an algorithm for computation of (initial) partially material-ized stream cube by popular-path approach.

ALGORITHM 1(**Popular-path-based stream cube computation**) Comput-ing initial stream cube, i.e., the cuboids along the *popular-path* between the m-layer and the o-layer, based on the currently collected set of input stream data.

Input. (1) multi-dimensional multi-level stream data (which consists of a set of tuples, each carrying the corresponding time stamps), (2) the m- and o-layer specifications, and (3) a given popular drilling path.

Output. All the aggregated cells of the cuboids along the popular path between the m- and o-layers.

Method.

1 Each tuple, which represents a minimal addressing unit of multi-dimensional multi-level stream data, is scanned once and generalized to the m-layer. The generalized tuple is then inserted into the corresponding path of the H-tree, increasing the count and aggregating the measure values of the corresponding leaf node in the corresponding slot of the tilted time frame.

2 Since each branch of the H-tree is organized in the same order as the specified popular path, aggregation for each corresponding slot in the tilted time frame is performed from the m-layer all the way up to the o-layer by aggregating along the popular path. The step-by-step aggregation is performed while inserting the new generalized tuples in the corresponding time slot.

3 The aggregated cells are stored in the nonleaf nodes in the H-tree, forming the computed cuboids along the popular path.

Analysis. The H-tree ordering is based on the popular drilling path given by users or experts. This ordering facilitates the computation and storage of the cuboids along the path. The aggregations along the drilling path from the m-layer to the o-layer are performed during the generalizing of the stream data to the m-layer, which takes only one scan of stream data. Since all the cells to be computed are the cuboids along the popular path, and the cuboids to be computed are the nonleaf nodes associated with the H-tree, both space and computation overheads are minimized. ◇

Second, we discuss how to perform incremental update of the stream data cube in the popular-path cubing approach. Here we deal with the "always-grow" nature of time-series stream data in an on-line, continuously growing manner.

The process is essentially an incremental computation method illustrated below, using the tilted time frame of Figure 6.1. Assuming that the memory contains the previously computed m- and o-layers, plus the cuboids along the popular path, and stream data arrives at every second. The new stream data is accumulated in the corresponding H-tree leaf nodes. Suppose the time granularity of the m-layer is minute. At the end of every minute, the accumulated data will be propagated from the leaf to the corresponding higher level cuboids. When reaching a cuboid whose time granularity is quarter, the rolled measure information remains in the corresponding minute slot until it reaches the full quarter (i.e., 15 minutes) and then it rolls up to even higher levels, and so on.

Notice in this process, the measure in the time interval of each cuboid will be accumulated and promoted to the corresponding coarser time granularity, when the accumulated data reaches the corresponding time boundary. For example, the measure information of every four quarters will be aggregated to one hour and be promoted to the hour slot, and in the mean time, the quarter slots will still retain sufficient information for quarter-based analysis. This design ensures that although the stream data flows in-and-out, measure always keeps up to the most recent granularity time unit at each layer.

Third, we examine how an online query can be answered with such a partially materialized popular-path data cube. If a query inquires on the information that is completely contained in the popular-path cuboids, it can be answered by directly retrieving the information stored in the popular-path cuboids. Thus our discussion will focus on the kind of queries that involve the aggregate cells not contained in the popular-path cuboids.

A multi-dimensional multi-level stream query usually provides a few instantiated constants and inquires information related to one or a small number of dimensions. Thus one can consider a query involving a set of instantiated dimensions, $\{D_{ci}, \ldots, D_{cj}\}$, and a set of inquired dimensions, $\{D_{ql}, \ldots, D_{qk}\}$. The set of relevant dimensions, D_r, is the union of the sets of instantiated dimensions and the inquired dimensions. For maximal use of the precomputed information available in the popular path cuboids, one needs to find the highest-level popular path cuboids that contains D_r. If one cannot find such a cuboid in the path, one will have to use the *base cuboid* at the m-layer to compute it. In either case, the remaining computation can be performed by fetching the relevant data set from the so-found cuboid and then computing the cuboid consisting of the inquired dimensions.

5.   Performance Study

To evaluate the effectiveness and efficiency of our proposed stream cube and OLAP computation methods, we performed an extensive performance study on synthetic datasets. Our result shows that the total memory and computation time taken by the proposed algorithms are small, in comparison with several other alternatives, and it is realistic to compute such a partially aggregated cube, incrementally update them, and perform fast OLAP analysis of stream data using such precomputed cube.

Besides our experiments on the synthetic datasets, the methods have also been tested on the real datasets in the MAIDS (Mining Alarming Incidents in Data Streams) project at NCSA ([10]). The multidimensional analysis engine of the MAID system is constructed based on the algorithms presented in this paper. The experiments demonstrate similar performance results as reported in this study.

Here we report our performance studies with synthetic data streams of various characteristics. The data stream is generated by a data generator similar in spirit to the IBM data generator ([5]) designed for testing data mining algorithms. The convention for the data sets is as follows: $D3L3C10T400K$ means there are 3 dimensions, each dimension contains 3 levels (from the m-layer to the o-layer, inclusive), the node fan-out factor (cardinality) is 10 (i.e., 10 children per node), and there are in total 400K merged m-layer tuples.

Notice that all the experiments are conducted in a static environment as a simulation of the online stream processing. This is because the cube computation, especially for full cube and top-k cube, may take much more time than

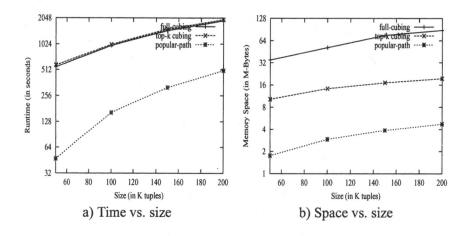

a) Time vs. size   b) Space vs. size

Figure 6.7. Cube computation: time and memory usage vs. # tuples at the m-layer for the data set $D5L3C10$

the stream flow allows. If this is performed in the online streaming environment, substantial amount of stream data could have been lost due to the slow computation of such data cubes. This simulation serves our purpose since it clear demonstrates the cost and the possible delays of stream cubing and indicates what could be the realistic choice if they were put in a dynamic streaming environment.

All experiments were conducted on a 2GHz Pentium PC with 1 GB main memory, running Microsoft Windows-XP Server. All the methods were implemented using Sun Microsystems' Java 1.3.1.

Our design framework has some **obvious performance advantages** over some alternatives in a few aspects, including (1) *tilted time frame* vs. *full non-tilted time frame*, (2) *using minimal interesting layer* vs. *examining stream data at the raw data layer*, and (3) *computing the cube up to the apex layer* vs. *computing it up to the observation layer*. Consequently, our feasibility study will not compare the design that does not have such advantages since they will be obvious losers.

Since a data analyst needs fast on-line response, and both space and time are critical in processing, we examine both time and space consumption. In our study, besides presenting the total time and memory taken to compute and store such a stream cube, we compare the two measures (time and space) of the *popular path* approach against two alternatives: (1) the *full-cubing* approach, i.e., materializing all the cuboids between the m- and o- layers, and (2) the *top-k cubing* approach, i.e., materializing only the top-k measured cells of the cuboids between the m- and o- layers, and we set top-k threshold to be 10%, i.e., only top 10% (in measure) cells will be stored at each layer (cuboid). Notice that top-k cubing cannot be used for incremental stream cubing. However, since people may like to pay attention only to top-k cubes, we still put it into our performance study (as initial cube computation). From the performance results, one can see that if top-k cubing cannot compete with the popular path approach, with its difficulty at handling incremental updating, it will not likely be a choice for stream cubing architecture.

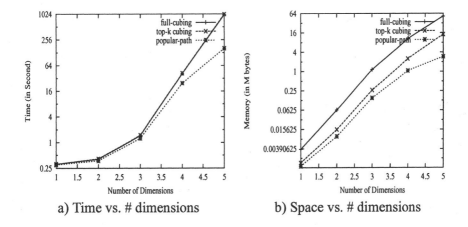

a) Time vs. # dimensions   b) Space vs. # dimensions

Figure 6.8. Cube computation: time and space vs. # of dimensions for the data set $L3C10T100K$

The performance results of stream data cubing (cube computation) are reported from Figure 6.7 to Figure 6.9.

Figure 6.7 shows the processing time and memory usage for the three approaches, with increasing size of the data set, where the size is measured as the number of tuples at the m-layer for the data set $D5L3C10$. Since *full-cubing* and *top-k cubing* compute all the cells from the m-layer all the way up to the o-layer, their total processing time is much higher than popular-path. Also, since *full-cubing* saves all the cube cells, its space consumption is much higher than popular-path. The memory usage of *top-k cubing* falls in between of the two approaches, and the concrete amount will depend on the k value.

Figure 6.8 shows the processing time and memory usage for the three approaches, with an increasing number of dimensions, for the data set $L3C10T100K$. Figure 6.9 shows the processing time and memory usage for the three approaches, with an increasing number of levels, for the data set $D5C10T50K$. The performance results show that popular-path is more efficient than both *full-cubing* and *top-k cubing* in computation time and memory usage. Moreover, one can see that increment of dimensions has much stronger impact on the computation cost (both time and space) in comparison with the increment of levels.

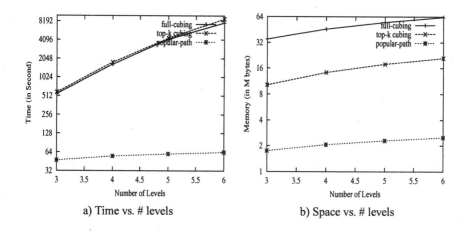

Figure 6.9.   Cube computation: time and space vs. # of levels for the data set $D5C10T50K$

Since incremental update of stream data cube carries the similar comparative costs for both popular-path and *full-cubing* approaches, and moreover, *top-k cubing* is inappropriate for incremental updating, we will not present this part of performance comparison. Notice that for incrementally computing the newly generated stream data, the computation time should be shorter than that shown here due to less number of cells involved in computation although the total memory usage may not reduce due to the need to store data in the layers along the popular path between two critical layers in the main memory.

Performance study has also been conducted on online query processing, which also shows the superior efficiency of the popular-path approach in comparison with other alternatives. Thus we conclude that popular-path is an efficient and feasible method for computing multi-dimensional, multi-level stream cubes.

6.   Related Work

Our work is related to *on-line analytical processing and mining in data cubes*, and *management and mining of stream data*. We briefly review previous research in these areas and point out the differences from our work.

In data warehousing and OLAP, much progress has been made on the efficient support of standard and advanced OLAP queries in data cubes, including selective materialization ([21]), cube computation ([7, 20, 30, 27]), cube gradient analysis ([23, 13]), exception ([26]), intelligent roll-up ([28]), and high-dimensional OLAP analysis ([24]). However, previous studies do not consider the support for stream data, which needs to handle huge amount of fast changing stream data and restricts that a data stream can be scanned only once. In contrast, our work considers complex measures in the form of stream data and studies OLAP and mining over partially materialized stream data cubes. Our data structure, to certain extent, extend the previous work on H-tree and H-cubing ([20]). However, instead of computing a materialized data cube as in H-cubing, we only use the H-tree structure to store a small number of cuboids along the popular path. This will save substantial amount of computation time and storage space and leads to high performance in both cube computation and query processing. We have also studied whether it is appropriate to use other cube structures, such as star-trees in StarCubing ([30]), dense-sparse partitioning in MM-cubing ([27]) and shell-fragments in high-dimensional OLAP ([24]). Our conclusion is that H-tree is still the most appropriate structure since most other structure needs to either scan data set more than once or know the sparse or dense part beforehand, which does not fit the single-scan and dynamic nature of data streams.

Recently, there have been intensive studies on the management and querying of stream data ([8, 17, 18, 16]), and data mining (classification and clustering) on stream data ([22, 19, 25, 29, 2, 15, 3, 4]). Although such studies lead to deep insight and interesting results on stream query processing and stream data mining, they do not address the issues of multidimensional, online analytical processing of stream data. Multidimensional stream data analysis is an essential step to understand the general statistics, trends and outliers as well as other data characteristics of online stream data and will play an essential role in stream data analysis. This study sets a framework and outlines an interesting approach to stream cubing and stream OLAP, and distinguishes itself from the previous works on stream query processing and stream data mining.

7. Possible Extensions

There are many potential extensions of the work towards comprehensive, high performance analysis of data streams. Here we outline a few.

- **Disk-based stream cube.** Although a stream cube usually retains in main memory for fast computation, updating, and accessing, it is important to have its important or substantial portion stored or mirrored on disk, which may enhance data reliability and system performance. There are several ways to do it. First, based on the design of the tilted time frame, the distant time portion in the data cube can be stored on disk. This may help reduce the total main memory requirement and the update overhead. The incremental propagation of data in such distant portion can be done by other processors using other memory space. Second, to ensure that the data is not lost in case of system error or power failure, it is important to keep a mirror copy of the stream data cube on disk. Such a mirroring process can be processed in parallel by other processors. Also, it is possible that a stream cube may miss a period of data due to software error, equipment malfunction, system failure, or other unexpected reasons. Thus a robust stream data cube should build the functionality to run despite the missing of a short period of data in the tilted time frame. The data so missed can be treated by special routines, like data smoothing, data cleaning, or other special handling so that the overall stream data can be interpreted correctly without interruption.

- **Computing complex measures in stream cubes.** Although we did not discuss the computation of complex measures in the data cube environment, it is obvious that complex measures, such as sum, avg, min, max, last, standard deviation, and many other measures can be handled for the stream data cube in the same manner as discussed in this study. Regression stream cubes can be computed efficiently as indicated in the study of ([12]). The distributed and algebraic measures of prediction cubes, as defined in ([9]), in principle, can be computed efficiently in the data stream environment. However, it is not clear how to efficiently handle holistic measures ([14]) in the stream data cubing environment. For example, it is still not clear that how some holistic measures, such as quantiles, rank, median, and so on, can be computed efficiently in this framework. This issue is left for future research.

- **Toward multidimensional online stream mining.** This study is on multidimensional OLAP stream data analysis. Many data mining tasks requires deeper analysis than simple OLAP analysis, such as classification, clustering and frequent pattern analysis. In principle, the general framework worked out in this study, including tilted time frame, minimal generalized layer and observation layers, as well as partial precomputation for powerful online analysis, will be useful for in-depth data mining methods. It is an interesting research theme on how to extend this framework towards online stream data mining.

8.　　Conclusions

In this paper, we have promoted on-line analytical processing of stream data, and proposed a feasible framework for on-line computation of multidimensional, multi-level stream cube.

We have proposed a general stream cube architecture and a stream data cubing method for on-line analysis of stream data. Our method uses a *tilted time frame*, explores *minimal interesting and observation layers*, and adopts a *popular path approach* for efficient computation and storage of stream cube to facilitate OLAP analysis of stream data. Our performance study shows that the method is cost-efficient and is a realistic approach based on the current computer technology. Recently, this stream data cubing methodology has been successfully implemented in the MAIDS project at NCSA (National Center for Supercomputing Applications) at the University of Illinois, and tested its effectiveness using online stream data sets ([10]).

Our proposed stream cube architecture shows a promising direction for realization of on-line, multi-dimensional analysis of data streams. There are a lot of issues to be explored further. In particular, it is important to further develop data mining methods to take advantage of stream cubes for on-line mining of multi-dimensional knowledge in stream data.

References

[1] S. Agarwal, R. Agrawal, P. M. Deshpande, A. Gupta, J. F. Naughton, R. Ramakrishnan, and S. Sarawagi. On the computation of multidimensional aggregates. In *Proc. 1996 Int. Conf. Very Large Data Bases (VLDB'96)*, pages 506–521, Bombay, India, Sept. 1996.

[2] C. C. Aggarwal, J. Han, J. Wang, and P. S. Yu. A framework for clustering evolving data streams. In *Proc. 2003 Int. Conf. Very Large Data Bases (VLDB'03)*, pages 81–92, Berlin, Germany, Sept. 2003.

[3] C. Aggarwal, J. Han, J. Wang, and P. S. Yu. A framework for projected clustering of high dimensional data streams. In *Proc. 2004 Int. Conf. Very Large Data Bases (VLDB'04)*, pages 852–863, Toronto, Canada, Aug. 2004.

[4] C. Aggarwal, J. Han, J. Wang, and P. S. Yu. On demand classification of data streams. In *Proc. 2004 ACM SIGKDD Int. Conf. Knowledge Discovery in Databases (KDD'04)*, pages 503–508, Seattle, WA, Aug. 2004.

[5] R. Agrawal and R. Srikant. Mining sequential patterns. In *Proc. 1995 Int. Conf. Data Engineering (ICDE'95)*, pages 3–14, Taipei, Taiwan, Mar. 1995.

[6] B. Babcock, S. Babu, M. Datar, R. Motwani, and J. Widom. Models and issues in data stream systems. In *Proc. 2002 ACM Symp. Principles of Database Systems (PODS'02)*, pages 1–16, Madison, WI, June 2002.

[7] K. Beyer and R. Ramakrishnan. Bottom-up computation of sparse and iceberg cubes. In *Proc. 1999 ACM-SIGMOD Int. Conf. Management of Data (SIGMOD'99)*, pages 359–370, Philadelphia, PA, June 1999.

[8] S. Babu and J. Widom. Continuous queries over data streams. *SIGMOD Record*, 30:109–120, 2001.

[9] B.-C. Chen, L. Chen, Y. Lin, and R. Ramakrishnan. Prediction cubes. In *Proc. 2005 Int. Conf. Very Large Data Bases (VLDB'05)*, pages 982–993, Trondheim, Norway, Aug. 2005.

[10] Y. D. Cai, D. Clutter, G. Pape, J. Han, M. Welge, and L. Auvil. MAIDS: Mining alarming incidents from data streams. In *Proc. 2004 ACM-SIGMOD Int. Conf. Management of Data (SIGMOD '04)*, pages 919–920, Paris, France, June 2004.

[11] S. Chaudhuri and U. Dayal. An overview of data warehousing and OLAP technology. *SIGMOD Record*, 26:65–74, 1997.

[12] Y. Chen, G. Dong, J. Han, B. W. Wah, and J. Wang. Multi-dimensional regression analysis of time-series data streams. In *Proc. 2002 Int. Conf. Very Large Data Bases (VLDB '02)*, pages 323–334, Hong Kong, China, Aug. 2002.

[13] G. Dong, J. Han, J. Lam, J. Pei, and K. Wang. Mining multi-dimensional constrained gradients in data cubes. In *Proc. 2001 Int. Conf. on Very Large Data Bases (VLDB '01)*, pages 321–330, Rome, Italy, Sept. 2001.

[14] J. Gray, S. Chaudhuri, A. Bosworth, A. Layman, D. Reichart, M. Venkatrao, F. Pellow, and H. Pirahesh. Data cube: A relational aggregation operator generalizing group-by, cross-tab and sub-totals. *Data Mining and Knowledge Discovery*, 1:29–54, 1997.

[15] C. Giannella, J. Han, J. Pei, X. Yan, and P. S. Yu. Mining frequent patterns in data streams at multiple time granularities. In H. Kargupta, A. Joshi, K. Sivakumar, and Y. Yesha, editors, *Data Mining: Next Generation Challenges and Future Directions*. AAAI/MIT Press, 2004.

[16] M. Greenwald and S. Khanna. Space-efficient online computation of quantile summaries. In *Proc. 2001 ACM-SIGMOD Int. Conf. Management of Data (SIGMOD '01)*, pages 58–66, Santa Barbara, CA, May 2001.

[17] A. C. Gilbert, Y. Kotidis, S. Muthukrishnan, and M. Strauss. Surfing wavelets on streams: One-pass summaries for approximate aggregate queries. In *Proc. 2001 Int. Conf. on Very Large Data Bases (VLDB '01)*, pages 79–88, Rome, Italy, Sept. 2001.

[18] J. Gehrke, F. Korn, and D. Srivastava. On computing correlated aggregates over continuous data streams. In *Proc. 2001 ACM-SIGMOD Int. Conf. Management of Data (SIGMOD '01)*, pages 13–24, Santa Barbara, CA, May 2001.

[19] S. Guha, N. Mishra, R. Motwani, and L. O'Callaghan. Clustering data streams. In *Proc. 2000 Symp. Foundations of Computer Science (FOCS '00)*, pages 359–366, Redondo Beach, CA, 2000.

[20] J. Han, J. Pei, G. Dong, and K. Wang. Efficient computation of iceberg cubes with complex measures. In *Proc. 2001 ACM-SIGMOD Int. Conf. Management of Data (SIGMOD '01)*, pages 1–12, Santa Barbara, CA, May 2001.

[21] V. Harinarayan, A. Rajaraman, and J. D. Ullman. Implementing data cubes efficiently. In *Proc. 1996 ACM-SIGMOD Int. Conf. Management of Data (SIGMOD'96)*, pages 205–216, Montreal, Canada, June 1996.

[22] G. Hulten, L. Spencer, and P. Domingos. Mining time-changing data streams. In *Proc. 2001 ACM SIGKDD Int. Conf. Knowledge Discovery in Databases (KDD'01)*, San Fransisco, CA, Aug. 2001.

[23] T. Imielinski, L. Khachiyan, and A. Abdulghani. Cubegrades: Generalizing association rules. *Data Mining and Knowledge Discovery*, 6:219–258, 2002.

[24] X. Li, J. Han, and H. Gonzalez. High-dimensional OLAP: A minimal cubing approach. In *Proc. 2004 Int. Conf. Very Large Data Bases (VLDB'04)*, pages 528–539, Toronto, Canada, Aug. 2004.

[25] G. Manku and R. Motwani. Approximate frequency counts over data streams. In *Proc. 2002 Int. Conf. Very Large Data Bases (VLDB'02)*, pages 346–357, Hong Kong, China, Aug. 2002.

[26] S. Sarawagi, R. Agrawal, and N. Megiddo. Discovery-driven exploration of OLAP data cubes. In *Proc. Int. Conf. of Extending Database Technology (EDBT'98)*, pages 168–182, Valencia, Spain, Mar. 1998.

[27] Z. Shao, J. Han, and D. Xin. MM-Cubing: Computing iceberg cubes by factorizing the lattice space. In *Proc. 2004 Int. Conf. on Scientific and Statistical Database Management (SSDBM'04)*, pages 213–222, Santorini Island, Greece, June 2004.

[28] G. Sathe and S. Sarawagi. Intelligent rollups in multidimensional OLAP data. In *Proc. 2001 Int. Conf. Very Large Data Bases (VLDB'01)*, pages 531–540, Rome, Italy, Sept. 2001.

[29] H. Wang, W. Fan, P. S. Yu, and J. Han. Mining concept-drifting data streams using ensemble classifiers. In *Proc. 2003 ACM SIGKDD Int. Conf. Knowledge Discovery and Data Mining (KDD'03)*, pages 226–235, Washington, DC, Aug. 2003.

[30] D. Xin, J. Han, X. Li, and B. W. Wah. Star-cubing: Computing iceberg cubes by top-down and bottom-up integration. In *Proc. 2003 Int. Conf. Very Large Data Bases (VLDB'03)*, pages 476–487, Berlin, Germany, Sept. 2003.

[31] Y. Zhao, P. Deshpande, and J. Naughton. An array-based algorithm for simultaneous multi-dimensional aggregates. In *Proc. ACM-SIGMOD International Conference on Management of Data*, pages 159-170, 1997.

Chapter 7

LOAD SHEDDING IN DATA STREAM SYSTEMS

Brian Babcock

Department of Computer Science
Stanford University
babcock@cs.stanford.edu

Mayur Datar

Google, Inc.
datar@cs.stanford.edu

Rajeev Motwani

Department of Computer Science
Stanford University
rajeev@cs.stanford.edu

Abstract   Systems for processing continuous monitoring queries over data streams must be adaptive because data streams are often bursty and data characteristics may vary over time. In this chapter, we focus on one particular type of adaptivity: the ability to gracefully degrade performance via "load shedding" (dropping unprocessed tuples to reduce system load) when the demands placed on the system cannot be met in full given available resources. Focusing on aggregation queries, we present algorithms that determine at what points in a query plan should load shedding be performed and what amount of load should be shed at each point in order to minimize the degree of inaccuracy introduced into query answers. We also discuss strategies for load shedding for other types of queries (set-valued queries, join queries, and classification queries).

Keywords: data streams, load shedding, adaptive query processing, sliding windows, autonomic computing

One of the main attractions of a streaming mode of data processing — as opposed to the more conventional approach for dealing with massive data sets, in which data are periodically collected and analyzed in batch mode — is the *timeliness* of the insights that are provided. In many cases, the ability to issue continuous queries and receive real-time updates to the query answers as new data arrives can be the primary motivation for preferring data stream processing technology to alternatives such as data warehousing. For this reason, it is important that data stream processing system be able to continue to provide timely answers even under difficult conditions, such as when temporary bursts in data arrival rates threaten to overwhelm the capabilities of the system.

Many data stream sources (for example, web site access patterns, transactions in financial markets, and communication network traffic) are prone to dramatic spikes in volume (e.g., spikes in traffic at a corporate web following the announcement of a new product or the spikes in traffic experienced by news web sites and telephone networks on September 11, 2001). Because peak load during a spike can be orders of magnitude higher than typical loads, fully provisioning a data stream monitoring system to handle the peak load is generally impractical. However, in many monitoring scenarios, it is precisely during bursts of high load that the function performed by the monitoring application is most critical. Therefore, it is particularly important for systems processing continuous monitoring queries over data streams to be able to automatically adapt to unanticipated spikes in input data rates that exceed the capacity of the system. An overloaded system will be unable to process all of its input data and keep up with the rate of data arrival, so *load shedding*, i.e., discarding some fraction of the unprocessed data, becomes necessary in order for the system to continue to provide up-to-date query responses. In this chapter, we consider the question of how best to perform load shedding: How many tuples should be dropped, and where in the query plan should they be dropped, so that the system is able to keep up with the rate of data arrival, while minimizing the degree of inaccuracy in the query answers introduced as a result of load shedding?

The answer to this question often differs depending on the type of queries being answered, since different classes of queries have different loss metrics for measuring the degradation in answer quality caused by load shedding. In the first part of the chapter, we perform a detailed study of the load shedding problem for one particular class of queries, sliding window aggregate queries. Afterwards, we consider several other classes of queries (set-valued queries with tuple-level utility functions, sliding-window join queries, and classification queries), and we briefly discuss load shedding techniques appropriate for each query class.

1.   Load Shedding for Aggregation Queries

The continuous monitoring queries that we consider in this section are *sliding window aggregate queries*, possibly including filters and foreign-key joins with stored relations, over continuous data streams. We restrict out attention to this query class, omitting monitoring queries involving joins between multiple streams, or non-foreign-key joins between streams and stored relations.

Overview of Approach In this section, we will describe a technique involving the introduction of load shedding operators, or *load shedders*, at various points in the query plan. Each load shedder is parameterized by a sampling rate p. The load shedder flips a coin for each tuple that passes through it. With probability p, the tuple is passed on to the next operator, and with probability $1-p$, the tuple is discarded. To compensate for the lost tuples caused by the introduction of load shedders, the aggregate values calculated by the system are scaled appropriately to produce unbiased approximate query answers.

The decisions about where to introduce load shedders and how to set the sampling rate for each load shedder are based on statistics about the data streams, including observed stream arrival rates and operator selectivities. We use statistical techniques similar to those used in approximate query processing systems to make these decisions in such a way as to achieve the best attainable accuracy given data input rates.

1.1   Problem Formulation

Preliminaries.   For our purposes, a *continuous data stream* S will be defined as a potentially unbounded sequence of tuples $\{s_1, s_2, s_3, \ldots\}$ that arrive over time and must be processed online as they arrive. A *sliding window aggregate* is an aggregation function applied over a sliding window of the most recently-arrived data stream tuples (for example, a moving average). The aggregation functions that we consider are SUM and COUNT, though the techniques described can be generalized to other functions such as AVG and MEDIAN. Sliding windows may be either *time-based*, meaning that the window consists of all tuples that have arrived within some time interval w of the present (e.g., the last 4 hours), or *tuple-based*, meaning that the window consists of the N most recently arrived tuples (e.g., the last 10,000 tuples). A *filter* is a local selection condition on tuples from a data stream.

This class of queries (sliding window aggregate queries) is important and useful for many data stream monitoring applications, including network traffic engineering, which we will use as an example application domain. Network analysts often monitor sliding window aggregates covering multiple timescales over packet traces from routers, typically filtering based on the internet protocol used, source and destination port numbers, and similar considerations. Foreign-key joins or semijoins with stored relations may be used in monitoring queries to perform filtering based on some auxiliary information that is not stored in the data stream itself (e.g., the industry grouping for a security in a financial

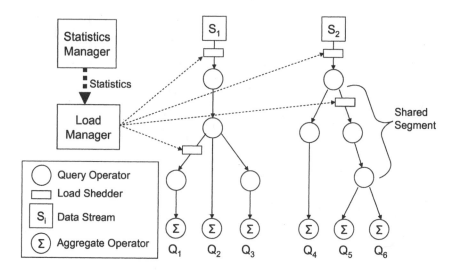

Figure 7.1. Data Flow Diagram

monitoring application). For our purposes, such joins have the same structure and effects as an expensive selection predicate or a user-defined function.

Most data stream monitoring scenarios involve multiple concurrent continuous queries. Sharing of common sub-expressions among queries is desirable to improve the scalability of the monitoring system. For this reason, it is important that a load shedding policy take into account the structure of operator sharing among query plans rather than attempting to treat each query as an isolated unit.

The input to the load shedding problem consists of a set of queries $q_1, \ldots, q_n$ over data streams $S_1, \ldots, S_m$, a set of query operators $O_1, \ldots, O_k$, and some associated statistics that are described below. The operators are arranged into a *data flow diagram* (similar to [3]) consisting of a directed acyclic graph with m source nodes representing the data streams, n sink nodes representing the queries, and k internal nodes representing the query operators. (Please refer to Figure 7.1.) The edges in the graph represent data flow between query operators. For each query q_i, there is a corresponding path in the data flow diagram from some data stream S_j though a set of query operators $O_{i_1}, O_{i_2}, \ldots, O_{i_p}$ to node q_i. This path represents the processing necessary to compute the answer to query q_i, and it is called the *query path* for query q_i. Because we do not consider joins between data streams, the data flow diagram can be thought of as being composed of a set of trees. The root node of each tree is a data stream S_j, and the leaf nodes of the tree are the queries that monitor stream S_j. Let $T(S_j)$ denote the tree of operators rooted at stream source S_j.

Every operator O_i in the data flow diagram is associated with two parameters: its selectivity s_i and its processing time per tuple t_i. The selectivity of an operator is defined as the ratio between the number of output tuples produced by the operator and the number of input tuples consumed by the operator. The processing time per tuple for an operator is defined as the average amount of time required by the operator to process each input tuple. The last operator along any query path is a windowed aggregate operator. The output of this operator is the final answer to the query and therefore not consumed by any other operator, so the selectivity of such an operator can be considered to be zero. Each SUM aggregate operator O_i is associated with two additional parameters, the mean μ_i and standard deviation σ_i of the values in the input tuples that are being aggregated. The final parameters to the load shedding problem are the rate parameters r_j, one for each data stream S_j. Rate parameter r_j represents the average rate of tuple arrival on stream S_j, measured in tuples per unit time.

Estimation of Input Parameters. Although we have described these input parameters (selectivity, stream rate, etc.) as known, fixed quantities, in reality their exact values will vary over time and cannot be known precisely in advance. In the data stream management system STREAM [8], there is a Statistics Manager module that estimates the values of these parameters. The query operators in STREAM are instrumented to report statistics on the number of tuples processed and output by the operator and the total processor time devoted to the operator. Based on these statistics, the Statistics Manager can estimate the selectivity and processing times of operators as well as the data stream arrival rates. During times when statistics gathering is enabled, the SUM aggregation operator additionally maintains statistics on the sum and sum-of-squares of the aggregate values of tuples that it processes, allowing the estimation of the mean and standard deviation of the values of the attribute being summed. As stream arrival rate and data characteristics change, the appropriate amount of load to shed and the right places to shed it may change as well. Therefore, in the STREAM system, estimates for the load shedding input parameters are periodically refreshed by the Statistics Manager, and load shedding decisions are periodically revisited.

Accuracy Metric. Let $A_1, A_2, \ldots, A_n$ be the answers to queries $q_1, q_2, \ldots, q_n$ at some point in time, and let $\widehat{A}_1, \widehat{A}_2, \ldots, \widehat{A}_n$ be the answers produced by the data stream monitoring system. If the input rates are high enough that load shedding becomes necessary, the data stream monitoring system may not be able to produce the correct query answers, i.e., $\widehat{A}_i \neq A_i$ for some or all queries q_i. The quality of a load shedding policy can be measured in terms of the deviation of the estimated answers produced by the system from the actual answers. Since the relative error in a query answer is generally more important than the absolute magnitude of the error, the goal of our load shedding policy will be to minimize the relative error for each query, defined as $\epsilon_i = |A_i - \widehat{A}_i|/|A_i|$. Moreover, as there are multiple queries, we aim to minimize the maximum error across all queries, $\epsilon_{max} = \max_{1 \leq i \leq n} \epsilon_i$.

Load Constraint.   The purpose of load shedding is to increase the throughput of the monitoring system so that the rate at which tuples are processed is at least as high as the rate at which new input tuples are arriving on the data streams. If this relationship does not hold, then the system will be unable to keep up with the arriving data streams, and input buffers and latency of responses will grow without bound. We capture this requirement in an equation, which we call the *load equation*, that acts as a constraint on load shedding decisions.

Before presenting the load equation, we will first introduce some additional notation. As mentioned earlier, each operator O_i is part of some tree of operators $T(S_j)$. Let U_i denote the set of operators "upstream" of O_i—that is, the operators that fall on the path from S_j to O_i in the data flow diagram. If some of the operators upstream of O_i are selective, the data input rate seen by operator O_i will be less than the data stream rate r_j at the stream source since some tuples are filtered out before reaching O_i. Furthermore, if load shedders are introduced upstream of O_i, they will also reduce the effective input rate seen by O_i. Let us define p_i as the sampling rate of the load shedder introduced immediately before operator O_i and let $p_i = 1$ when no such load shedder exists. Thus to measure the time spent in processing operator O_i, we are interested in the *effective input rate* for O_i, which we denote $r(O_i) = r_{src(i)} p_i \prod_{O_x \in U_i} s_x p_x$. (Here $src(i)$ denotes the index of the data stream source for operator O_i, i.e. $src(i) = j$ for $O_i \in T(S_j)$.) This leads to the load equation:

EQUATION 1.1 (LOAD EQUATION) *Any load shedding policy must select sampling rates p_i to ensure:*

$$\sum_{1 \leq i \leq k} \left(t_i r_{src(i)} p_i \prod_{O_x \in U_i} s_x p_x \right) \leq 1 \qquad (7.1)$$

The left hand side of Equation 1.1 gives the total amount of time required for the system to process the tuples that arrive during one time unit, assuming that the overhead introduced by load shedding is negligible. Clearly, this processing time can be at most one time unit, or else the system will be unable to keep up with the arriving data streams. The assumption that the cost of load shedding is small relative to the cost of query operators, and can therefore be safely ignored, is borne out by experimental evidence [2].

Problem Statement.   The formal statement of the load shedding problem is as follows: *Given a data flow diagram, the parameters $s_i, t_i, \mu_i, \sigma_i$ for each* operator O_i, and the rate parameters r_j for each data stream S_j, select load shedding sampling rates p_i to minimize the maximum relative error $\epsilon_{max} = \max_{1 \leq i \leq n} \epsilon_i$, subject to the constraint that the load equation, Equation 1.1, must be satisfied.

1.2 Load Shedding Algorithm

In this section, we describe our algorithm for determining the locations at which load shedding should be performed and setting the sampling rate parameters p_i. The algorithm has two steps:

1 Determine the effective sampling rates for each query that will distribute error evenly among all queries.

2 Determine where in the data flow diagram load shedding should be performed to achieve the appropriate rates and satisfy the load equation.

These two steps are described in detail below.

Allocation of Work Among Queries. Recall that the error metric we use to measure the accuracy of a query response is the relative error. It is impossible to precisely predict the relative error in query answers that will arise from a particular choice of a load shedding policy, because the data values in the discarded tuples are unknown. However, if we assume some knowledge about the distribution of data values, for example based on previously-seen tuples from the data streams, then we can use probabilistic techniques to get good estimates of what the relative error will be. There is some variability in the relative error, even if the data distribution is known exactly, because the approximate answers produced by the system depend on the outcomes of the random coin flips made by the load shedders. Therefore, to compare alternative load shedding policies, we do the following: for a fixed small constant δ (we use 0.01), we say that a load shedding policy achieves error ϵ if, for each query q_i, the relative error resulting from using the policy to estimate the answer to q_i exceeds ϵ with probability at most δ.

Relating Sampling Rate and Error Suppose the query path for a SUM query q_i consists of the sequence of operators $O_{i1}, O_{i2}, \ldots, O_{iz}$. Consider a load shedding policy that introduces load shedders along the query path with sampling rates $p_{i1}, p_{i2}, \ldots, p_{iz}$. Let τ be a tuple that would pass through all the query operators and contribute to the query answer in the absence of any load shedders. When load shedders are present, τ will contribute to the answer if and only if it passes through all the load shedders, which occurs with probability $P_i = p_{i1}p_{i2}\ldots p_{iz}$. We will refer to P_i as the *effective sampling rate* for query q_i.

Let $\mathcal{Q}_i$ denote the set of tuples from the current sliding window that would pass all selection conditions and contribute to the query answer in the absence of load shedders. Let N_i be the number of tuples in the set $\mathcal{Q}_i$. From the above discussion, it is clear that in the presence of load shedders, this aggregate query will be answered based on a sample of $\mathcal{Q}_i$ where each element gets included independently with probability P_i. For the tuples in the set $\mathcal{Q}_i$, let $v_1, v_2, \ldots, v_{N_i}$ denote the values of the attribute being summed, and let A_i be their sum. The approximate answer $\widehat{A}_i$ produced by the system will be the sum of v_i's for the tuples that get included in the sample, scaled by the inverse of the effective sampling rate $(1/P_i)$. The following proposition, which follows directly from a result due to Hoeffding (Theorem 2 in [6]), gives an upper bound on the probability that the relative error exceeds a certain threshold ϵ_i.

PROPOSITION 1.1 *Let* $X_1, X_2, \ldots, X_N$ *be* N *random variables, such that each random variable* X_j *takes the value* v_j/P *with probability* P *and the value zero otherwise. Let* $\widehat{A}_i$ *be the sum of these random variables and let* $A_i = \sum_{j=1}^{N} v_j$. *If we denote by* SS_i *the sum* $\sum_{j=1}^{N} v_j^2$, *then*

$$Pr\{|\widehat{A}_i - A_i| \geq \epsilon |A_i|\} \leq 2 \exp\left(-2P^2 \epsilon^2 A_i^2 / SS_i\right)$$

Thus, for a query q_i, to ensure that the probability that the relative error exceeds ϵ_i is at most δ, we must guarantee that $2 \exp\left(-2P_i^2 \epsilon_i^2 A_i^2 / SS_i\right) \leq \delta$, which occurs when $P_i \epsilon_i \geq C_i$, where we define $C_i = \sqrt{\frac{SS_i}{2A_i^2} \log \frac{2}{\delta}}$. Letting the mean and variance of the values $v_1, v_2, \ldots, v_{N_i}$ be denoted by $\mu_i = \sum_{j=1}^{N_i} v_j / N_i$ and $\sigma_i^2 = \sum_{j=1}^{N_i} (v_j - \mu_i)^2 / N_i$, respectively, the ratio SS_i/A_i^2 is equal to $(\sigma_i^2 + \mu_i^2)/(N_i \mu_i^2)$. Thus the right-hand side of the preceding inequality reduces to $C_i = \sqrt{\frac{\sigma_i^2 + \mu_i^2}{2N_i \mu_i^2} \log \frac{2}{\delta}}$.

If we want a load shedding policy to achieve relative error ϵ_i, we must guarantee that $P_i \geq C_i/\epsilon_i$. Thus, to set P_i correctly, we need to estimate C_i. Recall that we are given estimates for μ_i and σ_i (provided by the Statistics Manager) as inputs to the load shedding problem. The value of N_i can be calculated from the size of the sliding window, the estimated selectivities of the operators in the query path for q_i, and (in the case of time-based sliding windows) the estimated data stream rate r_j.

The larger the value of C_i, the larger the effective sampling rate P_i needs to be to achieve a fixed error ϵ_i with a fixed confidence bound δ. Clearly, C_i is larger for queries that are more selective, for queries over smaller sliding windows, and for queries where the distribution of values for the attribute being summed is more skewed. For a COUNT aggregate, $\mu_i = 1$ and $\sigma_i = 0$, so only the window size and predicate selectivity affect the effective sampling rate.

On the Effects of Parameter Estimation Errors   Since the values of the parameters that affect the effective sampling rate are known only approximately, and they are subject to change over time, using the estimated parameter values directly to calculate effective sampling rates may result in under-sampling for a particular query, causing higher relative error. For example, if the data characteristics change so that an attribute that previously had exhibited little skew suddenly becomes highly skewed, the relative error for a query which aggregates the attribute is likely to be higher than predicted.

The impact of a mistake in estimating parameters will be more pronounced for a query whose P_i is low than for a query with higher P_i. Therefore, for applications where rapid changes in data characteristics are of concern, a more conservative policy for setting effective sampling rates could be implemented by adding a constant "fudge factor" to the estimates for C_i for each query. In effect, this would result in resources being distributed among the queries somewhat more evenly than would be optimal based on the estimated parameter values. Such a modification would misallocate resources somewhat if the estimated parameters turn out to be correct, but it would be more forgiving in the case of significant errors in the estimates.

Choosing Target Errors for Queries   The objective that we seek to minimize is the maximum relative error ϵ_i across all queries q_i. It is easy to see that the optimal solution will achieve the same relative error ϵ for all queries.

OBSERVATION 1.2 *In the optimal solution, the relative error (ϵ_i) is equal for all queries for which load shedding is performed.*

PROOF: The proof is by contradiction. Suppose that $\epsilon_i < \epsilon_j$ for two queries q_i, q_j. Since $\epsilon_i = C_i/P_i < \epsilon_j$, we could reduce P_i to P_i' by introducing a load shedder before the final aggregation operator for q_i with effective sampling rate P_i'/P_i so that $\epsilon_i' = C_i/P_i' = \epsilon_j$. By doing so, we keep the maximum relative error unchanged but reduce the processing time, gaining some slack in the load equation. This slack can be distributed evenly across all queries by increasing all load shedder sampling rates slightly, reducing the relative error for all queries.

For an optimal solution, since the relative errors for all queries are the same, the effective sampling rate P_i for each query q_i will be proportional to the C_i value for that query, since $P_i = C_i/\epsilon_i = C_i/\epsilon_{max}$. Therefore, the problem of selecting the best load shedding policy reduces to determining the best achievable ϵ_{max} and inserting load shedders such that, for each query q_i, the effective sampling rate P_i, is equal to C_i/ϵ_{max}. In doing so we must guarantee that the modified query plan, after inserting load shedders, should satisfy the load equation (Equation 1.1).

Placement of Load Shedders.   For now, assume that we have guessed the
right value of ϵ_{max}, so that we know the exact effective sampling rate P_i for
each query. (In fact, this assumption is unnecessary, as we will explain below.)
Then our task is reduced to solving the following problem: *Given a data flow
diagram along with a set of target effective sampling rates P_i for each query q_i,
modify the diagram by inserting load shedding operators and set their sampling
rates so that the effective sampling rate for each query q_i is equal to P_i and the
total processing time is minimized.*

If there is no sharing of operators among queries, it is straightforward to
see that the optimal solution is to introduce a load shedder with sampling rate
$p_i = P_i$ before the first operator in the query path for each query q_i. Introducing
a load shedder as early in the query path as possible reduces the effective
input rate for all "downstream" operators and conforms to the general query
optimization principle of pushing selection conditions down.

Introducing load shedders and setting their sampling rates is more complicated when there is sharing among query plans. Suppose that two queries q_1 and
q_2 share the first portion of their query paths but have different effective sampling rate targets P_1 and P_2. Since a load shedder placed at the shared beginning
of the query path will affect the effective sampling rates for both queries, it is
not immediately clear how to simultaneously achieve both effective sampling
rate targets in the most efficient manner, though clearly any solution will necessarily involve the introduction of load shedding at intermediate points in the
query paths.

We will define a *shared segment* in the data flow diagram as follows: Suppose
we label each operator with the set of all queries that contain the operator in
their query paths. Then the set of all operators having the same label is a shared
segment.

OBSERVATION 1.3 *In the optimal solution, load shedding is only performed
at the start of shared segments.*

This observation is true for the same reason that load shedding should always
be performed at the beginning of the query plan when no sharing is present:
The effective sampling rates for all queries will be the same regardless of the
position of the load shedder on the shared segment, but the total execution time
will be smallest when the load shedding is performed as early as possible.

The preceding observation rules out some types of load shedding configura-
tions, but it is not enough to determine exactly where load shedding should be
performed. The following simple example will lead us to a further observation
about the structure of the optimal solution:

EXAMPLE 7.1 *Consider a simple data flow diagram with 3 operators as shown
in Figure 7.2. Suppose the query nodes q_1 and q_2 must have effective sampling*

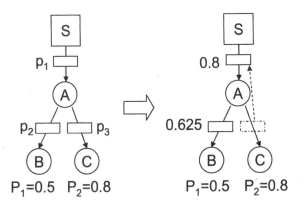

Figure 7.2. Illustration of Example 7.1

rates equal to 0.5 and 0.8 respectively. Each operator (A, B, and C) is in its own shared segment, so load shedding could potentially be performed before any operator. Imagine a solution that places load shedders before all three operators A, B, and C with sampling rates p_1, p_2, and p_3 respectively. Since $p_1 p_2 = 0.5$ and $p_1 p_3 = 0.8$, we know that the ratio $p_2/p_3 = 0.5/0.8 = 0.625$ in any solution. Consider the following modification to the solution: eliminate the load shedder before operator C and change the sampling rates for the other two load shedders to be $p'_1 = p_1 p_3 = 0.8$ and $p'_2 = p_2/p_3 = 0.625$. This change does not affect the effective sampling rates, because $p'_1 p'_2 = p_1 p_2 = 0.5$ and $p'_1 = p_1 p_3 = 0.8$, but the resulting plan has lower processing time per tuple. Effectively, we have pushed down the savings from load shedder p_3 to before operator A, thereby reducing the effective input rate to operator A while leaving all other effective input rates unchanged.

Let us define a *branch point* in a data flow diagram as a point where one shared segment ends by splitting into $k > 1$ new shared segments. We will call the shared segment terminating at a branch point the *parent segment* and the k shared segments originating at the branch point *child segments*. We can generalize the preceding example as follows:

OBSERVATION 1.4 *Let q_{max} be the query that has the highest effective sampling rate among all queries sharing the parent segment of a branch point B. In the optimal solution, the child segment of B that lies on the query path for q_{max} will not contain a load shedder. All other child segments of B will contain a load shedder with sampling rate P_{child}/P_{max}, where q_{child} is defined for each child segment as the query with the highest effective sampling rate among the queries sharing that child segment.*

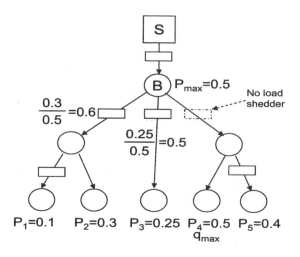

Figure 7.3. Illustration of Observation 1.4

Observation 1.4 is illustrated in Figure 7.3. The intuition underlying this observation is that, since all queries sharing the parent segment must shed at least a $(1 - P_{max})$-fraction of tuples, that portion of the load shedding should be performed as early as possible, no later than the beginning of the shared segment. The same intuition leads us to a final observation that completes our characterization of the optimal load shedding solution. Let us refer to a shared segment that originates at a data stream as an *initial segment*.

OBSERVATION 1.5 *Let q_{max} be the query that has the highest effective sampling rate among all queries sharing an initial segment S. In the optimal solution, S will contain a load shedder with sampling rate P_{max}.*

The combination of Observations 1.3, 1.4, and 1.5 completely specifies the optimal load shedding policy. This policy can be implemented using a simple top-down algorithm. If we collapse shared segments in the data flow diagram into single edges, the result is a set of trees where the root node for each tree is a data stream S_j, the internal nodes are branch points, and the leaf nodes are queries. We will refer to the resulting set of trees as the *collapsed tree representation* of the data flow diagram. For any internal node x in the collapsed tree representaiton, let P_x denote the maximum over all the effective sampling rates P_i corresponding to the leaves of the subtree rooted at this node.

The following definition will be useful in the proof of Theorem 1.7.

DEFINITION 1.6 *The **prefix path probability** of a node x in the collapsed tree representation is defined as the product of the sampling rates of all the load*

Algorithm 1 Procedure $SetSamplingRate(x, R_x)$

---

   if x is a leaf node **then**
      return
   end if
   Let $x_1, x_2, \ldots x_k$ be the children of x
   for $i = 1$ **to** k **do**
      if $P_{x_i} < R_x$ **then**
         Shed load with $p = P_{x_i}/R_x$ on edge (x, x_i)
      end if
      SetSamplingRate(x_i, P_{x_i})
   end for

Figure 7.4. Procedure $SetSamplingRate(x, R_x)$

shedders on the path from node x to the root of its tree. If there are no load shedders between the root and node x, then the prefix path probability of x is 1.

The pseudocode in Algorithm 7.4 operates over the collapsed tree representation to introduce load shedders and assign sampling rates starting with the call SetSamplingRate$(S_j, 1)$ for each data stream S_j.

THEOREM 1.7 *Among all possible choices for the placement of load shedders and their sampling rates which result in a given set of effective sampling rates for the queries, the solution generated by the* SetSamplingRate *procedure has the lowest processing time per tuple.*

PROOF: Note that in each recursive invocation of SetSamplingRate(x, R_x), the second parameter R_x is equal to the prefix path probability of node x. To prove the theorem, we first prove the claim that for each node x other than the root, the prefix path probability of x is equal to P_x.

The proof of the claim is by induction on the height of the tree. The base case consists of the root node and its children. The claim is trivially true for the root node. For a node n that is the child of the root, the top-level invocation of SetSamplingRate, with $R_{root} = 1$, places a load shedder with sampling rate $P_n/R_{root} = P_n$ at the beginning of edge $(root, n)$, so the prefix path probability of n is equal to P_n.

For the inductive case, consider any node b in the tree which is the child of some non-root node a. Assume that the claim holds for node a. When SetSamplingRate is called with a as an argument, it places a load shedder with sampling rate P_b/P_a at the beginning of edge (a, b). Thus, by the inductive hypothesis, the product of sampling rates of load shedders from the root to node b equals $P_a \times \frac{P_b}{P_a} = P_b$, proving the claim.

Thus we guarantee that the prefix path probability of any node is equal to the highest effective sampling rate of any query which includes that node in its query path. No solution could set a prefix path probability less than this value since it would otherwise violate the effective sampling rates for that query. Thus the effective input rate of each operator is the minimum that can be achieved subject to the constraint that prefix path probabilities at the leaf nodes should equal the specified effective sampling rates. This proves the optimality of the algorithm.

Determining the Value of ϵ_{max}   An important point to note about the algorithm is that except for the first load shedder that is introduced just after the root node, the sampling rates for all others depend only on the ratios between effective sampling rates (each sampling rate is equal to $P_i/P_j = C_i/C_j$ for some i, j) and not on the actual P_i values themselves. As a consequence, it is not actually necessary for us to know the value of ϵ_{max} in advance. Instead, we can express each effective sampling rate P_i as $C_i\lambda$, where $\lambda = 1/\epsilon_{max}$ is an unknown multiplier. On each query path, there is at most one load shedder whose sampling rate depends on λ, and therefore the load equation becomes a linear function of λ. After running Algorithm 7.4, we can easily solve Equation 1.1 for the resulting configuration to obtain the correct value of λ that makes the inequality in Equation 1.1 tight.

Another consequence of the fact that only load shedders on initial segments depend on the actual P_i values is that the load shedding structure remains stable as the data stream arrival rates r_j change. The effective sampling rate P_i for each query q_i over a given data stream S_j depends on the rate r_j in the same way. Therefore, changing r_j does not affect the ratio between the P_i values for these queries. The only impact that a small change to r_j will have is to modify the sampling rates for the load shedders on the initial segments.

When determining ϵ_{max} in situations when the system load is only slightly above system capacity, an additional consideration sometimes needs to be taken into account: When no load shedding is performed along the query path for a given query, the error on that query drops to zero. By contrast, for each query, there is a minimum error threshold (C_i) below which no error guarantees based on Proposition 1.1 can be given as long as *any* load shedding is performed along the query path. As the effective sampling rate P_i increases, the relative error ϵ_i decreases continuously while $P_i < 1$ then makes a discontinuous jump (from $\epsilon_i = C_i$ to $\epsilon_i = 0$) at $P_i = 1$. Our algorithm can be easily modified to incorporate this discontinuity, as described in the next paragraph.

In some cases, the value of λ that makes the inequality in Equation 1.1 tight may be greater than $1/C_{max}$, where C_{max} is the proportionality constant (derived using Proposition 1.1) of the query q_{max} with maximum target effective sampling rate. Such a value of λ corresponds to an infeasible target effective sampling rate for query q_{max}, since $P_{max} = C_{max}\lambda > 1$. It is not meaningful to have a load shedder with sampling rate greater than one, so the maximum possible effective sampling rate for any query is 1, which is attained when no load shedding is performed for that query. To handle this case, we set $P_{max} = 1$ and re-compute the placement of load shedders using the SetSamplingRate procedure (Algorithm 7.4). This re-computation may be need to be performed several times—each time forcing an additional query's target sampling rate equal to 1—until eventually $P_i \leq 1$ for all queries q_i.

1.3   Extensions

We briefly discuss how to extend our techniques to incorporate quality of services guarantees and a more general class of queries.

Quality of Service.   By taking as our objective the minimization of the maximum relative error across all queries, we have made the implicit assumption that all queries are equally important. In reality, in many monitoring applications some queries can be identified as being more critical than others. Our techniques can easily be adapted to incorporate varying quality of service requirements for different queries, either through the introduction of query weights, or query priorities, or both.

One modification would be to allow users to associate a weight or importance w_i with each query q_i. With weighted queries, the goal of the system is to minimize the maximum *weighted* relative error. When computing the effective sampling rate target for the queries, instead of ensuring that C_i/ϵ_{max} is equal for all queries q_i, we ensure that $C_i/(w_i\epsilon_{max})$ is equal. In other words, instead of $P_i \propto C_i$ we have $P_i \propto C_iw_i$.

An alternative way of specifying query importance is to assign a discrete priority level to each query. Then the goal of the system is to minimize the maximum relative error across all queries of the highest priority level. If all these queries can be answered exactly, then the system attempts to minimize the maximum relative error across queries with the second-highest priority level, and so on.

More General Query Classes.   We have discussed the load shedding problem in the context of a particular class of data stream monitoring queries, aggregation queries over sliding windows. However, the same techniques that we have developed can be applied to other classes of queries as well. One example is monitoring queries that have the same structure as the ones we have studied, except that they have set-valued answers instead of ending with an aggregation operator. In the case of set-valued queries, an approximate answer consists of a random sample of the tuples in the output set. The metric of relative error is not applicable to set-valued queries. Instead, we can measure error as the percentage of tuples from the query answer that are missing in the approximate answer. The goal of the system is to minimize the maximum value of this quantity across all queries, optionally with query weights or priorities. Our algorithm can be made to optimize for this objective by simply setting C_i for each query equal to 1.

Another class of queries that arises in data stream monitoring applications is aggregation queries with "group-bys". One can view a group-by query as multiple queries, one query for each group. However, all these queries share the entire query path and thus will have the same effective sampling rate. Consequently, the group with maximum relative error will be the one with the maximum C_i value. Since our error metric is the maximum relative error among all groups across queries, within each group-by query, the group with maximum C_i value will be the only group that counts in the design of our solution. Thus, we can treat the group with maximum C_i value as the representative group for that query.

Incorporating Load Shedding Overhead.   The results we have presented are based on the assumption that the cost (in terms of processing time) to perform load shedding is small relative to the the cost of query operators. In an actual system implementation, even simple query operators like basic selections generally have considerable overhead associated with them. A load shedder, on the other hand, involves little more than a single call to a random number generator and thus can be very efficiently implemented. In empirical tests using the STREAM system, we found that the processing time per tuple for a load shedding operator was only a small fraction of the total processing time per tuple even for a very simple query.

In some applications, however, the relative cost of load shedding may be larger, to the point where ignoring the overhead of load shedding when deciding on the placement of load shedders leads to inefficiencies. The same basic approach that we have described can be applied in such a context by associating a processing cost per tuple with load shedding operators. In this case, the best placement of load shedders can be found using dynamic programming [1].

2.   Load Shedding in Aurora

Similar to STREAM [8], Aurora [3] is a prototype of a data stream management system that has been designed to deal with a very large numbers of data streams. The query network in Aurora is a directed acyclic graph (DAG), with sources as data streams and sinks as query output nodes. Internal nodes represent one of seven primitive operators that process tuples, and edges represent queues that feed into these operators. The Aurora query-specification model differs from the one we have described earlier in two important aspects:

- The query network allows for binary operators that take input from two queues, e.g. (windowed) join of streams. Thus, the query network is not neccesarily a collection of trees.

- Aurora allows users to specify three types of *quality of service* (QoS) functions that capture the utility of the output to the user: utility as a function either of output latency, or of the percentage loss in tuples, or of the output value of tuples.

A paper by Tatbul et al. [9] discusses load shedding techniques used in the Aurora system. We highlight the similarities and differences between their approach and the one that we have described earlier. The query network structure in both systems is very similar, except for the provision for binary operators in Aurora. This leads to very similar equations for computing the load on the system, taking into the account the rates for the input streams, selectivity of operators and the time required to process each tuple by different operators. Both approaches use statistics gathered in the near past to estimate these quantities. In case of Aurora, the input rate into a binary operator is simply the sum of input rates of the individual input queues. The load equation is periodically computed to determine if the system is overloaded or not and whether we need to shed additional load or reverse any previously-introduced load shedding. Load shedding solutions by both approaches employ the *push load shedding upstream* mantra by virtue of which load shedders are always placed at the beginning of a shared segment.

The technique that we have described earlier focuses on the class of sliding-window aggregation queries, where the output at any instant is a single numeric value. The aim was to minimize the maximum (weighted) relative error for all queries. In contrast, the Aurora load-shedding paper focuses on set-valued (non-aggregate) queries. One could define different metrics when load-shedding in the context of set-valued queries. We have already described one such simple metric, namely the fraction of tuples lost for each query. The provision to be able to specify QoS functions leads to an interesting metric in the context of the Aurora system: minimize the loss in utility due to load shedding. The QoS functions that relate output value and utility let users specify relative importance of tuples as identified by their attribute values. This leads to a new type of load shedding operator, one that filters and drops tuples based on their value, as opposed to randomly dropping a fixed fraction of tuples. These are referred to as *semantic load shedders*. The load shedding algorithms in Aurora follow a greedy approach of introducing load shedders in the query plan that maximize the gain (amount of load reduced) and minimize the loss in utility as measured by QoS fuctions. For every potential location for a load shedder, a loss/gain ratio is computed which is the ratio of computing cycles that will be saved for all downstream operators to the loss in utility of all downstream queries, if we drop a fixed fraction of tuples at this location. In case of semantic load shedders, filters are introduced that first shed tuples with the least useful values. A plan that introduces drops at different locations along with amount of tuples dropped is called a Load Shedding Road Map (LSRM). A set of LSRMs is precomputed based on current statistics and at run-time the system picks the appropriate LSRM based on the current load on the system.

3.   Load Shedding for Sliding Window Joins

Queries that involve joins between two or more data streams present an interesting challenge for load shedding because of the complex interactions between load shedding decisions on the streams being joined. Joins between data streams are typically *sliding window joins*. A sliding window join with window size w introduces an implicit join predicate that restricts the difference between the timestamps of two joining tuples to be at most w. The implicit time-based predicate is in addition to the ordinary join predicate.

Kang, Naughton, and Viglas [7] study load shedding for sliding window join queries with the objective of maximizing the number of output tuples that are produced. They restrict their attention to queries consisting of a single sliding-window join operator and consider the question of how best to allocate resources between the two streams that are involved in a join. Their conclusion is that the maximum rate of output tuple production is achieved when the input rates of the two data streams being joined, adjusted for the effects of load shedding, are equal. In other words, if stream S_1 arrives at rate r_1 and stream S_2 arrives at rate r_2, and load shedders are placed on each stream upstream of the join, then the sampling rate of the load shedder on stream S_i should be proportional to $1/r_i$, with the constant of proportionality chosen such that the system is exactly able to keep up with the data arrival rates.

The paper by Das, Gehrke and Riedwald [5] also addresses the same problem, namely maximizing the join size in the context of load shedding for queries containing a single sliding window join. Additionally, they introduce a metric called the *Archive-metric* (ArM) that assumes that any tuples that are load-shed by the system can be archived to allow for computing the exact answer at a later time when the load on the system is less. The ArM metric measures the amount of work that will need to be done at a later time to compute the exact answer then. They also introduce new models, inspired by different application scenarios such as sensor networks, where they distinguish between the cases when the system is bounded in terms of its CPU speed versus when it is bounded by memory. In the latter case, the goal is to bound the size of the join state measured in terms of the number of tuples stored for join processing.

The Das et al. paper mainly differs from the Kang et al. paper in that it allows for semantic load shedding as opposed to just random load shedding. The ability to drop tuples based on their join attribute value leads to interesting problems. The one that is the focus of the paper arises from the bounded memory model. In this case, the problem translates to keeping M tuples at all times so as to maximize the join size, assuming that all incoming tuples are processed and joined with the partner tuples from other stream that are stored at that time as part of the M tuples. In the static case, when the streams are not really streams but relations, they provide an optimal dynamic programming solution for binary joins and show that for an m-relation join, they show that the static problem is NP-hard. For the offline case of join between two streams, where the arrival order of tuples on both streams is assumed to be known, they provide a polynomial-time (though impractical) solution that is based on reducing the problem to a max-flow computation. They also provide two heuristic solutions that can be implemented in a real system.

4.   Load Shedding for Classification Queries

Loadstar [4] is a system for executing classification queries over data streams. Data elements arrive on multiple data streams, and the system examines each data item as it arrives and attempts to assign it to one of a finite set of classes using a data mining algorithm. An example would be monitoring images from multiple security cameras and attempting to determine which person (if any) is displayed in each image. If the data arrival rates on the streams are too high for the system to keep up, then the system must discard certain data elements unexamined, but it must nonetheless provide a predicted classification for the discarded elements. The Loadstar system is designed to deal with cases where only a small fraction of the data elements can actually be examined, because examining a data element requires expensive feature extraction steps.

The designers of Loadstar introduce two main ideas that are used for load shedding in this context:

1. A *quality of decision* metric can be used to quantify the expected degradation in classification accuracy from failing to examine a data item. In general the quality of decision function will be different for different streams. (E.g., examining an image from a security camera in a poorly-lit or low-traffic area may not yield much improvement over always guessing "no person shown", whereas analyzing images from other cameras may allow them to be classfied with high accuracy.)

2. The features used in classification often exhibit a high degree of temporal correlation. Thus, if a data element from a particular stream has been examined in the recent past, it may be a reasonable assumption that future (unexamined) data elements have similar attribute values. As time passes, uncertainty about the attribute values increases.

The load shedding strategy used in Loadstar makes use of these two ideas to decide which data elements should be examined. Loadstar uses a quality of decision metric based on Bayesian decision theory and learns a Markov model for each stream to model the rate of dispersion of attribute values over time. By combining these two factors, the Loadstar system is able to achieve better classification accuracy than the naive approach that sheds an equal fraction of load from each data stream.

5.   Summary

It is important for computer systems to be able to adapt to changes in their operating environments. This is particularly true of systems for monitoring continuous data streams, which are often prone to unpredictable changes in data arrival rates and data characteristics. We have described a framework for one type of adaptive data stream processing, namely graceful performance degradation via load shedding in response to excessive system loads. In the context of data stream aggregation queries, we formalized load shedding as an optimization problem with the objective of minimizing query inaccuracy within the limits imposed by resource constraints. Our solution to the load shedding problem uses probabilistic bounds to determine the sensitivity of different queries to load shedding in order to perform load shedding where it will have minimum adverse impact on the accuracy of query answers. Different query classes have different measurements of answer quality, and thus require different techniques for load shedding; we described three additional query classes and summarized load shedding approaches for each.

References

[1] B. Babcock. *Processing Continuous Queries over Streaming Data With Limited System Resources*. PhD thesis, Stanford University, Department of Computer Science, 2005.

[2] B. Babcock, M. Datar, and R. Motwani. Load shedding for aggregation queries over data streams. In *Proceedings of the 2004 International Conference on Data Engineering*, pages 350–361, March 2004.

[3] D. Carney, U. Cetintemel, M. Cherniack, C. Convey, S. Lee, G. Seidman, M. Stonebraker, N. Tatbul, and S. Zdonik. Monitoring streams–a new class of data management applications. In *Proc. 28th Intl. Conf. on Very Large Data Bases*, August 2002.

[4] Y. Chi, P. S. Yu, H. Wang, and R. R. Muntz. Loadstar: A load shedding scheme for classifying data streams. In *Proceedings of the 2005 SIAM International Data Mining Conference*, April 2005.

[5] A. Das, J. Gehrke, and M. Riedwald. Approximate join processing over data streams. In *Proceedings of the 2003 ACM SIGMOD International Conf. on Management of Data*, pages 40–51, 2003.

[6] W. Hoeffding. Probability inequalities for sums of bounded random variables. In *Journal of the American Statistical Association*, volume 58, pages 13–30, March 1963.

[7] J. Kang, J. F. Naughton, and S. Viglas. Evaluating window joins over unbounded streams. In *Proceedings of the 2003 International Conference on Data Engineering*, March 2003.

[8] R. Motwani, J. Widom, A. Arasu, B. Babcock, S. Babu, M. Datar, G. Manku, C. Olston, J. Rosenstein, and R. Varma. Query processing, approximation, and resource management in a data stream management system. In *Proc. First Biennial Conf. on Innovative Data Systems Research (CIDR)*, January 2003.

[9] N. Tatbul, U. Cetintemel, S. Zdonik, M. Cherniack, and M. Stonebraker. Load shedding in a data stream manager. In *Proceedings of the 2003 International Conference on Very Large Data Bases*, pages 309–320, September 2003.

Chapter 8

THE SLIDING-WINDOW COMPUTATION MODEL AND RESULTS*

Mayur Datar
Google, Inc.
datar@cs.stanford.edu

Rajeev Motwani
Department of Computer Science
Stanford University
rajeev@cs.stanford.edu

Abstract   The sliding-window model of computation is motivated by the assumption that, in certain data-stream processing applications, recent data is more useful and pertinent than older data. In such cases, we would like to answer questions about the data only over the last N most recent data elements (N is a parameter). We formalize this model of computation and answer questions about how much space and computation time is required to solve certain problems under the sliding-window model.

Keywords: sliding-window, exponential histograms, space lower bounds

Sliding-Window Model: Motivation

In this chapter we present some results related to small space computation over sliding windows in the data-stream model. Most research in the data-stream model (e.g. , see [1, 10, 15, 11, 13, 14, 19]), including results presented in some of the other chapters, assume that all data elements seen so far in the stream are equally important and synopses, statistics or models that are built should reflect the entire data set. However, for many applications this

---

*Material in this chapter also appears in **Data Stream Management: Processing High-Speed Data Streams**, edited by *Minos Garofalakis, Johannes Gehrke and Rajeev Rastogi*, published by *Springer-Verlag*.

assumption is not true, particularly those that ascribe more importance to recent data items. One way to discount old data items and only consider recent ones for analysis is the *sliding-window model*: Data elements arrive at every instant; each data element expires after exactly N time steps; and, the portion of data that is relevant to gathering statistics or answering queries is the set of last N elements to arrive. The sliding window refers to the window of active data elements at a given time instant and window size refers to N.

0.1   Motivation and Road Map

Our aim is to develop algorithms for maintaining statistics and models that use space sublinear in the window size N. The following example motivates why we may not be ready to tolerate memory usage that is linear in the size of the window. Consider the following network-traffic engineering scenario: a high speed router working at 40 gigabits per second line speed. For every packet that flows through this router we do a prefix match to check if it originates from the stanford.edu domain. At every instant, we would like to know how many packets, of the last 10^{10} packets, belonged to the stanford.edu domain. The above question can be rephrased as the following simple problem:

PROBLEM 0.1 (BASICCOUNTING) *Given a stream of data elements, consisting of 0's and 1's, maintain at every time instant the count of the number of 1's in the last N elements.*

A data element equals one if it corresponds to a packet from the stanford.edu domain and is zero otherwise. A trivial solution[1] exists for this problem that requires N bits of space. However, in such a scenario as the high-speed router, where on-chip memory is expensive and limited, and particularly when we would like to ask multiple (thousands) such continuous queries, it is prohibitive to use even $N = 10^{10}$ (window size) bits of memory for each query. Unfortunately, it is easy to see that the trivial solution is the best we can do in terms of memory usage, unless we are ready to settle for approximate answers, i.e. an exact solution to BASICCOUNTING requires $\Theta(N)$ bits of memory. We will present a solution to the problem that uses no more than $O(\frac{1}{\epsilon} \log^2 N)$ bits of memory (i.e., $O(\frac{1}{\epsilon} \log N)$ words of memory) and provides an answer at each instant that is accurate within a factor of $1 \pm \epsilon$. Thus, for $\epsilon = 0.1$ (10% accuracy) our solution will use about 300 words of memory for a window size of 10^{10}.

Given our concern that derives from working with limited space, it is natural to ask "Is this the best we can do with respect with memory utilization?" We answer this question by demonstrating a matching space lower bound, i.e. we show that any approximation algorithm (deterministic or randomized) for BA-

---

[1]Maintain a FIFO queue and update counter.

sicCounting with relative error ϵ must use $\Omega(\frac{1}{\epsilon} \log^2 N)$ bits of memory. The lower bound proves that the above mentioned algorithm is optimal, to within constant factors, in terms of memory usage.

Besides maintaining simple statistics like a bit count, as in BASICCOUNTING, there are various applications where we would like to maintain more complex statistics. Consider the following motivating example:

A fundamental operation in database systems is a join between two or more relations. Analogously, one can define a join between multiple streams, which is primarily useful for correlating events across multiple data sources. However, since the input streams are unbounded, producing join results requires unbounded memory. Moreover, in most cases, we are only interested in those join results where the joining tuples exhibit temporal locality. Consequently, in most data-stream applications, a relevant notion of joins that is often employed is sliding-window joins, where tuples from each stream only join with tuples that belong to a sliding window over the other stream. The semantics of such a join are clear to the user and also such joins can be processed in a non-blocking manner using limited memory. As a result, sliding-window joins are quite popular in most stream applications.

In order to improve join processing, database systems maintain "join statistics" for the relations participating in the join. Similarly, in order to efficiently process sliding-window joins, we would like to maintain statistics over the sliding windows, for streams participating in the join. Besides being useful for the exact computation of sliding-window joins, such statistics could also be used to approximate them. Sliding-window join approximations have been studied by Das, Gehrke and Riedwald [6] and Kang, Naughton and Viglas [16]. This further motivates the need to maintain various statistics over sliding windows, using small space and update time.

This chapter presents a general technique, called the Exponential Histogram (EH) technique, that can be used to solve a wide variety of problems in the sliding-window model; typically problems that require us to maintain statistics. We will showcase this technique through solutions to two problems: the BASICCOUNTING problem above and the SUM problem that we will define shortly. However, our aim is not to solely present solutions to these problems, rather to explain the EH technique itself, such that the reader can appropriately modify it to solve more complex problems that may arise in various applications. Already, the technique has been applied to various other problems, of which we will present a summary in Section 4.

The road map for this chapter is as follows: After presenting an algorithm for the BASICCOUNTING problem and the associated space lower bound in sections 1 and 2 respectively, we present a modified version of the algorithm in Section 3 that solves the following generalization of the BASICCOUNTING problem:

PROBLEM 0.2 (SUM) *Given a stream of data elements that are positive integers in the range* $[0 \ldots R]$, *maintain at every time instant the sum of the last N elements.*

A summary of other results in the sliding-window model is given in Section 4, before concluding in Section 8.1

1.   A Solution to the BASICCOUNTING Problem

It is instructive to observe why naive schemes do not suffice for producing approximate answers with a low memory requirement. For instance, it is natural to consider random sampling as a solution technique for solving the problem. However, maintaining a uniform random sample of the window elements will result in poor accuracy in the case where the 1's are relatively sparse.

Another approach is to maintain histograms. While the algorithm that we present follows this approach, it is important to note why previously known histogram techniques from databases are not effective for this problem. A histogram technique is characterized by the policy used to maintain the bucket boundaries. We would like to build time-based histograms in which every bucket summarizes a contiguous time interval and stores the number of 1's that arrived in that interval. As with all histogram techniques, when a query is presented we may have to interpolate in some bucket to estimate the answer, because some of the bucket's elements may have expired. Let us consider some schemes of bucketizing and see why they will not work. The first scheme that we consider is that of dividing into k equi-width (width of time interval) buckets. The problem is that the distribution of 1's in the buckets may be nonuniform. We will incur large error when the interpolation takes place in buckets with a majority of the 1's. This observation suggests another scheme where we use buckets of nonuniform width, so as to ensure that each bucket has a near-uniform number of 1's. The problem is that total number of 1's in the sliding window could change dramatically with time, and current buckets may turn out to have more or less than their fair shares of 1's as the window slides forward. The solution we present is a form of histogram that avoids these problems by using a set of well-structured and nonuniform bucket sizes. It is called the Exponential Histogram (EH) for reasons that will be clear later. Before getting into the details of the solution we introduce some notation.

We follow the conventions illustrated in Figure 8.1. In particular, we assume that new data elements are coming from the right and the elements at the left are ones already seen. Note that each data element has an *arrival time* which increments by one at each arrival, with the leftmost element considered to have arrived at time 1. But, in addition, we employ the notion of a *timestamp* which corresponds to the position of an *active* data element in the current window. We timestamp the active data elements from right to left, with the most recent element being at position 1. Clearly, the timestamps change with every new arrival and we do not wish to make explicit updates. A simple solution is to record the arrival times in a wraparound counter of $\log N$ bits and then the timestamp can be extracted by comparison with counter value of the current arrival. As mentioned earlier, we concentrate on the 1's in the data stream. When we refer to the k-th 1, we mean the k-th most recent 1 encountered in the data stream.

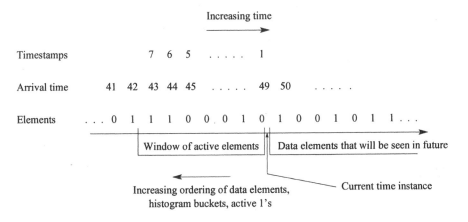

Figure 8.1. Sliding window model notation

For an illustration of this notation, consider the situation presented in Figure 8.1. The current time instant is 49 and the most recent arrival is a zero. The element with arrival time 48 is the most recent 1 and has timestamp 2 since it is the second most recent arrival in the current window. The element with arrival time 44 is the second most recent 1 and has timestamp 6.

We will maintain histograms for the active 1's in the data stream. For every bucket in the histogram, we keep the timestamp of the most recent 1 (called *timestamp* for the bucket), and the number of 1's (called *bucket size*). For example, in our figure, a bucket with timestamp 2 and size 2 represents a bucket that contains the two most recent 1's with timestamps 2 and 6. Note that timestamp of a bucket increases as new elements arrive. When the timestamp of a bucket expires (reaches $N+1$), we are no longer interested in data elements contained in it, so we drop that bucket and reclaim its memory. If a bucket is still active, we are guaranteed that it contains at least a single 1 that has not expired. Thus, at any instant there is at most one bucket (the last bucket) containing 1's that may have expired. At any time instant we may produce an estimate of the number of active 1's as follows. For all but the last bucket, we add the number of 1's that are in them. For the last bucket, let C be the count of the number of 1's in that bucket. The actual number of active 1's in this bucket could be anywhere between 1 and C, so we estimate it to be $C/2$. We obtain the following:

FACT 1.1 *The absolute error in our estimate is at most $C/2$, where C is the size of the last bucket.*

Note that, for this approach, the window size does not have to be fixed a-priori at N. Given a window size S ($S \leq N$), we do the same thing as before except that the last bucket is the bucket with the largest timestamp less than S.

1.1   The Approximation Scheme

We now define the Exponential Histograms and present a technique to maintain them, so as to guarantee count estimates with relative error at most ϵ, for any $\epsilon > 0$. Define $k = \lceil \frac{1}{\epsilon} \rceil$, and assume that $\frac{k}{2}$ is an integer; if $\frac{k}{2}$ is not an integer we can replace $\frac{k}{2}$ by $\lceil \frac{k}{2} \rceil$ without affecting the basic results.

As per Fact 1.1, the absolute error in the estimate is $C/2$, where C is the size of the last bucket. Let the buckets be numbered from right to left with the most recent bucket being numbered 1. Let m denote the number of buckets and C_i denote the size of the i-th bucket. We know that the true count is at least $1 + \sum_{i=1}^{m-1} C_i$, since the last bucket contains at least one unexpired 1 and the remaining buckets contribute exactly their size to total count. Thus, the relative estimation error is at most $(C_m/2)/(1 + \sum_{i=1}^{m-1} C_i)$. We will ensure that the relative error is at most $1/k$ by maintaining the following invariant:

INVARIANT 1.2 *At all times, the bucket sizes $C_1, \ldots, C_m$ are such that: For all $j \leq m$, we have $C_j / \langle 2(1 + \sum_{i=1}^{j-1} C_i) \leq \frac{1}{k} \rangle$.*

Let $N' \leq N$ be the number of 1's that are active at any instant. Then the bucket sizes must satisfy $\sum_{i=1}^{m} C_i \geq N'$. Our goal is to satisfy this property and Invariant 1.2 with as few buckets as possible. In order to achieve this goal we maintain buckets with exponentially increasing sizes so as to satisfy the following second invariant.

INVARIANT 1.3 *At all times the bucket sizes are nondecreasing, i.e., $C_1 \leq C_2 \leq \cdots \leq C_{m-1} \leq C_m$. Further, bucket sizes are constrained to the following: $\{1, 2, 4, \ldots, 2^{m'}\}$, for some $m' \leq m$ and $m' \leq \log \frac{2N}{k} + 1$. For every bucket size other than the size of the first and last bucket, there are at most $\frac{k}{2} + 1$ and at least $\frac{k}{2}$ buckets of that size. For the size of the first bucket, which is equal to one, there are at most $k + 1$ and at least k buckets of that size. There are at most $\frac{k}{2}$ buckets with size equal to the size of the last bucket.*

Let $C_j = 2^r$ ($r > 0$) be the size of the j-th bucket. If the size of the last bucket is 1 then there is no error in estimation since there is only data element in that bucket for which we know the timestamp exactly. If Invariant 1.3 is satisfied, then we are guaranteed that there are at least $\frac{k}{2}$ buckets each of sizes $2, 4, \ldots, 2^{r-1}$ and at least k buckets of size 1, which have indexes less than j. Consequently, $C_j < \frac{2}{k}(1 + \sum_{i=1}^{j-1} C_i)$. It follows that if Invariant 1.3 is satisfied then Invariant 1.2 is automatically satisfied, at least with respect to buckets that have sizes greater than 1. If we maintain Invariant 1.3, it is easy to see that to cover all the active 1's, we would require no more than $m \leq (\frac{k}{2} + 1)(\log(\frac{2N}{k}) + 2)$ buckets. Associated with each bucket is its size and a timestamp. The bucket size takes at most $\log N$ values, and hence we can maintain them using $\log \log N$ bits. Since a timestamp requires $\log N$ bits, the total memory requirement of each bucket is $\log N + \log \log N$ bits. Therefore, the total memory requirement (in bits) for an EH is $O(\frac{1}{\epsilon} \log^2 N)$. It is implied that by maintaining Invariant 1.3, we are guaranteed the desired relative error and memory bounds.

The query time for EH can be made $O(1)$ by maintaining two counters, one for the size of the last bucket (LAST) and one for the sum of the sizes of all buckets (TOTAL). The estimate itself is TOTAL minus half of LAST. Both counters can be updated in $O(1)$ time for every data element. See the box below for a detailed description of the update algorithm.

Algorithm (Insert):

   1 When a new data element arrives, calculate the new expiry time. If the timestamp of the last bucket indicates expiry, delete that bucket and update the counter LAST containing the size of the last bucket and the counter TOTAL containing the total size of the buckets.

   2 If the new data element is 0 ignore it; else, create a new bucket with size 1 and the current timestamp, and increment the counter TOTAL.

   3 Traverse the list of buckets in order of increasing sizes. If there are $\frac{k}{2} + 2$ buckets of the same size ($k + 2$ buckets if the bucket size equals 1), merge the oldest two of these buckets into a single bucket of double the size. (A merger of buckets of size 2^r may cause the number of buckets of size 2^{r+1} to exceed $\frac{k}{2} + 1$, leading to a cascade of such mergers.) Update the counter LAST if the last bucket is the result of a new merger.

EXAMPLE 8.1 *We illustrate the execution of the algorithm for 10 steps, where at each step the new data element is 1. The numbers indicate the bucket sizes from left to right, and we assume that $\frac{k}{2} = 1$.*

```
32, 32, 16, 8, 8, 4, 2, 1,1
32, 32, 16, 8, 8, 4, 4, 2, 1, 1, 1 (new 1 arrived)
32, 32, 16, 8, 8, 4, 4, 2, 1, 1, 1, 1 (new 1 arrived)
32, 32, 16, 8, 8, 4, 4, 2, 2, 1, 1 (merged the older 1's)
32, 32, 16, 8, 8, 4, 4, 2, 2, 1, 1, 1 (new 1 arrived)
32, 32, 16, 8, 8, 4, 4, 2, 2, 1, 1, 1, 1 (new 1 arrived)
32, 32, 16, 8, 8, 4, 4, 2, 2, 2, 1, 1 (merged the older 1's)
32, 32, 16, 8, 8, 4, 4, 4, 2, 1, 1 (merged the older 2's)
```

```
32, 32, 16, 8, 8, 8, 4, 2, 1, 1 (merged the older 4's)
32, 32, 16, 16, 8, 4, 2, 1, 1 (merged the older 8's)
```

Merging two buckets corresponds to creating a new bucket whose size is equal to the sum of the sizes of the two buckets and whose timestamp is the timestamp of the more recent of the two buckets, i.e. the timestamp of the bucket that is to the right. A merger requires $O(1)$ time. Moreover, while cascading may require $\Theta(\log \frac{2N}{k})$ mergers upon the arrival of a single new element, a simple argument, presented in the next proof, allows us to argue that the amortized cost of mergers is $O(1)$ per new data element. It is easy to see that the above algorithm maintains Invariant 1.3. We obtain the following theorem:

THEOREM 1.4 *The EH algorithm maintains a data structure that gives an estimate for the* BASICCOUNTING *problem with relative error at most ϵ using at most $(\frac{k}{2}+1)(\log(\frac{2N}{k})+2)$ buckets, where $k = \lceil \frac{1}{\epsilon} \rceil$. The memory requirement is $\log N + \log \log N$ bits per bucket. The arrival of each new element can be processed in $O(1)$ amortized time and $O(\log N)$ worst-case time. At each time instant, the data structure provides a count estimate in $O(1)$ time.*

PROOF: The EH algorithm above, by its very design, maintains Invariant 1.3. As noted earlier, an algorithm that maintains Invariant 1.3, requires no more than $(\frac{k}{2}+1)(\log(\frac{2N}{k})+2)$ buckets to cover all the active 1's. Furthermore, the invariant also guarantees that our estimation procedure has a relative error no more than $1/k \le \epsilon$.

Each bucket maintains a timestamp and the size for that bucket. Since we maintain timestamps using wraparound arrival times, they require no more than $\log N$ bits of memory. As per Invariant 1.3, bucket sizes can take only one of the $\log \frac{2N}{k} + 1$ unique values, and can be represented using $\log \log N$ bits. Thus, the total memory requirement of each bucket is no more than $\log N + \log \log N$ bits.

On the arrival of a new element, we may perform a cascading merge of buckets, that takes time proportional to the number of buckets. Since there are $O(\log N)$ buckets, this gives a worst case update time of $O(\log N)$. Whenever two buckets are merged, the size of the merged bucket is double the size of those that are merged. The cost of the merging can be amortized among all the 1's that fall in the merged bucket. Thus, an element that belongs to a bucket of size 2^p, pays an amortized cost $1 + 1/2 + 1/4 + \cdots + 1/2^p \le 2$. This is because, whenever it gets charged, the size of the bucket it belongs to doubles and consequently the charge it incurs halves. Thus, we get that the amortized cost of merging buckets is $O(1)$ per new element, in fact $O(1)$ per new element that has value 1.

We maintain counters TOTAL and LAST, which can be updated in $O(1)$ time for each new element, and which enable us to give a count estimate in $O(1)$ time whenever a query is asked.

If instead of maintaining a timestamp for every bucket, we maintain a timestamp for the most recent bucket and maintain the difference between the timestamps for the successive buckets then we can reduce the total memory requirement to $O(k \log^2 \frac{N}{k})$.

2. Space Lower Bound for BASICCOUNTING Problem

We provide a lower bound which verifies that the algorithms is optimal in its memory requirement. We start with a deterministic lower bound of $\Omega(k \log^2 \frac{N}{k})$. We omit proofs for lack of space, and refer the reader to [8].

THEOREM 2.1 *Any deterministic algorithm that provides an estimate for the* BASICCOUNTING *problem at every time instant with relative error less than* $\frac{1}{k}$ *for some integer* $k \leq 4\sqrt{N}$ *requires at least* $\frac{k}{16} \log^2 \frac{N}{k}$ *bits of memory.*

The proof argument goes as follows: At any time instant, the space utilized by any algorithm, is used to summarize the contents of the current active window. For a window of size N, we can show that there are a large number of possible input instances, i.e. arrangements of 0's and 1's, such that any deterministic algorithm which provides estimates with small relative error (i.e. less than $\frac{1}{k}$) has to differentiate between every pair of these arrangements. The number of memory bits required by such an algorithm must therefore exceed the logarithm of the number of arrangements. The above argument is formalized by the following lemma.

LEMMA 2.2 *For* $k/4 \leq B \leq N$, *there exist* $L = \binom{B}{k/4}^{\lfloor \log \frac{N}{B} \rfloor}$ *arrangements of 0's and 1's of length* N *such that any deterministic algorithm for* BASIC-COUNTING *with relative error less than* $\frac{1}{k}$ *must differentiate between any two of the arrangements.*

To prove Theorem 2.1, observe that if we choose $B = \sqrt{Nk}$ in the lemma above then $\log L \geq \frac{k}{16} \log^2 \frac{N}{k}$. While the lower bound above is for a deterministic algorithm, a standard technique for establishing lower bounds for randomized algorithms, called the *minimax principle* [18], lets us extend this lower bound on the space complexity to randomized algorithms.

As a reminder, a *Las Vegas algorithm* is a randomized algorithm that always produces the correct answer, although the running time or space requirement of the algorithm may vary with the different random choices that the algorithm makes. On the other hand, a *Monte Carlo algorithm* is a randomized algorithm that sometimes produces an incorrect solution. We obtain the following lower bounds for these two classes of algorithms.

THEOREM 2.3 *Any randomized Las Vegas algorithm for* BASICCOUNTING *with relative error less than* $\frac{1}{k}$, *for some integer* $k \leq 4\sqrt{N}$, *requires an expected memory of at least* $\frac{k}{16} \log^2 \frac{N}{k}$ *bits.*

THEOREM 2.4 *Any randomized Monte Carlo algorithm for* BASICCOUNT-ING *problem with relative error less than* $\frac{1}{k}$, *for some integer* $k \leq 4\sqrt{N}$, *with probability at least* $1 - \delta$ *(for* $\delta < \frac{1}{2}$*) requires an expected memory of at least* $\frac{k}{32} \log^2 \frac{N}{k} + \frac{1}{2} \log(1 - 2\delta)$ *bits.*

3.   Beyond 0's and 1's

The BASICCOUNTING problem, discussed in the last two sections, is one of the basic operations that one can define over sliding windows. While the problem in its original form has various applications, it is natural to ask "What are the other problems, in the sliding-window model, that can be solved using small space and small update time?". For instance, instead of the data elements being binary values, namely 0 and 1, what if they were positive integers in the range $[0 \dots R]$? Could we efficiently maintain the sum of these numbers in the sliding-window model? We have already defined this problem, in Section 8, as the SUM problem.

We will now present a modification of the algorithm from Section 1, that solves the SUM problem. In doing so, we intend to highlight the characteristic elements of the solution technique, so that readers may find it easy to adapt the technique to other problems. Already, the underlying technique has been successfully applied to many problems, some of which will be listed in the following section.

One way to solve the SUM problem would be to maintain separately a sliding window sum for each of the $\log R$ bit positions using an EH from Section 1.1. As before, let $k = \lceil \frac{1}{\epsilon} \rceil$. The memory requirement for this approach would be $O(k \log^2 N \log R)$ bits. We will present a more direct approach that uses less memory. In the process we demonstrate how the EH technique introduced in Section 1 can be generalized to solving a bigger class of problems.

Typically, a problem in the sliding-window model requires us to maintain a function f defined over the elements in the sliding window. Let $f(B)$ denote the function value restricted to the elements in a bucket B. For example, in case of the SUM problem, the function f equals the sum of the positive integers that fall inside the sliding-window. In case of BASIC COUNTING the function f is simply the number of 1's that fall inside the sliding-window. We note the following central ingredients of the EH technique from Section 1 and adapt them for the SUM problem :

1 **Size of a Bucket:** The *size* of each bucket is defined as the value of the function f that we are estimating (over the sliding window), restricted to that bucket B, i.e., $f(B)$. In the earlier case *size* was simply the count of 1's falling inside the bucket. For SUM, we define *size* analogously as the sum of integers falling inside the bucket.

2 **Procedure to Merge Buckets or Combination Rule:** Whenever the algorithm decides to merge two adjacent buckets, a new bucket is created with timestamp equal to that of the newer bucket or the bucket to the right. The *size* of this new bucket is computed using the sizes of the individual buckets (i.e., , using $f(B)$'s for the buckets that are merged) and any additional information that may be stored with the buckets.[6] Clearly, for the problem of maintaining the sum of data elements, which are either 0's and 1's or positive integers, no additional information is required. By definition, the size of the new merged bucket is simply the sum of the sizes of buckets being merged .

3 **Estimation:** Whenever a query is asked, we need to estimate the answer at that moment based on the sizes of all the buckets and any additional information that we may have kept. In order to estimate the answer, we may be required to "interpolate" over the last bucket that is part inside and part outside the sliding window, i.e., the "straddling" bucket.

   Typically, this is done by computing the function value f over all buckets other than the last bucket. In order to do this, we use the same procedure as in the Merge step. To this value we may add the interpolated value of the function f from the last bucket.

   Again, for the problem of maintaining the sum of positive integers this task is relatively straightforward. We simply add up the sizes of all the buckets that are completely inside the sliding window. To this we add the "interpolated" value from the last bucket, which is simply half the size of the last bucket.

4 **Deleting the Oldest Bucket:** In order to reclaim memory, the algorithm deletes the oldest bucket when its timestamp reaches $N + 1$. This step is same irrespective of the function f we are estimating.

The technique differs for different problems in the particulars of how the steps above are executed and the rules for when to merge old buckets and

---

[6]There are problems for which just knowing the sizes of the buckets that are merged is not sufficient to compute the size of the new merged bucket. For example, if the function f is the variance of numbers, in addition to knowing the variance of the buckets that are merged, we also need to know the number of elements in each bucket and mean value of the elements from each bucket, in order to compute the variance for the merged bucket. See [4] for details.

create new ones, as new data elements get inserted. The goal is to maintain as few buckets as possible, i.e., merge buckets whenever possible, while at the same time making sure that the error due to the estimation procedure, which interpolates for the last bucket, is bounded. Typically, this goal is achieved by maintaining that the bucket sizes grow exponentially from right to left (new to old) and hence the name Exponential Histograms (EH). It is shown in [8] that the technique can be used to estimate a general class of functions f, called *weakly-additive* functions, over sliding windows. In the following section, we list different problems over sliding windows, that can be solved using the EH technique.

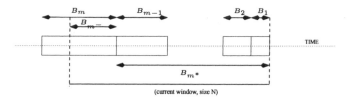

Figure 8.2.   An illustration of an Exponential Histogram (EH).

We need some more notation to demonstrate the EH technique for the SUM problem. Let the buckets in the histogram be numbered $B_1, B_2, \ldots, B_m$, starting from most recent (B_1) to oldest (B_m); further, $t_1, t_2, \ldots, t_m$ denote the bucket timestamps. See Figure 8.2 for an illustration. In addition to the buckets maintained by the algorithm, we define another set of *suffix buckets*, denoted $B_{1*}, \ldots, B_{j*}$, that represent suffixes of the data stream. Bucket B_{i*} represents all elements in the data stream that arrived after the elements of bucket B_i, that is, $B_{i*} = \bigcup_{l=1}^{i-1} B_l$. We do not explicitly maintain the suffix buckets. Let S_i denote the size of bucket B_i. Similarly, let S_{i*} denote the size of the suffix bucket B_{i*}. Note, for the **Sum** problem $S_{i*} = \sum_{l=1}^{i-1} S_l$. Let $B_{i,i-1}$ denote the bucket that would be formed by merging buckets i and $i - 1$, and $S_{i,i-1}$ ($S_{i,i-1} = S_i + S_{i-1}$) denote the size of this bucket. We maintain the following two invariants that guarantee a small relative error ϵ in estimation and small number of buckets:

INVARIANT 3.1 *For every bucket B_i, $\frac{1}{2\epsilon} S_i \le S_{i*}$.*

INVARIANT 3.2 *For each $i > 1$, for every bucket B_i,*

$$\frac{1}{2\epsilon} S_{i,i-1} > S_{i-1*}.$$

It follows from Invariant 3.1 that the relative error in estimation is no more than $\frac{S_m}{2S_{m*}} \le \epsilon$. Invariant 3.2 guarantees that the number of buckets is no more than $O(\frac{1}{\epsilon}(\log N + \log R))$. It is easy to see the proof of this claim. Since $S_{i*} = \sum_{l=1}^{i-1} S_l$, i.e., the bucket sizes are additive, after every $\lceil 1/\epsilon \rceil$ buckets (rather $\lceil 1/2\epsilon \rceil$ pairs of buckets) the value of S_{i*} doubles. As a result, after $O(\frac{1}{\epsilon}(\log N + \log R))$ buckets the value of S_{i*} exceeds NR, which is the maximum value that can be achieved by S_{i*}. We now present a simple insert algorithm that maintains the two invariants above as new elements arrive.

Algorithm (Insert): x_t denotes the most recent element.

1. If $x_t = 0$, then do nothing. Otherwise, create a new bucket for x_t. The new bucket becomes B_1 with $S_1 = x_t$. Every old bucket B_i becomes B_{i+1}.

2. If the oldest bucket B_m has timestamp greater than N, delete the bucket. Bucket B_{m-1} becomes the new oldest bucket.

3. **Merge step:** Let $k = \frac{1}{2\epsilon}$. While there exists an index $i > 2$ such that $kS_{i,i-1} \leq S_{i-1*}$, find the smallest such i and combine buckets B_i and B_{i-1} using the combination rule described earlier. Note that S_{i*} value can be computed incrementally by adding S_{i-1} and S_{i-1*}, as we make the sweep.

Note, Invariant 3.1 holds for buckets that have been formed as a result of the merging of two or more buckets because the merging condition assures that it holds for the merged bucket. Addition of new elements in the future does not violate the invariant, since the right-hand side of the invariant can only increase by addition of the new elements. However, the invariant may not hold for a bucket that contains a singleton nonzero element and was never merged. The fact that the invariant does not hold for such a bucket, does not affect the error bound for the estimation procedure because, if such a bucket were to become the last bucket, we know the exact timestamp for the only non zero element in the bucket. As a result there is no interpolation error in that case.

Analogously to the variables TOTAL and LAST in Section 1.1, we can maintain $S_m + S_{m*}$ and S_m that enable us to answer queries in $O(1)$ time. The algorithm for insertion requires $O(\frac{1}{\epsilon}(\log N + \log R))$ time per new element. Most of the time is spent in Step 3, where we make the sweep to combine buckets. This time is proportional to number of buckets, $(O(\frac{1}{\epsilon}(\log N + \log R)))$. A simple trick, to skip Step 3 until we have seen $\Theta(\frac{1}{\epsilon}(\log N + \log R))$ data points, ensures that the running time of the algorithm is amortized $O(1)$. While we may violate Invariant 3.2 temporarily, we restore it after seeing $\Theta(\frac{1}{\epsilon}(\log N + \log R))$ data points, by executing Step 3, which ensures that the number of buckets is $O(\frac{1}{\epsilon}(\log N + \log R))$. The space requirement for each bucket (memory needed to maintain timestamp and size) is $\log N + \log R$ bits. If we assume that a word is at least $\log N + \log R$ bits long, equivalently the size required to count up to NR, which is the maximum value of the answer, we get that the total memory requirement is $O(\frac{1}{\epsilon}(\log N + \log R))$ words or $O(\frac{1}{\epsilon}(\log N + \log R)^2)$ bits. Please refer to [8] for a more complex procedure that has similar time requirements and space requirement $O(\frac{1}{\epsilon} \log N (\log N + \log R))$ bits. To summarize, we get the following theorem:

THEOREM 3.3 *The sum of positive integers in the range $[0 \dots R]$ can be estimated over sliding windows with relative error at most ϵ using $O(\frac{1}{\epsilon}(\log N + \log R))$ words of memory. The time to update the underlying EH is worst case $O(\frac{1}{\epsilon}(\log N + \log R))$ and amortized $O(1)$.*

Similar to the space lower bound that we presented in Section 2, one can show a space lower bound of $\Omega(\frac{1}{\epsilon}(\log N + \log R)(\log N))$ bits for the SUM problem. See [8] for details. This is asymptotically equal to the upper bound for the algorithm in [8] that we mentioned earlier.

It is natural to ask the question: What happens if we do not restrict data elements to positive integers and are interested in estimating the sum over sliding windows. We show that even if we restrict the set of unique data elements to $\{1, 0, -1\}$, to approximate the sum within a constant factor requires $\Omega(N)$ bits of memory. Moreover, it is easy to maintain the sum by storing the last N integers which requires $O(N)$ bits of memory. We assume that the storage required for every integer is a constant independent of the window size N. With this assumption, we have that the complexity of the problem in the general case (allowing positive and negative integers) is $\Theta(N)$.

We now argue the lower bound of $\Omega(N)$. Consider an algorithm A that provides a constant-factor approximation to the problem of maintaining the general sum. Given a bit vector of size $N/2$ we present the algorithm A with the pair $(-1, 1)$ for every 1 in the bit vector and the pair $(1, -1)$ for every 0. Consider the state (time instance) after we have presented all the $N/2$ pairs to the algorithm. We claim that we can completely recover the original bit vector by presenting a sequence of 0's henceforth and querying the algorithm on every odd time instance. If the current time instance is T (after having presented the $N/2$ pairs) then it is easy to see that the correct answer at time instance $T + 2i - 1$ ($1 \le i \le N/2$) is 1 iff the ith bit was 1 and -1 iff the ith bit was 0. Since the algorithm A gives a constant factor approximation its estimate would be positive if the correct answer is 1 and negative if the correct answer was -1. Since the state of the algorithm after feeding the $N/2$ pairs enables us to recover the bit vector exactly for any arbitrary bit vector it must be using at least $N/2$ bits of memory to encode it. This proves the lower bound. We can state the following theorem:

THEOREM 3.4 *The space complexity of any algorithm that gives a constant factor approximation, at every instant, to the problem of maintaining the sum of last N integers (positive or negative) that appear as stream of data elements is equal to $\Theta(N)$.*

4. References and Related Work

The EH technique, that we demonstrate through solutions to the BASIC-COUNTING and SUM problem, is by Datar, Gionis, Indyk and Motwani [8]. The space lower bounds, presented above, are also from that paper. In the same paper, the authors characterize a general class of *weakly additive* functions that can be efficiently estimated over sliding windows, using the EH technique. Also see, Datar's PhD thesis [7] for more details.

As we have seen in other chapters from this book, it is often the case that input data streams are best visualized as a high dimensional vector. A standard operation is to compute the l_p norm, for $0 < p \leq 2$, of these vectors or the l_p norm of the difference between two vectors. In Chapter, we have seen *sketching* techniques to estimate these l_p norms using small space. It turns out that, when each data element in the data stream represents an increment to some dimension of the underlying high dimensional vector, the l_p norm of a vector belongs to the class of *weakly additive* functions mentioned above. Consequently, for the restricted case when the increments are positive, the EH technique in conjunction with the sketching technique, can be adapted to the estimate l_p norms over the sliding windows. See [8, 7] for details.

Babcock, Datar, Motwani and O'Callaghan [4] showed that the variance of real numbers with maximum absolute value R, can be estimated over sliding windows with relative error at most ϵ using $O(\frac{1}{\epsilon^2}(\log N + \log R))$ words of memory. The update time for the data structure is worst case $O(\frac{1}{\epsilon^2}(\log N + \log R))$ and amortized $O(1)$. In the same paper, the authors look at the problem of maintaining k-medians clustering of points over a sliding window. They present an algorithm that uses $O(\frac{k}{\tau^4}N^{2\tau}\log^2 N)$ memory[9] and presents k centers, for which the objective function value is within a constant factor $(2^{O(1/\tau)})$ of optimal, where $\tau < 1/2$ is a parameter which captures the trade-off between the space bound and the approximation ratio. The update time for the data structure is worst case $\tilde{O}(\frac{k^2}{\tau^3}N^{2\tau})$ and amortized $\tilde{O}(k)$. Both these algorithms are an adaptation of the EH technique, presented in Section 3 above.

In this chapter, we have focussed on the sliding-window model, that assumes that the pertinent data set is the last N data elements, i.e., we focus on *sequence*-based sliding-window model. In other words, we assumed that data items arrive at regular time intervals and arrival time increases by one with every new data item that we have seen. Such regularity in arrival of data items is seldom true for most real life applications, for which arrival rates of data items may be bursty. Often, we would like to define the sliding window based on real time. It is easy

---

[9] The space required to hold a single data point, which in this case is a point from some metric space, is assumed to be $O(1)$ words.

to adapt the EH technique to such a *time-based* sliding-window model. See [8, 7] for details.

One may argue that the sliding-window model is not the right model to discount old data, in the least not the only model. If our aim is to assign a smaller weight to older elements so that they contribute less to any statistics or models we maintain, we may want to consider other monotonically decreasing functions (time decayed functions) for assigning weights to elements other than the step function (1 for the last N elements and 0 beyond) that is implicit in the sliding-window model. A natural decay function is the exponentially decreasing weight function that was considered by Gilbert et al. [12] in maintaining *aged aggregates*: For a data stream $\ldots, x_{(-2)}, x_{(-1)}, x_{(0)}$, where $x_{(0)}$ is the most recently seen data element, λ-aging aggregate is defined as $\lambda x_{(0)} + \lambda(1 - \lambda)x_{(-1)} + \lambda(1 - \lambda)^2 x_{(-2)} + \ldots$. Exponentially decayed statistics as above are easy to maintain, although one may argue that exponential decay of weights is not suited for all applications or is too restrictive. We may desire a richer class of decay functions, e.g. polynomially decaying weight functions instead of exponential decay. Cohen and Strauss [5] show how to maintain statistics efficiently for a general class of time decaying functions. Their solutions use the EH technique as a building block or subroutine, there by demonstrating the applicability of the EH technique to a wider class of models that allow for time decay, besides the sliding-window model that we have considered.

See [7] for solutions to other problems in the sliding-window model, that do not rely on the EH technique. These problems include maintaining a uniform random sample(See also [3]), maintaining the min/max of real numbers, estimating the ratio of *rare*[11] elements to the number of distinct elements(See also [9]), and estimating the similarity between two data streams measured according to the Jaccard coefficient for set similarity between two sets A, B: $|A \bigcap B|/|A \bigcup B|$(See also [9]).

Maintaining approximate counts of high frequency elements and maintaining approximate quantiles, are important problems that have been studied in database research as maintaining end-biased histograms and maintaining equidepth histograms. These problems are particularly useful for sliding-window join processing; they provide the necessary join statistics and can also be used for approximate computation of joins. A solution to these problems, in the sliding-window model, is presented by Arasu and Manku [2] and Lu et al. [17].

5.   Conclusion

In this chapter we have studied algorithms for two simple problems, BASIC-COUNTING and SUM, in the sliding-window model; a natural model to discount

---

[11] An element is termed rare if it occurs only once (or a small number of times) in the sliding window

Problem	Space requirement requirement (in words)	Space lower bound (in words) (when available)	Amortized update time	Worst case update time
BASIC COUNTING	$O(\frac{1}{\epsilon}\log N)$	$\Omega(\frac{1}{\epsilon}\log N)$	$O(1)$	$\Omega(\frac{1}{\epsilon}\log N)$
SUM	$O(\frac{1}{\epsilon}\log N)$	$\Omega(\frac{1}{\epsilon}\log N)$	$O(1)$	$\Omega(\frac{1}{\epsilon}(\log N + \log R))$
Variance	$O(\frac{1}{\epsilon^2}(\log N + \log R))$	$\Omega(\frac{1}{\epsilon}\log N)$	$O(1)$	$O(\frac{1}{\epsilon^2}(\log N + \log R))$
l_p norm sketches	$O(\log N)$		$O(1)$	$O(\log N)$
k-median $(2^{O(1/\tau)}$-approx.)	$O(\frac{k}{\tau^4}N^{2\tau} \cdot \log^2 N)$	$\Omega(\frac{1}{\epsilon}\log N)$	$\tilde{O}(k)$	$\tilde{O}(\frac{k^2}{\tau^3}N^{2\tau})$
Min/Max	$O(N)$	$\Omega(N)$	$O(\log N)$	$O(\log N)$
Similarity	$O(\log N)$		$O(\log\log N)$ (w.h.p)	$O(\log\log N)$ (w.h.p.)
Rarity	$O(\log N)$		$O(\log\log N)$ (w.h.p)	$O(\log\log N)$ (w.h.p.)
Approx. counts	$O(\frac{1}{\epsilon}\log^2\frac{1}{\epsilon})$	$\Omega(1/\epsilon)$	$O(\log(1/\epsilon))$	$O(1/\epsilon)$
Quantiles	$O(\frac{1}{\epsilon}\log\frac{1}{\epsilon}\log N)$	$\Omega(\frac{1}{\epsilon}\log(\epsilon N/\log(1/\epsilon)))$	$O(\log(1/\epsilon)\log(\epsilon N/\log(1/\epsilon)))$	$O(1/\epsilon\log(1/\epsilon))$

Table 8.1. Summary of results for the sliding-window model.

stale data that only considers the last N elements as being pertinent. Our aim was to showcase the Exponential Histogram (EH) technique that has been used to efficiently solve various problems over sliding windows. We also presented space lower bounds for the two problems above. See Table 8.1 for a summary of results in the sliding-window model. Note, for this summary, we measure memory in words, where a word is assumed large enough to hold the answer or one unit of answer. For example, in the case of BASICCOUNTING a word is assumed to be $\log N$ bits long, for SUM word is assumed to be $\log N + \log R$ bits long, for l_p norm sketches we assume that sketches can fit in a word, for clustering a word is assumed large enough to be able to hold a single point from the metric space, and so on. Similarly, we assume it is possible to do a single word operation in one unit of time while measuring time requirements.

References

[1] N. Alon, Y. Matias, and M. Szegedy. The space complexity of approximating the frequency moments. In *Proc. of the 1996 Annual ACM Symp. on Theory of Computing*, pages 20–29, 1996.

[2] A. Arasu and G. Manku. Approximate counts and quantiles over sliding windows,. In *Proc. of the 2004 ACM Symp. Principles of Database Systems*, pages 286–296, June 2004.

[3] B. Babcock, M. Datar, and R. Motwani. Sampling from a moving window over streaming data. In *Proc. of the 2002 Annual ACM-SIAM Symp. on Discrete Algorithms*, pages 633–634, 2002.

[4] B. Babcock, M. Datar, R. Motwani, and L. O'Callaghan. Maintaining variance and k-medians over data stream windows. In *Proc. of the 2003 ACM Symp. on Principles of Database Systems*, pages 234–243, June 2003.

[5] E. Cohen and M. Strauss. Maintaining time-decaying stream aggregates. In *Proc. of the 2003 ACM Symp. on Principles of Database Systems*, pages 223–233, June 2003.

[6] A. Das, J. Gehrke, and M. Riedwald. Approximate join processing over data streams. In *Proc. of the 2003 ACM SIGMOD Intl. Conf. on Management of Data*, pages 40–51, 2003.

[7] M. Datar. *Algorithms for Data Stream Systems*. PhD thesis, Stanford University, Stanford, CA, USA, December 2003.

[8] M. Datar, A. Gionis, P. Indyk, and R. Motwani. Maintaining stream statistics over sliding windows. *SIAM Journal on Computing*, 31(6):1794–1813, 2002.

[9] M. Datar and S. Muthukrishnan. Estimating rarity and similarity over data stream windows. In *Proc. of the 2002 Annual European Symp. on Algorithms*, pages 323–334, September 2002.

[10] J. Feigenbaum, S. Kannan, M. Strauss, and M. Viswanathan. An approximate 11-difference algorithm for massive data streams. In *Proc. of the 1999 Annual IEEE Symp. on Foundations of Computer Science*, pages 501–511, 1999.

[11] A. Gilbert, S. Guha, P. Indyk, Y. Kotidis, S. Muthukrishnan, and M. Strauss. Fast, small-space algorithms for approximate histogram maintenance. In *Proc. of the 2002 Annual ACM Symp. on Theory of Computing*, 2002.

[12] A. Gilbert, Y. Kotidis, S. Muthukrishnan, and M. Strauss. Surfing wavelets on streams: One-pass summaries for approximate aggregate queries. In *Proc. of the 2001 Intl. Conf. on Very Large Data Bases*, pages 79–88, 2001.

[13] M. Greenwald and S. Khanna. Space-efficient online computation of quantile summaries. In *Proc. of the 2001 ACM SIGMOD Intl. Conf. on Management of Data*, pages 58–66, 2001.

[14] S. Guha, N. Mishra, R. Motwani, and L. O'Callaghan. Clustering data streams. In *Proc. of the 2000 Annual IEEE Symp. on Foundations of Computer Science*, pages 359–366, November 2000.

[15] P. Indyk. Stable distributions, pseudorandom generators, embeddings and data stream computation. In *Proc. of the 2000 Annual IEEE Symp. on Foundations of Computer Science*, pages 189–197, 2000.

[16] J. Kang, J. F. Naughton, and S. Viglas. Evaluating window joins over unbounded streams. In *Proc. of the 2003 Intl. Conf. on Data Engineering*, March 2003.

[17] X. Lin, H. Lu, J. Xu, and J. X. Yu. Continuously maintaining quantile summaries of the most recent n elements over a data stream. In *Proc. of the 2004 Intl. Conf. on Data Engineering*, March 2004.

[18] R. Motwani and P. Raghavan. *Randomized Algorithms*. Cambridge University Press, 1995.

[19] J.S. Vitter. Random sampling with a reservoir. *ACM Trans. on Mathematical Software*, 11(1):37–57, 1985.

Chapter 9

A SURVEY OF SYNOPSIS CONSTRUCTION IN DATA STREAMS

Charu C. Aggarwal
IBM T. J. Watson Research Center
Hawthorne, NY 10532
charu@us.ibm.com

Philip S. Yu
IBM T. J. Watson Research Center
Hawthorne, NY 10532
psyu@us.ibm.com

Abstract

The large volume of data streams poses unique space and time constraints on the computation process. Many query processing, database operations, and mining algorithms require efficient execution which can be difficult to achieve with a fast data stream. In many cases, it may be acceptable to generate *approximate solutions* for such problems. In recent years a number of *synopsis structures* have been developed, which can be used in conjunction with a variety of mining and query processing techniques in data stream processing. Some key synopsis methods include those of sampling, wavelets, sketches and histograms. In this chapter, we will provide a survey of the key synopsis techniques, and the mining techniques supported by such methods. We will discuss the challenges and tradeoffs associated with using different kinds of techniques, and the important research directions for synopsis construction.

1.   Introduction

Data streams pose a unique challenge to many database and data mining applications because of the computational and storage costs associated with the large volume of the data stream. In many cases, synopsis data structures and statistics can be constructed from streams which are useful for a variety of applications. Some examples of such applications are as follows:

- **Approximate Query Estimation:** The problem of query estimation is possibly the most widely used application of synopsis structures [11]. The problem is particularly important from an efficiency point of view, since queries usually have to be resolved in online time. Therefore, most synopsis methods such as sampling, histograms, wavelets and sketches are usually designed to be able to solve the query estimation problem.

- **Approximate Join Estimation:** The efficient estimation of join size is a particularly challenging problem in streams when the domain of the join attributes is particularly large. Many methods [5, 26, 27] have recently been designed for efficient join estimation over data streams.

- **Computing Aggregates:** In many data stream computation problems, it may be desirable to compute aggregate statistics [40] over data streams. Some applications include estimation of frequency counts, quantiles, and heavy hitters [13, 18, 72, 76]. A variety of synopsis structures such as sketches or histograms can be useful for such cases.

- **Data Mining Applications:** A variety of data mining applications such as change detection do not require to use the individual data points, but only require a temporal synopsis which provides an overview of the behavior of the stream. Methods such as clustering [1] and sketches [88] can be used for effective change detection in data streams. Similarly, many classification methods [2] can be used on a supervised synopsis of the stream.

The design and choice of a particular synopsis method depends on the problem being solved with it. Therefore, the synopsis needs to be constructed in a way which is friendly to the needs of the particular problem being solved. For example, a synopsis structure used for query estimation is likely to be very different from a synopsis structure used for data mining problems such as change detection and classification. In general, we would like to construct the synopsis structure in such a way that it has wide applicability across broad classes of problems. In addition, the applicability to data streams makes the efficiency issue of space and time-construction critical. In particular, the desiderata for effective synopsis construction are as follows:

- **Broad Applicability:** Since synopsis structures are used for a variety of data mining applications, it is desirable for them to have as broad an applicability as possible. This is because one may desire to use the underlying data stream for as many different applications. If synopsis construction methods have narrow applicability, then a different structure will need to be computed for each application. This will reduce the time and space efficiency of synopsis construction.

- **One Pass Constraint:** Since data streams typically contain a large number of points, the contents of the stream cannot be examined more than once during the course of computation. Therefore, all synopsis construction algorithms are designed under a one-pass constraint.

- **Time and Space Efficiency:** In many traditional synopsis methods on static data sets (such as histograms), the underlying dynamic programming methodologies require super-linear space and time. This is not acceptable for a data stream. For the case of space efficiency, it is not desirable to have a complexity which is more than linear in the size of the stream. In fact, in some methods such as sketches [44], the space complexity is often designed to be logarithmic in the *domain-size* of the stream.

- **Robustness:** The error metric of a synopsis structure needs to be designed in a robust way according to the needs of the underlying application. For example, it has often been observed that some wavelet based methods for approximate query processing may be optimal from a global perspective, but may provide very large error on some of the points in the stream [65]. This is an issue which needs the design of robust metrics such as the maximum error metric for stream based wavelet computation.

- **Evolution Sensitive:** Data Streams rarely show stable distributions, but rapidly evolve over time. Synopsis methods for static data sets are often not designed to deal with the rapid evolution of a data stream. For this purpose, methods such as clustering [1] are used for the purpose of synopsis driven applications such as classification [2]. Carefully designed synopsis structures can also be used for forecasting futuristic queries [3], with the use of evolution-sensitive synopsis.

There are a variety of techniques which can be used for synopsis construction in data streams. We summarize these methods below:

- **Sampling methods:** Sampling methods are among the most simple methods for synopsis construction in data streams. It is also relatively easy to use these synopsis with a wide variety of application since their representation is not specialized and uses the same multi-dimensional representation as the original data points. In particular reservoir based sampling methods [92] are very useful for data streams.

- **Histograms:** Histogram based methods are widely used for static data sets. However most traditional algorithms on static data sets require super-linear time and space. This is because of the use of dynamic programming techniques for optimal histogram construction. Their extension to the data stream case is a challenging task. A number of recent techniques [37] discuss the design of histograms for the dynamic case.

- **Wavelets:** Wavelets have traditionally been used in a variety of image and query processing applications. In this chapter, we will discuss the issues and challenges involved in dynamic wavelet construction. In particular, the dynamic maintenance of the dominant coefficients of the wavelet representation requires some novel algorithmic techniques.

- **Sketches:** Sketch-based methods derive their inspiration from wavelet techniques. In fact, sketch based methods can be considered a randomized version of wavelet techniques, and are among the most space-efficient of all methods. However, because of the difficulty of intuitive interpretations of sketch based representations, they are sometimes difficult to apply to arbitrary applications. In particular, the generalization of sketch methods to the multi-dimensional case is still an open problem.

- **Micro-cluster based summarization:** A recent micro-clustering method [1] can be used be perform synopsis construction of data streams. The advantage of micro-cluster summarization is that it is applicable to the multi-dimensional case, and adjusts well to the evolution of the underlying data stream. While the empirical effectiveness of the method is quite good, its heuristic nature makes it difficult to find good theoretical bounds on its effectiveness. Since this method is discussed in detail in another chapter of this book, we will not elaborate on it further.

In this chapter, we will provide an overview of the different methods for synopsis construction, and their application to a variety of data mining and database problems. This chapter is organized as follows. In the next section, we will discuss the sampling method and its application to different kinds of data mining problems. In section 3, we will discuss the technique of wavelets for data approximation. In section 4, we will discuss the technique of sketches for data stream approximation. The method of histograms is discussed in section 4. Section 5 discusses the conclusions and challenges in effective data stream summarization.

2.   Sampling Methods

Sampling is a popular tool used for many applications, and has several advantages from an application perspective. One advantage is that sampling is easy and efficient, and usually provides an *unbiased* estimate of the underlying data with *provable error guarantees*. Another advantage of sampling methods is that since they use the original representation of the records, they are easy to use with any data mining application or database operation. In most cases, the error guarantees of sampling methods generalize to the mining behavior of the underlying application. Many synopsis methods such as wavelets, histograms, and sketches are not easy to use for the multi-dimensional cases. The random sampling technique is often the only method of choice for high dimensional applications.

Before discussing the application to data streams, let us examine some properties of the random sampling approach. Let us assume that we have a database $\mathcal{D}$ containing N points which are denoted by $X_1 \ldots X_N$. Let us assume that the function $f(\mathcal{D})$ represents an operation which we wish to perform on the database $\mathcal{D}$. For example $f(\mathcal{D})$ may represent the mean or sum of one of the attributes in database $\mathcal{D}$. We note that a random sample S from database $\mathcal{D}$ defines a random variable $f(S)$ which is (often) closely related to $f(\mathcal{D})$ for many commonly used functions. It is also possible to estimate the standard deviation of $f(S)$ in many cases. In the case of *aggregation based* functions in linear separable form (eg. sum, mean), the law of large numbers allows us to approximate the random variable $f(S)$ as a normal distribution, and characterize the value of $f(\mathcal{D})$ probabilistically. However, not all functions are aggregation based (eg. min, max). In such cases, it is desirable to estimate the mean μ and standard deviation σ of $f(S)$. These parameters allows us to design *probabilistic bounds* on the value of $f(S)$. This is often quite acceptable as an alternative to characterizing the entire distribution of $f(S)$. Such probabilistic bounds can be estimated using a number of inequalities which are also often referred to as *tail bounds*.

The markov inequality is a weak inequality which provides the following bound for the random variable X:

$$P(X > a) \leq E[X]/a = \mu/a \tag{9.1}$$

By applying the Markov inequality to the random variable $(X - \mu)^2/\sigma^2$, we obtain the Chebychev inequality:

$$P(|X - \mu| > a) \leq \sigma^2/a^2 \tag{9.2}$$

While the Markov and Chebychev inequalities are farily general inequalities, they are quite loose in practice, and can be tightened when the distribution of the random variable X is known. We note that the Chebychev inequality is derived by applying the Markov inequality on a function of the random variable X. Even tighter bounds can be obtained when the random variable X shows a specific form, by applying the Markov inequality to parameterized functions of X and optimizing the parameter using the particular characteristics of the random variable X.

The Chernoff bound [14] applies when X is the sum of several independent and identical Bernoulli random variables, and has a lower tail bound as well as an upper tail bound:

$$P(X < (1 - \delta)\mu) \le e^{-\mu\delta^2/2} \tag{9.3}$$

$$P(X > (1 + \delta)\mu) \le \max\{2^{-\delta\mu}, e^{-\mu\delta^2/4}\} \tag{9.4}$$

Another kind of inequality often used in stream mining is the Hoeffding inequality. In this inequality, we bound the sum of k *independent bounded* random variables. For example, for a set of k independent random variables lying in the range $[a, b]$, the sum of these k random variables X satisfies the following inequality:

$$P(|X - \mu| > \delta) \le 2e^{-2k \cdot \delta^2/(b-a)^2} \tag{9.5}$$

We note that the Hoeffding inequality is slightly more general than the Chernoff bound, and both bounds have similar form for overlapping cases. These bounds have been used for a variety of problems in data stream mining such as classification, and query estimation [28, 58]. In general, the method of random sampling is quite powerful, and can be used for a variety of problems such as order statistics estimation, and distinct value queries [41, 72].

In many applications, it may be desirable to pick out a sample (reservoir) from the stream with a pre-decided size, and apply the algorithm of interest to this sample in order to estimate the results. One key issue in the case of data streams is that we are not sampling from a fixed data set with *known size* N. Rather, the value of N is unknown in advance, and the sampling must be performed dynamically as data points arrive. Therefore, in order to maintain an unbiased representation of the underlying data, the probability of including a point in the random sample should not be fixed in advance, but should change with progression of the data stream. For this purpose, reservoir based sampling methods are usually quite effective in practice.

2.1   Random Sampling with a Reservoir

Reservoir based methods [92] were originally proposed in the context of one-pass access of data from magnetic storage devices such as tapes. As in the case of streams, the number of records N are not known in advance and the sampling must be performed dynamically as the records from the tape are read.

Let us assume that we wish to obtain an unbiased sample of size n from the data stream. In this algorithm, we maintain a reservoir of size n from the data stream. The first n points in the data streams are added to the reservoir for initialization. Subsequently, when the $(t + 1)$th point from the data stream is received, it is added to the reservoir with probability $n/(t + 1)$. In order to make room for the new point, any of the current points in the reservoir are sampled with equal probability and subsequently removed.

The proof that this sampling approach maintains the unbiased character of the reservoir is straightforward, and uses induction on t. The probability of the $(t + 1)$th point being included in the reservoir is $n/(t + 1)$. The probability of any of the last t points being included in the reservoir is defined by the sum of the probabilities of the events corresponding to whether or not the $(t + 1)$th point is added to the reservoir. From the inductive assumption, we know that the first t points have equal probability of being included in the reservoir and have probability equal to n/t. In addition, since the points remain in the reservoir with equal probability of $(n - 1)/n$, the conditional probability of a point (among the first t points) remaining in the reservoir given that the $(t + 1)$ point is added is equal to $(n/t) \cdot (n-1)/n = (n-1)/t$. By summing the probability over the cases where the $(t+1)$th point is added to the reservoir (or not), we get a total probability of $((n/(t+1)) \cdot (n-1)/t + (1-(n/(t+1))) \cdot (n/t) = n/(t+1)$. Therefore, the inclusion of all points in the reservoir has equal probability which is equal to $n/(t + 1)$. As a result, at the end of the stream sampling process, all points in the stream have equal probability of being included in the reservoir, which is equal to n/N.

In many cases, the stream data may evolve over time, and the corresponding data mining or query results may also change over time. Thus, the results of a query over a more recent window may be quite different from the results of a query over a more distant window. Similarly, the entire history of the data stream may not relevant for use in a repetitive data mining application such as classification. Recently, the reservoir sampling algorithm was adapted to sample from a moving window over data streams [8]. This is useful for data streams, since only a small amount of recent history is more relevant that the entire data stream. However, this can sometimes be an extreme solution, since one may desire to sample from varying lengths of the stream history. While recent queries may be more frequent, it is also not possible to completely disregard queries over more distant horizons in the data stream. A method in [4] designs methods for *biased reservoir sampling*, which uses a bias function to regulate the sampling from the stream. This bias function is quite effective since it regulates the sampling in a smooth way so that queries over recent horizons are more accurately resolved. While the design of a reservoir for arbitrary bias function is extremely difficult, it is shown in [4], that certain classes of bias functions (exponential bias functions) allow the use of a straightforward replacement algorithm. The advantage of a bias function is that it can smoothly regulate the sampling process so that acceptable accuracy is retained for more distant queries. The method in [4] can also be used in data mining applications so that the quality of the results do not degrade very quickly.

2.2   Concise Sampling

The effectiveness of the reservoir based sampling method can be improved further with the use of concise sampling. We note that the size of the reservoir is sometimes restricted by the available main memory. It is desirable to increase the sample size within the available main memory restrictions. For this purpose, the technique of concise sampling is quite effective.

The method of concise sampling exploits the fact that the number of *distinct* values of an attribute is often significantly smaller than the size of the data stream. This technique is most applicable while performing univariate sampling along a single dimension. For the case of multi-dimensional sampling, the simple reservoir based method discussed above is more appropriate. The repeated occurrence of the same value can be exploited in order to increase the sample size beyond the relevant space restrictions. We note that when the number of distinct values in the stream is smaller than the main memory limitations, the entire stream can be maintained in main memory, and therefore sampling may not even be necessary. For current desktop systems in which the memory sizes may be of the order of several gigabytes, very large sample sizes can be main memory resident, as long as the number of distinct values does not exceed the memory constraints. On the other hand, for more challenging streams with an unusually large number of distinct values, we can use the following approach.

The sample is maintained as a set S of <value, count> pairs. For those pairs in which the value of count is one, we do not maintain the count explicitly, but we maintain the value as a *singleton*. The number of elements in this representation is referred to as the footprint and is bounded above by n. We note that the footprint size is always smaller than or equal to than the true sample size. If the count of any distinct element is larger than 2, then the footprint size is strictly smaller than the sample size. We use a *threshold parameter* τ which defines the probability of successive sampling from the stream. The value of τ is initialized to be 1. As the points in the stream arrive, we add them to the current sample with probability $1/\tau$. We note that if the corresponding value-count pair is already included in the set S, then we only need to increment the count by 1. Therefore, the footprint size does not increase. On the other hand, if the value of the current point is distinct from all the values encountered so far, or it exists as a singleton then the foot print increases by 1. This is because either a singleton needs to be added, or a singleton gets converted to a value-count pair with a count of 2. The increase in footprint size may potentially require the removal of an element from sample S in order to make room for the new insertion. When this situation arises, we pick a new (higher) value of the threshold τ', and apply this threshold to the footprint in repeated passes. In each pass, we reduce the count of a value with probability τ/τ', until at least one value-count pair reverts to a singleton or a singleton is removed. Subsequent

Granularity (Order k)	Averages Φ values	DWT Coefficients ψ values
$k = 4$	$(8, 6, 2, 3, 4, 6, 6, 5)$	-
$k = 3$	$(7, 2.5, 5, 5.5)$	$(1, -0.5, -1, 0.5)$
$k = 2$	$(4.75, 5.25)$	$(2.25, -0.25)$
$k = 1$	(5)	(-0.25)

Table 9.1. An Example of Wavelet Coefficient Computation

points from the stream are sampled with probability $1/\tau'$. As in the previous case, the probability of sampling reduces with stream progression, though we have much more flexibility in picking the threshold parameters in this case. More details on the approach may be found in [41].

One of the interesting characteristics of this approach is that the sample S continues to remain an unbiased representative of the data stream irrespective of the choice of τ. In practice, τ' may be chosen to be about 10% larger than the value of τ. The choice of different values of τ provides different tradeoffs between the average (true) sample size and the computational requirements of reducing the footprint size. In general, the approach turns out to be quite robust across wide ranges of the parameter τ.

3. Wavelets

Wavelets [66] are a well known technique which is often used in databases for hierarchical data decomposition and summarization. A discussion of applications of wavelets may be found in [10, 66, 89]. In this chapter, we will discuss the particular case of the *Haar Wavelet*. This technique is particularly simple to implement, and is widely used in the literature for hierarchical decomposition and summarization. The basic idea in the wavelet technique is to create a decomposition of the data characteristics into a set of wavelet functions and basis functions. The property of the wavelet method is that the higher order coefficients of the decomposition illustrate the broad trends in the data, whereas the more localized trends are captured by the lower order coefficients.

We assume for ease in description that the length q of the series is a power of 2. This is without loss of generality, because it is always possible to decompose a series into segments, each of which has a length that is a power of two. The Haar Wavelet decomposition defines 2^{k-1} coefficients of order k. Each of these 2^{k-1} coefficients corresponds to a contiguous portion of the time series of length $q/2^{k-1}$. The ith of these 2^{k-1} coefficients corresponds to the segment in the series starting from position $(i-1) \cdot q/2^{k-1} + 1$ to position $i * q/2^{k-1}$. Let us denote this coefficient by ψ_k^i and the corresponding time series segment by S_k^i. At the same time, let us define the average value of the first half of the S_k^i by

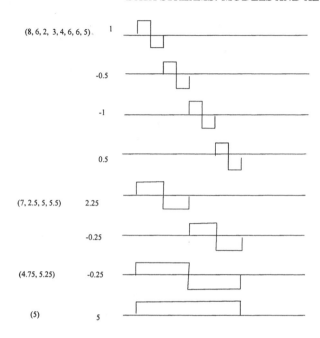

Figure 9.1.   Illustration of the Wavelet Decomposition

a_k^i and the second half by b_k^i. Then, the value of ψ_k^i is given by $(a_k^i - b_k^i)/2$. More formally, if Φ_k^i denote the average value of the S_k^i, then the value of ψ_k^i can be defined recursively as follows:

$$\psi_k^i = (\Phi_{k+1}^{2 \cdot i - 1} - \Phi_{k+1}^{2 \cdot i})/2 \qquad (9.6)$$

The set of Haar coefficients is defined by the Ψ_k^i coefficients of order 1 to $\log_2(q)$. In addition, the global average Φ_1^1 is required for the purpose of perfect reconstruction. We note that the coefficients of different order provide an understanding of the major trends in the data at a particular level of granularity. For example, the coefficient ψ_k^i is half the quantity by which the first half of the segment S_k^i is larger than the second half of the same segment. Since larger values of k correspond to geometrically reducing segment sizes, one can obtain an understanding of the basic trends at different levels of granularity. We note that this definition of the Haar wavelet makes it very easy to compute by a sequence of averaging and differencing operations. In Table 9.1, we have illustrated how the wavelet coefficients are computed for the case of the sequence $(8, 6, 2, 3, 4, 6, 6, 5)$. This decomposition is illustrated in graphical form in Figure 9.1. We also note that each value can be represented as a sum of $\log_2(8) = 3$ linear decomposition components. In general, the entire decomposition may be represented as a tree of depth 3, which represents the

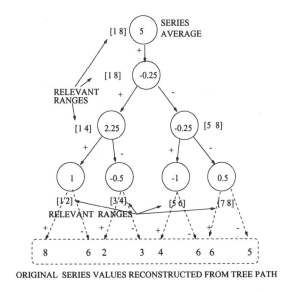

Figure 9.2. The Error Tree from the Wavelet Decomposition

hierarchical decomposition of the entire series. This is also referred to as the *error tree*, and was introduced in [73]. In Figure 9.2, we have illustrated the error tree for the wavelet decomposition illustrated in Table 9.1. The nodes in the tree contain the values of the wavelet coefficients, except for a special *super-root* node which contains the series average. This super-root node is not necessary if we are only considering the relative values in the series, or the series values have been normalized so that the average is already zero. We further note that the number of wavelet coefficients in this series is 8, which is also the length of the original series. The original series has been replicated just below the error-tree in Figure 9.2, and it can be reconstructed by adding or subtracting the values in the nodes along the path leading to that value. We note that each coefficient in a node should be added, if we use the left branch below it to reach to the series values. Otherwise, it should be subtracted. This natural decomposition means that an entire contiguous range along the series can be reconstructed by using only the portion of the error-tree which is relevant to it. Furthermore, we only need to retain those coefficients whose values are significantly large, and therefore affect the values of the underlying series. In general, we would like to minimize the reconstruction error by retaining only a fixed number of coefficients, as defined by the space constraints.

We further note that the coefficients represented in Figure 9.1 are un-normalized. For a time series T, let $\overline{W_1} \ldots \overline{W_t}$ be the corresponding basis vectors of length t. In Figure 9.1, each component of these basis vectors is 0, +1, or -1. The list of basis vectors in Figure 9.1 (in the same order as the corresponding wavelets illustrated) are as follows:

(1 -1 0 0 0 0 0 0)
(0 0 1 -1 0 0 0 0)
(0 0 0 0 1 -1 0 0)
(0 0 0 0 0 0 1 -1)
(1 1 -1 -1 0 0 0 0)
(0 0 0 0 1 1 -1 -1)
(1 1 1 1 -1 -1 -1 -1)

The most detailed coefficients have only one +1 and one -1, whereas the most coarse coefficient has $t/2$ +1 and -1 entries. Thus, in this case, we need $2^3 - 1 = 7$ wavelet vectors. In addition, the vector (11111111) is needed to represent the special coefficient which corresponds to the series average. Then, if $a_1 \ldots a_t$ be the wavelet coefficients for the wavelet vectors $\overline{W_1} \ldots \overline{W_t}$, the time series T can be represented as follows:

$$T = \sum_{i=1}^{t} a_i \cdot \overline{W_i} \qquad (9.7)$$

$$= \sum_{i=1}^{t} (a_i \cdot |\overline{W_i}|) \cdot \frac{\overline{W_i}}{|\overline{W_i}|} \qquad (9.8)$$

While a_i is the un-normalized value from Figure 9.1, the values $a_i \cdot |\overline{W_i}|$ represent normalized coefficients. We note that the values of $|\overline{W_i}|$ are different for coefficients of different orders, and may be equal to either $\sqrt{2}$, $\sqrt{4}$ or $\sqrt{8}$ in this particular example. For example, in the case of Figure 9.1, the broadest level un-normalized coefficient is -0.25, whereas the corresponding normalized value is $-0.25 \cdot \sqrt{8}$. After normalization, the basis vectors $\overline{W_1} \ldots \overline{W_t}$ are orthonormal, and therefore, the sum of the squares of the corresponding (normalized) coefficients is equal to the energy in the time series T. Since the normalized coefficients provide a new coordinate representation after axis rotation, euclidian distances between time series are preserved in this new representation.

The total number of coefficients is equal to the length of the data stream. Therefore, for very large time series or data streams, the number of coefficients is also large. This makes it impractical to retain the entire decomposition throughout the computation. The wavelet decomposition method provides a natural method for dimensionality reduction, by retaining only the coefficients with large absolute values. All other coefficients are implicitly approximated to zero. This makes it possible to approximately represent the series with a small number of coefficients. The idea is to retain only a pre-defined number of coefficients from the decomposition, so that the error of the reduced representation is minimized. Wavelets are used extensively for efficient and approximate query processing of different kinds of data [11, 93]. They are particularly useful for range queries, since contiguous ranges can easily be reconstructed with a small number of wavelet coefficients. The efficiency of the query processing arises from the reduced representation of the data. At the same time, since only the small coefficients are discarded the results are quite accurate.

A key issue for the accuracy of the query processing is the choice of coefficients which should be retained. While it may be tempting to choose only the coefficients with large absolute values, this is not always the best choice, since a more judicious choice of coefficients can lead to minimizing specific error criteria. Two such metrics are the minimization of the mean square error or the maximum error metric. The mean square error minimizes the L_2 error in approximation of the wavelet coefficients, whereas maximum error metrics minimize the maximum error of any coefficient. Another related metric is the relative maximum error which normalizes the maximum error with the absolute coefficient value.

It has been shown in [89] that the choice of largest B (normalized) coefficients minimizes the mean square error criterion. This should also be evident from the fact that the normalized coefficients render an orthonormal decomposition, as a result of which the energy in the series is equal to the sum of the squares of the coefficients. However, the use of the mean square error metric is not without its disadvantages. A key disadvantage is that a global optimization criterion implies that the local behavior of the approximation is ignored. Therefore, the approximation arising from reconstruction can be arbitrarily poor for certain regions of the series. This is especially relevant in many streaming applications in which the queries are performed only over recent time windows. In many cases, the maximum error metric provides much more robust guarantees. In such cases, the errors are spread out over the different coefficients more evenly. As a result, the worst-case behavior of the approximation over different queries is much more robust.

Two such methods for minimization of maximum error metrics are discussed in [38, 39]. The method in [38] is probabilistic, but its application of probabilistic expectation is questionable according to [53]. One feature of the method in [38] is that the space is bounded only in expectation, and the variance in space usage is large. The technique in [39] is deterministic and uses dynamic programming in order to optimize the maximum error metric. The key idea in [39] is to define a recursion over the nodes of the tree in top down fashion. For a given internal node, we compute the least maximum error over the two cases of either keeping or not keeping a wavelet coefficient of this node. In each case, we need to recursively compute the maximum error for its two children over all possible space allocations among two children nodes. While the method is quite elegant, it is computationally intensive, and it is therefore not suitable for the data stream case. We also note that the coefficient is defined according to the wavelet coefficient definition i.e. half the difference between the left hand and right hand side of the time series. While this choice of coefficient is optimal for the L_2 metric, this is not the case for maximum or arbitrary L_p error metrics.

Another important topic in wavelet decomposition is that of the use of multiple measures associated with the time series. The problem of multiple measures refers to the fact that many quantities may simultaneously be tracked in a given time series. For example, in a sensor application, one may simultaneously track many variables such as temperature, pressure and other parameters at each time instant. We would like to perform the wavelet decomposition over multiple measures simultaneously. The most natural technique [89] is to perform the decomposition along the different measures separately and pick the largest coefficients for each measure of the decomposition. This can be inefficient, since a coordinate needs to be associated with each separately stored coefficient and it may need to be stored multiple times. It would be more efficient to amortize the storage of a coordinate across multiple measures. The trade-off is that while a given coordinate may be the most effective representation for a particular measure, it may not simultaneously be the most effective representation across all measures. In [25], it has been proposed to use an extended wavelet representation which simultaneously tracks multi-measure coefficients of the wavelet representation. The idea in this technique is use a bitmap for each coefficient set to determine which dimensions are retained, and store all coefficients for this coordinate. The technique has been shown to significantly outperform the methods discussed in [89].

3.1 Recent Research on Wavelet Decomposition in Data Streams

The one-pass requirement of data streams makes the problem of wavelet decomposition somewhat more challenging. However, the case of optimizing the mean square error criterion is relatively simple, since a choice of the largest coefficients can preserve the effectiveness of the decomposition. Therefore, we only need to dynamically construct the wavelet decomposition, and keep track of the largest B coefficients encountered so far.

As discussed in [65], these methods can have a number of disadvantages in many situations, since many parts of the time series may be approximated very poorly. The method in [39] can effectively perform the wavelet decomposition with maximum error metrics. However, since the method uses dynamic programming, it is computationally intensive, it is quadratic in the length of the series. Therefore, it cannot be used effectively for the case of data streams, which require a one-pass methodology in linear time. in [51], it has been shown that all weighted L_m measures can be solved in a space-efficient manner using only $O(n)$ space. In [65], methods have been proposed for one-pass wavelet synopses with the maximum error metric. It has been shown in [65], that by using a number of intuitive thresholding techniques, it is possible to approximate the effectiveness of the technique discussed in [39]. A set of independent results obtained in [55] discuss how to minimize non-euclidean and relative error with the use of wavelet synopses. This includes metrics such as the L_p error or the relative error. Both the works of [65] and [55] were obtained independently and at a similar time. While the method in [65] is more deeply focussed on the use of maximum error metrics, the work in [55] also provides some worst case bounds on the quality of the approximation. The method of [65] depends on experimental results to illustrate the quality of the approximation. Another interesting point made in [55] is that most wavelet approximation methods solve a restricted version of the problem in which the wavelet coefficient for the basis is defined to be half the difference between the left hand and right hand side of the basis vectors. Thus, the problem is only one of picking the best B coefficients out of these pre-defined set of coefficients. While this is an intuitive method for computation of the wavelet coefficient, and is optimal for the case of the Euclidean error, it is not necessarily optimal for the case of the L_m-metric. For example, consider the time series vector $(1, 4, 5, 6)$. In this case, the wavelet transform is $(4, -1.5, -1.5, -0.5)$. Thus, for $B = 1$, the optimal coefficient picked is $(4, 0, 0, 0)$ for any L_m-metric. However, for the case of L_∞-metric, the optimal solution should be $(3.5, 0, 0, 0)$, since 3.5 represents the average between the minimum and maximum value. Clearly, any scheme which restricts itself only to wavelet coefficients defined in a particular way will not even consider this solution [55]. Almost all methods for non-euclidean wavelet computation tend to use this approach, possibly as a legacy from the Haar method of wavelet decomposition. This restriction has been removed in [55] and proposes a method for determining the optimal *synopsis coefficients* for the case of the weighted L_m metric. We distinguish between synopsis coefficients and wavelet coefficients, since the latter are defined by the simple subtractive methodology of the Haar decomposition. A related method was also proposed by Matias and Urieli [75] which discusses a near linear time optimal algorithm for the weighted L_m-error. This method is offline, and chooses a basis vector which depends upon the weights.

An interesting extension of the wavelet decomposition method is one in which *multiple measures* are associated with the time series. A natural solution is to treat each measure separately, and store the wavelet decomposition. However, this can be wasteful, since a coordinate needs to be stored with each coefficient, and we can amortize this storage by storing the same coordinate across multiple measures. A technique in [25] proposes the concept of *extended wavelets* in order to amortize the coordinate storage across multiple measures. In this representation, one or more coefficients are stored with each coordinate. Clearly, it can be tricky to determine which coordinates to store, since different coordinates will render larger coefficients across different measures. The technique in [25] uses a dynamic programming method to determine the optimal extended wavelet decomposition. However, this method is not time and space efficient. A method in [52] provides a fast algorithm whose space requirement is linear in the size of the synopsis and logarithmic in the size of the data stream.

Another important point to be noted is that the choice of the best wavelet decomposition is not necessarily pre-defined, but it depends upon the particular workload on which the wavelet decomposition is applied. Some interesting papers in this direction [77, 75] design methods for workload aware wavelet synopses of data streams. While this line of work has not been extensively researched, we believe that it is likely to be fruitful in many data stream scenarios.

4. Sketches

The idea of sketches is essentially an extension of the random projection technique [64] to the time series domain. The idea of using this technique for determining representative trends in the time series domain was first observed in [61]. In the method of random projection, we can reduce a data point of dimensionality d to an axis system of dimensionality k by picking k random vectors of dimensionality d and calculating the dot product of the data point with each of these random vectors. Each component of the k random vectors is drawn from the normal distribution with zero mean and unit variance. In addition, the random vector is normalized to one unit in magnitude. It has been shown in [64] that proportional L_2 distances between the data points are approximately preserved using this transformation. The accuracy bounds of the distance values are dependent on the value of k. The larger the chosen value of k, the greater the accuracy and vice-versa.

This general principle can be easily extended to the time series domain, by recognizing the fact that the length of a time series may be treated as its dimensionality, and correspondingly we need to compute a random vector of length equal to the time series, and use it for the purpose of sketch computation. If desired, the same computation can be performed over a sliding window of a given length by choosing a random vector of appropriate size. As proposed in [61], the following approximation bounds are preserved:

LEMMA 9.1 *Let L be a set of vectors of length l, for fixed $\epsilon < 1/2$, and $k = 9 \cdot log|L|/\epsilon^2$. Consider a pair of vectors $\overline{u}$, $\overline{w}$ in L, such that the corresponding sketches are denoted by $S(\overline{u})$ and $S(\overline{w})$ respectively. Then, we have:*

$$(1 - \epsilon) \cdot ||\overline{u} - \overline{w}||^2 \leq ||S(\overline{u}) - S(\overline{w})|| \leq (1 + \epsilon) \cdot ||\overline{u} - \overline{w}||^2 \qquad (9.9)$$

with probability $1/2$. Here $||U - V||^2$ is the L_2 distance between two vectors U and V.

The generalization to time series is fairly straightforward, and the work in [61] makes two primary contributions in extending the sketch methodology to finding time series trends.

4.1   Fixed Window Sketches for Massive Time Series

In this case, we wish to determine sliding window sketches with a fixed window length l. For each window of length l, we need to perform $l \cdot k$ operations for a sketch of size k. Since there are $O(n - l)$ sliding windows, this will require $O(n \cdot l \cdot k)$ computations. When l is large, and is of the same order of magnitude as the time series, the computation may be quadratic in the size of the series. This can be prohibitive for very large time series, as is usually the case with data streams. The key observation in [61], is that all such sketches can be viewed as the problem of computing the polynomial convolution of the random vector of appropriate length with the time series. Since the problem of polynomial convolution can be computed efficiently using fast fourier transform, this also means that the sketches may be computed efficiently. The problem of polynomial convolution is defined as follows:

DEFINITION 9.2 *Given two vectors $A[1 \ldots a]$ and $B[1 \ldots b]$, $a \geq b$, their convolution is the vector $C[1 \ldots a + b]$ where $C[k] = \sum_{i=1}^{b} A[k - i] \cdot B[i]$ for $k \in [2, a + b]$, with any out of range references assumed to be zero.*

The key point here is that the above polynomial convolution can be computed using FFT, in $O(a \cdot \log(b))$ operations rather than $O(a \cdot b)$ operations. This effectively means the following:

LEMMA 9.3 *Sketches of all subvectors of length l can be computed in time $O(n \cdot k \cdot \log(l))$ using polynomial convolution.*

4.2   Variable Window Sketches of Massive Time Series

The method in the previous subsection discussed the problem of sketch computation for a fixed window length. The more general version of the problem is one in which we wish to compute the sketch for any subvector between length l and u. In the worst-case this comprises $O(n^2)$ subvectors, most of which will have length $O(n)$. Therefore, the entire algorithm may require $O(n^3)$ operations, which can be prohibitive for massive time series streams.

The key idea in [61] is to store a *pool* of sketches. The size of this pool is significantly smaller than the entire set of sketches needed. However, it is carefully chosen so that the sketch of any sub-vector in the original vector can be computed in $O(1)$ time fairly accurately. In fact, it can be shown that the approximate sketches computed using this approach satisfy a slightly relaxed version of Lemma 9.1. We refer details to [61].

4.3   Sketches and their applications in Data Streams

In the previous sections we discussed the application of sketches to the problem of massive time series. Some of the methods such as fixed window sketch computation are inherently offline. This does not suffice in many scenarios in which it is desirable to continuously compute the sketch over the data stream. Furthermore, in many cases, it is desirable to efficiently use this sketch in order to work with a variety of applications such as query estimation. In this subsection, we will discuss the applications of sketches in the data stream scenario. Our earlier discussion corresponds to a sketch of the *time series* itself, and entails the storage of the random vector required for sketch generation. While such a technique can be used effectively for massive time series, it cannot always be used for time series data streams.

However, in certain other applications, it may be desirable to track the *frequencies* of the distinct values in the data stream. In this case, if $(u_1 \ldots u_N)$ be the frequencies of N distinct values in the data stream, then the sketch is defined by the dot product of the vector $(u_1 \ldots u_N)$ with a random vector of size N. As in the previous case, the number of distinct items N may be large, and therefore the size of the corresponding random vector will also be large. A natural solution is to pre-generate a set of k random vectors, and whenever the ith item is received, we add r_i^j to the jth sketch component. Therefore, the k random vectors may need to be pre-stored in order to perform the computation. However, the explicit storage of the random vector will defeat the purpose of the sketch computation, because of the high space complexity.

The key here is to store the random vectors implicitly in the form of a seed, which can be used to dynamically generate the vector. The key idea discussed in [6] is that it is possible to generate the random vectors from a seed of size $O(\log(N))$ provided that one is willing to work with the restriction that the values of $r_i^j \in \{-1, +1\}$ are only 4-wise independent. We note that having a seed of small size is critical in terms of the space-efficiency of the method. Furthermore, it has been shown in [6] that the theoretical results only require 4-wise independence. In [44], it has also been shown how to use Reed-Muller codes in order to generate 7-wise independent random numbers. These method suffices for the purpose of wavelet decomposition of the frequency distribution of different items.

Some key properties of the pseudo-random number generation approach and the sketch representation are as follows:

- A given component r_i^j can be generated in poly-logarithmic time from the seed.

- The dot-product of two vectors can be approximately computed using only their sketch representations. This follows from the fact that the dot product of two vectors is closely related to the Euclidean distance, a quantity easily approximated by the random projection approach [64]. Specifically, if $\overline{U}$ and $\overline{V}$ be two (normalized) vectors, then the euclidean distance and dot product are related as follows:

$$||\overline{U} - \overline{V}||^2 = ||\overline{U}||^2 + ||\overline{V}||^2 - 2 \cdot \overline{U} \cdot \overline{V} \qquad (9.10)$$

$$(9.11)$$

This relationship can be used to establish bounds on the quality of the dot product approximation of the sketch vectors. We refer to [44] for details of the proof.

The first property ensures that the sketch components can be updated and maintained efficiently. Whenever the ith value is received, we only need to add r_i^j to the jth component of the sketch vector. Since the quantity r_i^j can be efficiently computed, it follows that the update operations can be performed efficiently as well. In the event that the data stream also incorporates frequency counts with the arriving items (item i is associated with frequency count $f(i)$), then we simply need to add $f(i) \cdot r_i^j$ to the jth sketch component. We note that the efficient and accurate computation of the dot product of a given time series with the random vector is a key primitive which can be used to compute many properties such as the wavelet decomposition. This is because each wavelet coefficient can be computed as the dot product of the wavelet basis with the corresponding time series data stream; an approximation may be determined by using only their sketches. The key issue here is that we also need the sketch representation of the wavelet basis vectors, each of which may take $O(N)$ time in the worst case. In general, this can be time consuming; however the work in [44] shows how to do this in poly-logarithmic time for the special case in which the vectors are Haar-basis vectors. Once the coefficients have been computed, we only need to retain the B coefficients with the highest energy.

We note that one property of the results in [44] is that it uses the sketch representation of the frequency distribution of the *original stream* in order to derive the wavelet coefficients. A recent result in [16] works *directly* with the sketch representation of the wavelet coefficients rather than the sketch representation of the original data stream. Another advantage of the work in [16] is that the query times are much more efficient, and the work extends to the multi-dimensional domain. We note that while the wavelet representation in [44] is space efficient, the entire synopsis structure may need to be touched for updates and every wavelet coefficient must be touched in order to find the best B coefficients. The technique in [16] reduces the time and space efficiency for both updates and queries.

The method of sketches can be effectively used for second moment and join estimation. First, we discuss the problem of second moment estimation [6] and illustrate how it can be used for the problem of estimating the size of self joins. Consider a set of n quantitative values $U = (u_1 \ldots u_N)$. We would like to estimate the second moment $|U|^2$. Then, as before generate the random vectors $\overline{r^1} \ldots r^k$, (each of size N), and compute the dot product of these random vectors with U to create the sketch components denoted by $S_1 \ldots S_k$. Then, it can be shown that the expected value of S_i^2 is equal to the second moment. In fact, the approximation can be bounded with high probability.

LEMMA 9.4 *By selecting the median of* $O(log(1/\delta))$ *averages of* $O(1/\epsilon^2)$ *copies of* S_i^2, *it is possible to guarantee the accuracy of the sketch based approximation to within* $1 \pm \epsilon$ *with probability at least* $1 - \delta$.

In order to prove the above result, the first step is to show that the expected value of S_i^2 is equal to the second moment, and the variance of the variable S_i^2 is at most twice the square of the expected value. The orthogonality of the random projection vectors can be used to show the first result and the 4-wise independence of the values of r_i^j can be used to show the second. The relationship between the expected values and variance imply that the Chebychev inequality can be used to prove that the average of $O(1/\epsilon^2)$ copies provides the ϵ bound with a constant probability which is at least $7/8$. This constant probability can be tightened to at least $1 - \delta$ (for any small value of δ) with the use of the median of $O(log(1/\delta))$ independent copies of these averages. This is because the median would lie outside the ϵ-bound only if more than $log(1/\delta)/2$ copies (minimum required number of copies) lie outside the ϵ bound. However, the expected number of copies which lie outside the ϵ-bound is only $log(1/\delta)/8$, which is less than above-mentioned required number of copies by $3 \cdot log(1/\delta)/8$. The Chernoff tail bounds can then be applied on the random variable representing the number of copies lying outside the ϵ-bound. This can be used to show that the probability of more than half the $log(1/\delta)$ copies lying outside the ϵ-bound is at most δ. Details of the proof can be found in [6].

We note that the second moment immediately provides an estimation for self-joins. If u_i be the number of items corresponding to the ith value, then the second moment estimation is exactly the size of the self-join. We further note that the dot product function is not the only one which can be estimated from the sketch. In general, many functions such as the dot product, the L_2 distance, or the maximum frequency items can be robustly estimated from the sketch. This is essentially because the sketch simply projects the time series onto a new set of (expected) orthogonal vectors. Therefore many rotational invariant properties such as the L_2 distance, dot product, or second moment are approximately preserved by the sketch.

A number of interesting techniques have been discussed in [5, 26, 27] in order to perform the estimation more effectively over general joins and multi-joins. Consider the multi-join problem on relations $R1$, $R2$, $R3$, in which we wish to join attribute A of $R1$ with attribute B of $R2$, and attribute C of $R2$ with attribute D of $R3$. Let us assume that the join attribute on $R1$ with $R2$ has N distinct values, and the join attribute of $R2$ with $R3$ has M distinct values. Let $f(i)$ be the number of tuples in $R1$ with value i for attribute A. Let $g(i, j)$ be the number of tuples in $R2$ with values i and j for attributes B and C respectively. Let $h(j)$ be the number of tuples in $R3$ with value j for join attribute C. Then, the total estimated join size J is given by the following:

$$J = \sum_{i=1}^{N} \sum_{j=1}^{M} f(i) \cdot g(i,j) \cdot h(j) \tag{9.12}$$

In order to estimate the join size, we create *two independently* generated families of random vectors $\overline{r^1} \ldots \overline{r^k}$ and $\overline{s^1} \ldots \overline{s^k}$. We dynamically maintain the following quantities, as the stream points are received:

$$Z_1^j = \sum_{i=1}^{N} f(i) \cdot r_i^j \tag{9.13}$$

$$Z_2^j = \sum_{i=1}^{N} \sum_{k=1}^{M} g(i,k) \cdot r_k^j \cdot s_k^j \tag{9.14}$$

$$Z_3^j = \sum_{k=1}^{M} h(k) \cdot s_k^j \tag{9.15}$$

It can be shown [5], that the quantity $Z_1^j \cdot Z_2^j \cdot Z_3^j$ estimates the join size. We can use the multiple components of the sketch (different values of j) in order to improve the accuracy. It can be shown that the variance of this estimate is equal to the *product of the self-join sizes* for the three different relations. Since the tail bounds use the variance in order to provide quality estimates, a large value of the variance can reduce the effectiveness of such bounds. This is particularly a problem if the composite join has a small size, whereas the product of the self-join sizes is very large. In such cases, the errors can be very large in relation to the size of the result itself. Furthermore, the product of self-join sizes increases with the number of joins. This degrades the results. We further note that the error bound results for sketch based methods are proved with the use of the Chebychev inequality, which depends upon a low ratio of the variance to result size. A high ratio of variance to result size makes this inequality ineffective, and therefore the derivation of worst-case bounds requires a greater number of sketch components.

An interesting observation in [26] is that of *sketch partitioning*. In this technique, we intelligently partition the join attribute domain-space and use it in order to compute separate sketches of each partition. The resulting join estimate is computed as the sum over all partitions. The key observation here is that intelligent domain partitioning reduces the variance of the estimate, and is therefore more accurate for practical purposes. This method has also been discussed in more detail for the problem of multi-query processing [27].

Another interesting trick for improving join size estimation is that of *sketch skimming* [34]. The key insight is that the variance of the join estimation is highly affected by the most frequent components, which are typically small in number. A high variance is undesirable for accurate estimations. Therefore, we treat the frequent items in the stream specially, and can separately track them. A skimmed sketch can be constructed by subtracting out the sketch components of these frequent items. Finally, the join size can be estimated as a 4-wise addition of the join estimation across two pairs of partitions. It has been shown that this approach provides superior results because of the reduced variance of the estimations from the skimmed sketch.

4.4   Sketches with p-stable distributions

In our earlier sections, we did not discuss the effect of the distribution from which the random vectors are drawn. While the individual components of the random vector were drawn from the normal distribution, this is not the only possibility for sketch generation. In this section, we will discuss a special set of distributions for the random vectors which are referred to as p-stable distributions. A distribution $\mathcal{L}$ is said to be p-stable, if it satisfies the following property:

DEFINITION 9.5 *For any set of N i.i.d. random variables $X_1 \ldots X_N$ drawn from a p-stable distribution $\mathcal{L}$, and any set of real numbers $a_1 \ldots a_N$, the random variable $(\sum_{i=1}^{N} a_i \cdot X_i)/(\sum_{i=1}^{N} a_i^p)^{(1/p)}$ is drawn from $\mathcal{L}$.*

A classic example of the p-stable distribution is the normal distribution with $p = 2$. In general p-stable distributions can be defined for $p \in (0, 2]$.

The use of p-stable distributions has implications in the construction of sketches. Recall, that the ith sketch component is of the form $\sum_{i=1}^{N} u_j \cdot r_i^j$, where u_i is the frequency of the ith distinct value in the data stream. If each r_i^j is drawn from a p-stable distribution, then the above sum is also a (scaled) p-stable distribution, where the scale coefficient is given by $(\sum_{i=1}^{N} u_i^p)^{(1/p)}$. The ability to use the *exact distribution* of the sketch provides much stronger results than just the use of mean and variance of the sketch components. We note that the use of only mean and variance of the sketch components often restricts us to the use of generic tail bounds (such as the Chebychev inequality) which may not always be tight in practice. However, the knowledge of the sketch distribution can potentially provide very tight bounds on the behavior of each sketch component.

An immediate observation is that the scale coefficient $(\sum_{i=1}^{N} u_i^p)^{(1/p)}$ of each sketch component is simply the L_p-norm of the frequency distribution of the incoming items in the data stream. By using $O(\log(1/\delta)/\epsilon^2)$ independent sketch components, it is possible to approximate the L_p norm within ϵ with probability at least $1 - \delta$. We further note that the use of the L_0 norm provides the number of distinct values in the data stream. It has been shown in [17] that by using $p \to 0$ (small values of p), it is possible to closely approximate the number of distinct values in the data stream.

Other Applications of Sketches.   The method of sketches can be used for a variety of other applications. Some examples of such applications include the problem of *heavy hitters* [13, 18, 76, 21], a problem in which we determine the most frequent items over data streams. Other problems include those of finding significant network differences over data streams [19] and finding quantiles [46, 50] over data streams. Another interesting application is that of significant differences between data streams [32, 33], which has applications in numerous change detection scenarios. Another recent application to sketches has been to XML and tree-structured data [82, 83, 87]. In many cases, these synopses can be used for efficient resolution of the structured queries which are specified in the XQuery pattern-specification language.

Recently sketch based methods have found considerable applications to efficient communication of signals in sensor networks. Since sensors are battery constrained, it is critical to reduce the communication costs of the transmission. The space efficiency of the sketch computation approach implies that it can also be used in the sensor network domain in order to minimize the communication costs over different processors. In [22, 67, 50], it has been shown how to extend the sketch method to distributed query tracking in data streams. A particularly interesting method is the technique in [22] which reduces the communication costs further by using sketch skimming techniques [34], in order to reduce communication costs further. The key idea is to use models to estimate the future behavior of the sketch, and make changes to the sketch only when there are significant changes to the underlying model.

4.5   The Count-Min Sketch

One interesting variation of the sketching method for data streams is the *count-min sketch*, which uses a hash-based sketch of the stream. The broad ideas in the count-min sketch were first proposed in [13, 29, 30]. Subsequently, the method was enhanced with pairwise-independent hash functions, formalized, and extensively analyzed for a variety of applications in [20].

In the count-min sketch, we use $\lceil \ln(1/\delta) \rceil$ pairwise independent hash functions, each of which map on to uniformly random integers in the range $[0, e/\epsilon]$, where e is the base of the natural logarithm. Thus, we maintain a total of $\lceil \ln(1/\delta) \rceil$ hash tables, and there are a total of $O(\ln(1/\delta)/\epsilon)$ hash cells. This *apparently* provides a better space complexity than the $O(\ln(1/\delta)/\epsilon^2)$ bound of AMS sketches in [6]. We will discuss more on this point later.

We apply each hash function to any incoming element in the data stream, and add the count of the element to each of the corresponding $\lceil \ln(1/\delta) \rceil$ positions in the different hash tables. We note that because of collisions, the hash table counts will not exactly correspond to the count of any element in the incoming data stream. When incoming frequency counts are non-negative, the hash table counts will over-estimate the true count, whereas when the incoming frequency counts are either positive or negative (deletions), the hash table count could be either an over-estimation or an under-estimation. In either case, the use of the median count of the hash position of a given element among the $O(\ln(1/\delta))$ counts provided by the different hash functions provides a estimate which is within a $3 \cdot \epsilon$ factor of the L_1-norm of element counts with probability at least $1 - \delta^{1/4}$ [20]. In other words, if the frequencies of the N different items are $f_1 \ldots f_N$, then the estimated frequency of the item i lie between $f_i - 3 \cdot \epsilon \cdot \sum_{i=1}^{N} |f_i|$ and $f_i + 3 \cdot \epsilon \cdot \sum_{i=1}^{N} |f_i|$ with probability at least $1 - \delta^{1/4}$. The proof of this result relies on the fact that the expected inaccuracy of a given entry j is at most $\epsilon \cdot \sum_{i=1}^{N} |f_i|/e$, if the hash function is sufficiently uniform. This is because we expect the count of other (incorrect) entries which map onto the position of j to be $\sum_{i \in [1,N], i \neq j} f_i \cdot \epsilon/e$ for a sufficiently uniform hash function with $\lceil e/\epsilon \rceil$ entries. This is at most equal to $\epsilon \cdot \sum_{i=1}^{N} |f_i|/e$. By the Markov inequality, the probability of this number exceeding $3 \cdot \epsilon \cdot \sum_{i=1}^{N} |f_i|$ is less than $1/(3 \cdot e) < 1/8$. By using the earlier Chernoff bound trick (as in AMS sketches) in conjunction with the median selection operation, we get the desired result.

In the case of non-negative counts, the *minimum count* of any of the $\ln(1/\delta)$ possibilities provides a tighter ϵ-bound (of the L_1-norm) with probability at least $1 - \delta$. In this case, the estimated frequency of item i lies between f_i and $f_i + \epsilon \cdot \sum_{i=1}^{N} f_i$ with probability at least $1 - \delta$. As in the previous case, the expected inaccuracy is $\epsilon \cdot \sum_{i=1}^{N} f_i/e$. This is less than the maximum bound by a factor of e. By applying the Markov inequality, it is clear that the probability that the bound is violated for a given entry is $1/e$. Therefore, the probability that it is violated by all $\log(1/\delta)$ entries is at most $(1/e)^{\log(1/\delta)} = \delta$.

For the case of non-negative vectors, the dot product can be estimated by computing the dot product on the corresponding entries in the hash table. Each of the $\lceil \ln(1/\delta) \rceil$ such dot products is an over estimate, and the minimum of these provides an ϵ bound with probability at least $1 - \delta$. The dot product result immediately provides bounds for join size estimation. Details of extending the method to other applications such as heavy hitters and quantiles may be found in [20]. In many of these methods, the time and space complexity is bounded above by $O(\ln(1/\delta)/\epsilon)$, which is again apparently superior to the AMS sketch.

As noted in [20], the ϵ-bound in the count-min sketch cannot be directly compared with that of the AMS sketch. This is because the AMS sketch provides the ϵ-bound as a function of the L_2-norm, whereas the method in [20] provides the ϵ-bound only in terms of the L_1-norm. The L_1-norm can be *quadratically* larger (than the L_2-norm) in the most challenging case of non-skewed distributions, and the ratio between the two may be as large as $\sqrt{N}$. Therefore, the *equivalent* value of ϵ in the count-min sketch can be smaller than that in the AMS sketch by a factor of $\sqrt{N}$. Since N is typically large, and is in fact the motivation of the sketch-based approach, the *worst-case* time and space complexity of a truly equivalent count-min sketch may be *significantly* larger for practical values of ϵ. While this observation has been briefly mentioned in [20], there seems to be some confusion on this point in the current literature. This is because of the overloaded use of the parameter ϵ, which has different meaning for the AMS and count-min sketches. For the skewed case (which is quite common), the ratio of the L_1-norm to the L_2-norm reduces. However, since this case is less challenging, the general methods no longer remain relevant, and a number of other specialized methods (eg. sketch skimming [34]) exist in order to improve the experimental and worst-case effectiveness of both kinds of sketches. It would be interesting to experimentally compare the count-min and AMS methods to find out which is superior in different kinds of skewed and non-skewed cases. Some recent results [91] seem to suggest that the count-min sketch is experimentally superior to the AMS sketch in terms of maintaining counts of elements. On the other hand, the AMS sketch seems to be superior in terms of estimating aggregate functions such as the L_2-norm. Thus, the count-min sketch does seem to have a number of practical advantages in many scenarios.

4.6 Related Counting Methods: Hash Functions for Determining Distinct Elements

The method of sketches is a probabilistic counting method whereby a randomized function is applied to the data stream in order to perform the counting in a space-efficient way. While sketches are a good method to determine *large aggregate signals*, they are not very useful for counting infrequently occurring items in the stream. For example, problems such as the determination of the number of distinct elements cannot be performed with sketches. For this purpose, hash functions turn out to be a useful choice.

Consider a hash function that renders a mapping from a given word to an integer in the range $[0, 2^L - 1]$. Therefore, the binary representation of that integer will have length L. The position (least significant and rightmost bit is counted as 0) of the rightmost 1-bit of the binary representation of that integer is tracked, and the largest such value is retained. This value is logarithmically related to the number of distinct elements [31] in the stream.

The intuition behind this result is quite simple. For a sufficiently uniformly distributed hash function, the probability of the ith bit on the right taking on the first 1-value is simply equal to 2^{-i-1}. Therefore, for N distinct elements, the expected number of records taking on the ith bit as the first 1-value is $2^{-i-1} \cdot N$. Therefore, when i is picked larger than $\log(N)$, the expected number of such bitstrings falls off exponentially less than 1. It has been rigorously shown [31] that the expected position of the rightmost bit $E[R]$ is logarithmically related to the number of distinct elements as follows:

$$E[R] = \log_2(\phi N), \quad \phi = 0.77351 \tag{9.16}$$

The standard deviation $\sigma(R) = 1.12$. Therefore, the value of R provides an estimate for the number of distinct elements N.

The hash function technique is very useful for those estimations in which non-repetitive elements have the same level of importance as repetitive elements. Some examples of such functions are those of finding distinct values [31, 43], mining inverse distributions [23], or determining the cardinality of set expressions [35]. The method in [43] uses a technique similar to that discussed in [31] in order to obtain a random sample of the distinct elements. This is then used for estimation. In [23], the problem of inverse distributions is discussed, in which it is desirable to determine the elements in the stream with a particular frequency level. Clearly such an inverse query is made difficult by the fact that a query for an element with very low frequency is equally likely to that of an element with very high frequency. The method in [23] solves this problem using a hash based approach similar to that discussed in [31]. Another related problem is that of finding the number of *distinct elements* in a join after eliminating duplicates. For this purpose, a join-distinct sketch (or JD-Sketch) was proposed in [36], which uses a 2-level adaptation of the hash function approach in [31].

4.7   Advantages and Limitations of Sketch Based Methods

One of the key advantages of sketch based methods is that they require space which is sublinear in the data size being considered. Another advantage of sketch based methods that it is possible to maintain sketches in the presence of deletions. This is often not possible with many synopsis methods such as random samples. For example, when the ith item with frequency $f(i)$ is deleted, the jth component of the sketch can be updated by subtracting $f(i) \cdot r_i^j$ from it. Another advantage of using sketch based methods is that they are extraordinarily space efficient, and require space which is logarithmic in the *number of distinct items* in the stream. Since the number of distinct items is significantly smaller than the size of the stream itself, this is an extremely low space requirement.

We note that sketches are based on the Lipshitz embeddings, which preserve a number of aggregate measures such as the L_p norm or the dot product. However, the entire distribution on the data (including the local temporal behavior) are not captured in the sketch representation, unless one is willing to work with a much larger space requirement.

Most sketch methods are based on analysis along a single dimensional stream of data points. Many problems in the data stream scenario are inherently multi-dimensional, and may in fact involve hundreds or thousands of independent and simultaneous data streams. In such cases, it is unclear whether sketch based methods can be easily extended. While some recent work in [16] provides a few limited methods for multi-dimensional queries, these are not easily extensible for more general problems. This problem is not however unique to sketch based methods. Many other summarization methods such as wavelets or histograms can be extended in a limited way to the multi-dimensional case, and do not work well beyond dimensionalities of 4 or 5.

While the concept of sketches is potentially powerful, one may question whether sketch based methods have been used for the right problems in the data stream domain. Starting with the work in [6], most work on sketches focuses on the *aggregate frequency* behavior of individual items rather than the temporal characteristics of the stream. Some examples of such problems are those of finding the frequent items, estimation of *frequency moments*, and *join size estimation*. The underlying assumption of these methods is an extremely large *domain size* of the data stream. The actual problems solved (aggregate frequency counts, join size estimation, moments) are relatively simple for modest domain sizes in many practical problems over very fast data streams. In these cases, temporal information in terms of sequential arrival of items is aggregated and therefore lost. Some sketch-based techniques such as those in [61] perform temporal analysis over specific time windows. However, this method has much larger space requirements. It seems to us that many of the existing sketch based methods can be easily extended to the temporal representation of the stream. It would be interesting to explore how these methods compare with other synopsis methods for temporal stream representation.

We note that the problem of aggregate frequency counts is made difficult only by the assumption of *very large domain sizes*, and not by the speed of the stream itself. It can be argued that in most practical applications, the data stream itself may be very fast, but the number of distinct items in the stream may be of manageable size. For example, a motivating application in [44] uses the domain of call frequencies of phone records, an application in which the number of distinct items is bounded above by the number of phone numbers of a particular phone company. With modern computers, it may even be possible to hold the frequency counts of a few million distinct phone numbers in a main memory array. In the event that main memory is not sufficient, many efficient disk based index structures may be used to index and update frequency counts. We argue that many applications in the sketch based literature which attempts to find specific properties of the frequency counts (eg. second moments, join size estimation, heavy hitters) may in fact be implemented trivially by using simple main memory data structures, and the ability to do this will only increase over time with hardware improvements. There are however a number of applications in which hardware considerations make the applications of sketch based methods very useful. In our view, the most fruitful applications of sketch based methods lie in its recent application to the sensor network domain, in which in-network computation, storage and communication are greatly constrained by power and hardware considerations [22, 67, 68]. Many distributed applications such as those discussed in [9, 24, 70, 80] are particularly suited to this approach.

5.   Histograms

Another key method for data summarization is that of histograms. In the method of histograms, we essentially divide the data along any attribute into a set of ranges, and maintain the count for each bucket. Thus, the space requirement is defined by the number of buckets in the histogram. A naive representation of a histogram would discretize the data into partitions of equal length (equi-width partitioning) and store the frequencies of these buckets. At this point, we point out a simple connection between the histogram representation and Haar wavelet coefficients. If we construct the wavelet representation of the *frequency distribution* of a data set along any dimension, then the (non-normalized) Haar coefficients of any order provide the difference in relative frequencies in equi-width histogram buckets. Haar coefficients of different orders correspond to buckets of different levels of granularity.

It is relatively easy to use the histogram for answering different kinds of queries such as range queries, since we only need to determine the set of buckets which lie within the user specified ranges [69, 81]. A number of strategies can be devised for improved query resolution from the histogram [69, 81, 84, 85].

The key source of inaccuracy in the use of histograms is that the distribution of the data points within a bucket is not retained, and is therefore assumed to be uniform. This causes inaccuracy because of extrapolation at the query boundaries which typically contain only a fractional part of a histogram. Thus, an important design consideration in the construction of histograms is the determination of how the buckets in the histogram should be designed. For example, if each range is divided into equi-width partitions, then the number of data points would be distributed very unequally across different buckets. If such buckets include the range boundary of a query, this may lead to inaccurate query estimations.

Therefore, a natural choice is to pick equi-depth buckets, in which each range contains an approximately equal number of points. In such cases, the maximum inaccuracy of a query is equal to twice the count in any bucket. However, in the case of a stream, the choice of ranges which would result in equi-depth partitions is not known a-priori. We note that the design of equi-depth buckets is exactly the problem of quantile estimation in data streams, since the equi-depth partitions define different quantiles in the data.

A different choice for histogram construction is that of minimizing the frequency variance of the different values within a bucket, so that the uniform distribution assumption is approximately held for queries. This minimizes the boundary error of extrapolation in a query. Thus, if a bucket B with count $C(B)$ contains the frequency of $l(B)$ elements, then average frequency of each element in the bucket is $C(B)/l(B)$. Let $f_1 \ldots f_{l(B)}$ be the frequencies of the l values within the bucket. Then, the total variance $v(B)$ of the frequencies from the average is given by:

$$v(B) = \sum_{i=1}^{l} (f_i - C(B)/l(B))^2 \qquad (9.17)$$

Then, the total variance V across all buckets is given by the following:

$$V = \sum_{B} v(B) \qquad (9.18)$$

Such histograms are referred to as *V-Optimal histograms*. A different way of looking at the V-optimal histogram is as a least squares fit to the frequency distribution in the data. Algorithms for V-Optimal histogram construction have been proposed in [60, 63]. We also note that the objective function to be optimized has the form of an L_p-difference function between two vectors whose cardinality is defined by the number of distinct values. In our earlier observations, we noted that sketches are particularly useful in tracking such aggregate functions. This is particularly useful in the multi-dimensional case, where the number of buckets can be very large as a result of the combination of a large number of dimensions. Therefore sketch-based methods can be used for the multi-dimensional case. We will discuss this in detail slightly later. We note that a number of other objective functions also exist for optimizing histogram construction [86]. For example, one can minimize the difference in the area between the original distribution, and the corresponding histogram fit. Since the space requirement is dictated by the number of buckets, it is also desirable to minimize it. Therefore, the dual problem of minimizing the number of buckets, for a given threshold on the error has been discussed in [63, 78].

One problem with the above definitions is that they use they use absolute errors in order to define the accuracy. It has been pointed out in [73] that the use of absolute error may not always be a good representation of the error. Therefore, some methods for optimizing relative error have been proposed in [53]. While this method is quite efficient, it is not designed to be a data stream algorithm. Therefore, the design of relative error histogram construction for the stream case continues to be an open problem.

5.1   One Pass Construction of Equi-depth Histograms

In this section, we will develop algorithms for one-pass construction of equi-depth histograms. The simplest method for determination of the relevant quantiles in the data is that of sampling. In sampling, we simply compute the estimated quantile $q(S) \in [0, 1]$ of the true quantile $q \in [0, 1]$ on a random sample S of the data. Then, the Hoeffding inequality can be used to show that $q(S)$ lies in the range $(q - \epsilon, q + \epsilon)$ with probability at least $1 - \delta$, if the sample size S is chosen larger than $O(\log(\delta)/\epsilon^2)$. Note that this sample size is a constant, and is independent of the size of the underlying data stream.

Let v be the value of the element at quantile q. Then the probability of including an element in S with value less than v is a Bernoulli trial with probability q. Then the expected number of elements less than v is $q \cdot |S|$, and this number lies in the interval $(q \pm \epsilon)$ with probability at least $2 \cdot e^{-2 \cdot |S| \cdot \epsilon^2}$ (Hoeffding inequality). By picking a value of $|S| = O(\log(\delta)/\epsilon^2)$, the corresponding results may be easily proved. A nice analysis of the effect of sample sizes on histogram construction may be found in [12]. In addition, methods for incremental histogram maintenance may be found in [42]. The $O(\log(\delta)/\epsilon^2)$ space-requirements have been tightened to $O(\log(\delta)/\epsilon)$ in a variety of ways. For example, the algorithms in [71, 72] discuss probabilistic algorithms for tightening this bound, whereas the method in [49] provides a deterministic algorithm for the same goal.

5.2   Constructing V-Optimal Histograms

An interesting offline algorithm for constructing V-Optimal histograms has been discussed in [63]. The central idea in this approach is to set up a dynamic programming recursion in which the partition for the last bucket is determined. Let us consider a histogram drawn on the N ordered distinct values $[1 \ldots N]$. Let $Opt(k, N)$ be the error of the V-optimal histogram for the first N values, and k buckets. Let $Var(p, q)$ be the variances of values indexed by p through q in $(1 \ldots N)$. Then, if the last bucket contains values $r \ldots N$, then the error of the V-optimal histogram would be equal to the sum of the error of the $(k - 1)$-bucket V-optimal histogram for values up to $r - 1$, added to the error of the last bucket (which is simply the variance of the values indexed by r through N). Therefore, we have the following dynamic programming recursion:

$$Opt(k, N) = \min_r \{Opt(k - 1, r - 1) + Var(r, N)\} \qquad (9.19)$$

We note that there are $O(N \cdot k)$ entries for the set $Opt(k, N)$, and each entry can be computed in $O(N)$ time using the above dynamic programming recursion. Therefore, the total time complexity is $O(N^2 \cdot k)$.

While this is a neat approach for offline computation, it does not really apply to the data stream case because of the quadratic time complexity. In [54], a method has been proposed to construct $(1 + \epsilon)$-optimal histograms in $O(N \cdot k^2 \cdot \log(N)/\epsilon)$ time and $O(k^2 \cdot \log(N)/\epsilon)$ space. We note that the number of buckets k is typically small, and therefore the above time complexity is quite modest in practice. The central idea behind this approach is that the dynamic programming recursion of Equation 9.19 is the sum of a monotonically increasing and a monotonically decreasing function in r. This can be leveraged to reduce the amount of search in the dynamic programming recursion, if one is willing to settle for a $(1 + \epsilon)$-approximation. Details may be found in [54]. Other algorithms for V-optimal histogram construction may be found in [47, 56, 57].

5.3   Wavelet Based Histograms for Query Answering

Wavelet Based Histograms are a useful tool for selectivity estimation, and were first proposed in [73]. In this approach, we construct the Haar wavelet decomposition on the cumulative distribution of the data. We note that for a dimension with N distinct values, this requires N wavelet coefficients. As is usually the case with wavelet decomposition, we retain the B Haar coefficients with the largest absolute (normalized) value. The cumulative distribution $\theta(b)$ at a given value b can be constructed as the sum of $O(\log(N))$ coefficients on the error-tree. Then for a range query $[a, b]$, we only need to compute $\theta(b) - \theta(a)$.

In the case of data streams, we would like to have the ability to maintain the wavelet based histogram dynamically. In this case, we perform the maintenance with frequency distributions rather than cumulative distributions. We note that when a new data stream element x arrives, the frequency distribution along a given dimension gets updated. This can lead to the following kinds of changes in the maintained histogram:

- Some of the wavelet coefficients may change and may need to be updated. An important observation here is that only the $O(\log(N))$ wavelet coefficients whose ranges include x may need to be updated. We note that many of these coefficients may be small and may not be included in the histogram in the first place. Therefore, only those coefficients which are already included in the histogram need to be updated. For a coefficient including a range of length $l = 2^q$ we update it by adding or subtracting $1/l$. We first update all the wavelet coefficients which are currently included in the histogram.

- Some of the wavelet coefficients which are currently not included in the histogram may become large, and may therefore need to be added to it. Let c_{min} be the minimum value of any coefficient currently included in the histogram. For a wavelet coefficient with range $l = 2^q$, which is not currently included in the histogram, we add it to be histogram with probability $1/(l * c_{min})$. The initial value of the coefficient is set to c_{min}.

- The addition of new coefficients to the histogram will increase the total number of coefficients beyond the space constraint B. Therefore, after each addition, we delete the minimum coefficient in the histogram.

The correctness of the above method follows from the probabilistic counting results discussed in [31]. It has been shown in [74] that this probabilistic method for maintenance is effective in practice.

5.4   Sketch Based Methods for Multi-dimensional Histograms

Sketch based methods can also be used to construct V-optimal histograms in the multi-dimensional case [90]. This is a particularly useful application of sketches since the number of possible buckets in the N^d space increases exponentially with d. Furthermore, the objective function to be optimized has the form of an L_2-distance function over the different buckets. This can be approximated with the use of the Johnson-Lindenstrauss result [64].

We note that each d-dimensional vector can be sketched over N^d-space using the same method as the AMS sketch. The only difference is that we are associating the 4-wise independent random variables with d-dimensional items. The Johnson-Lindenstrauss Lemma implies that the L_2-distances in the sketched representation (optimized over $O(b \cdot d \cdot \log(N)/\epsilon^2)$ possibilities) are within a factor $(1 + \epsilon)$ of the L_2-distances in the original representation for a b-bucket histogram.

Therefore, if we can pick the buckets so that L_2-distances are optimized in the sketched representation, this would continue to be true for the original representation within factor $(1 + \epsilon)$. It turns out that a simple greedy algorithm is sufficient to achieve this. In this algorithm, we pick the buckets greedily, so that the L_2 distances in the sketched representation are optimized in each step. It can be shown [90], that this simple approach provides a near optimal histogram with high probability.

6.   Discussion and Challenges

In this paper, we provided an overview of the different methods for synopsis construction in data streams. We discussed random sampling, wavelets, sketches and histograms. In addition, many techniques such as clustering can also be used for synopses construction. Some of these methods are discussed in more detail in a different chapter of this book. Many methods such as wavelets and histograms are closely related to one another. This chapter explores the basic methodology of each technique and the connections between different techniques. Many challenges for improving synopsis construction methods remain:

- While many synopses construction methods work effectively in individual scenarios, it is as yet unknown how well the different methods compare with one another. A thorough performance study needs to be conducted in understanding the relative behavior of different synopsis methods. One important point to be kept in mind is that the "trusty-old" sampling method provides the most effective results in many practical situations, where space is not constrained by specialized hardware considerations (such as a distributed sensor network). This is especially true for multi-dimensional data sets with inter-attribute correlations, in which methods such as histograms and wavelets become increasingly ineffective. Sampling is however ineffective in counting measures which rely on *infrequent* behavior of the underlying data set. Some examples are distinct element counting and join size estimation. Such a study may reveal the importance and robustness of different kinds of methods in a wide variety of scenarios.

- A possible area of research is in the direction of designing *workload aware* synopsis construction methods [75, 78, 79]. While many methods for synopsis construction optimize average or worst-case performance, the real aim is to provide optimal results for *typical* workloads. This requires methods for modeling the workload as well as methods for leveraging these workloads for accurate solutions.

- Most synopsis structures are designed in the context of quantitative or categorical data sets. It would be interesting to examine how synopsis methods can be extended to the case of different kinds of domains such as string, text or XML data. Some recent work in this direction has designed methods for XCluster synopsis or sketch synopsis for XML data [82, 83, 87].

- Most methods for synopsis construction focus on construction of optimal synopsis over the *entire data stream*. In many cases, data streams may evolve over time, as a result of which it may be desirable to construct optimal synopsis over specific time windows. Furthermore, this window may not be known in advance. This problem may be quite challenging to solve in a space-efficient manner. A number of methods for maintaining exponential histograms and time-decaying stream aggregates [15, 48] try to account for evolution of the data stream. Some recent work on *biased reservoir sampling* [4] tries to extend such an approach to sampling methods.

We believe that there is considerable scope for extension of the current synopsis methods to domains such as sensor mining in which the hardware requirements force the use of space-optimal synopsis. However, the objective of constructing a given synopsis needs to be carefully calibrated in order to take the specific hardware requirements into account. While the broad theoretical foundations of this field are now in place, it remains to carefully examine how these methods may be leveraged for applications with different kinds of hardware, computational power, or space constraints.

References

[1] Aggarwal C., Han J., Wang J., Yu P. (2003) A Framework for Clustering Evolving Data Streams. *VLDB Conference.*

[2] Aggarwal C, Han J., Wang J., Yu P. (2004). On-Demand Classification of Data Streams. *ACM KDD Conference.*

[3] Aggarwal C. (2006) On Futuristic Query Processing in Data Streams. *EDBT Conference.*

[4] Aggarwal C. (2006) On Biased Reservoir Sampling in the Presence of Stream Evolution. *VLDB Conference.*

[5] Alon N., Gibbons P., Matias Y., Szegedy M. (1999) Tracking Joins and Self Joins in Limited Storage. *ACM PODS Conference.*

[6] Alon N., Matias Y., Szegedy M. (1996) The Space Complexity of Approximating the Frequency Moments. *ACM Symposium on Theory of Computing,* pp. 20–29/

[7] Arasu A., Manku G. S. Approximate quantiles and frequency counts over sliding windows. *ACM PODS Conference,* 2004.

[8] Babcock B., Datar M. Motwani R. (2002) Sampling from a Moving Window over Streaming Data. *ACM SIAM Symposium on Discrete Algorithms.*

[9] Babcock B., Olston C. (2003) Distributed Top-K Monitoring. *ACM SIGMOD Conference 2003.*

[10] Bulut A., Singh A. (2003) Hierarchical Stream summarization in Large Networks. *ICDE Conference.*

[11] Chakrabarti K., Garofalakis M., Rastogi R., Shim K. (2001) Approximate Query Processing with Wavelets. *VLDB Journal,* 10(2-3), pp. 199–223.

[12] Chaudhuri S., Motwani R., Narasayya V. (1998) Random Sampling for Histogram Construction: How much is enough? *ACM SIGMOD Conference.*

[13] Charikar M., Chen K., Farach-Colton M. (2002) Finding Frequent items in data streams. *ICALP.*

[14] Chernoff H. (1952) A measure of asymptotic efficiency for tests of a hypothesis based on the sum of observations. *The Annals of Mathematical Statistics*, 23:493–507.

[15] Cohen E., Strauss M. (2003). Maintaining Time Decaying Stream Aggregates. *ACM PODS Conference.*

[16] Cormode G., Garofalakis M., Sacharidis D. (2006) Fast Approximate Wavelet Tracking on Streams. *EDBT Conference.*

[17] Cormode G., Datar M., Indyk P., Muthukrishnan S. (2002) Comparing Data Streams using Hamming Norms. *VLDB Conference.*

[18] Cormode G., Muthukrishnan S. (2003) What's hot and what's not: Tracking most frequent items dynamically. *ACM PODS Conference.*

[19] Cormode G., Muthukrishnan S. (2004) What's new: Finding significant differences in network data streams. *IEEE Infocom.*

[20] Cormode G., Muthukrishnan S. (2004) An Improved Data Stream Summary: The Count-Min Sketch and Its Applications. *LATIN* pp. 29-38.

[21] Cormode G., Muthukrishnan S. (2004) Diamond in the Rough; Finding Hierarchical Heavy Hitters in Data Streams. *ACM SIGMOD Conference.*

[22] Cormode G., Garofalakis M. (2005) Sketching Streams Through the Net: Distributed approximate Query Tracking. *VLDB Conference.*

[23] Cormode G., Muthukrishnan S., Rozenbaum I. (2005) Summarizing and Mining Inverse Distributions on Data Streams via Dynamic Inverse Sampling. *VLDB Conference.*

[24] Das A., Ganguly S., Garofalakis M. Rastogi R. (2004) Distributed Set-Expression Cardinality Estimation. *VLDB Conference.*

[25] Degligiannakis A., Roussopoulos N. (2003) Extended Wavelets for multiple measures. *ACM SIGMOD Conference.*

[26] Dobra A., Garofalakis M., Gehrke J., Rastogi R. (2002) Processing complex aggregate queries over data streams. *SIGMOD Conference*, 2002.

[27] Dobra A., Garofalakis M. N., Gehrke J., Rastogi R. (2004) Sketch-Based Multi-query Processing over Data Streams. *EDBT Conference.*

[28] Domingos P., Hulten G. (2000) Mining Time Changing Data Streams. *ACM KDD Conference.*

[29] Estan C., Varghese G. (2002) New Directions in Traffic Measurement and Accounting, *ACM SIGCOMM*, 32(4), *Computer Communication Review.*

[30] Fang M., Shivakumar N., Garcia-Molina H., Motwani R., Ullman J. (1998) Computing Iceberg Cubes Efficiently. *VLDB Conference.*

[31] Flajolet P., Martin G. N. (1985) Probabilistic Counting for Database Applications. *Journal of Computer and System Sciences*, 31(2) pp. 182–209.

[32] Feigenbaum J., Kannan S., Strauss M. Viswanathan M. (1999) An Approximate L_1-difference algorithm for massive data streams. *FOCS Conference*.

[33] Fong J., Strauss M. (2000) An Approximate L_p-difference algorithm for massive data streams. *STACS Conference*.

[34] Ganguly S., Garofalakis M., Rastogi R. (2004) Processing Data Stream Join Aggregates using Skimmed Sketches. *EDBT Conference*.

[35] Ganguly S., Garofalakis M, Rastogi R. (2003) Processing set expressions over continuous Update Streams. *ACM SIGMOD Conference*

[36] Ganguly S., Garofalakis M., Kumar N., Rastogi R. (2005) Join-Distinct Aggregate Estimation over Update Streams. *ACM PODS Conference*.

[37] Garofalakis M., Gehrke J., Rastogi R. (2002) Querying and mining data streams: you only get one look (a tutorial). *SIGMOD Conference*.

[38] Garofalakis M., Gibbons P. (2002) Wavelet synopses with error guarantees. *ACM SIGMOD Conference*.

[39] Garofalakis M, Kumar A. (2004) Deterministic Wavelet Thresholding with Maximum Error Metrics. *ACM PODS Conference*.

[40] Gehrke J., Korn F., Srivastava D. (2001) On Computing Correlated Aggregates Over Continual Data Streams. SIGMOD Conference.

[41] Gibbons P., Mattias Y. (1998) New Sampling-Based Summary Statistics for Improving Approximate Query Answers. *ACM SIGMOD Conference Proceedings*.

[42] Gibbons P., Matias Y., and Poosala V. (1997) Fast Incremental Maintenance of Approximate Histograms. *VLDB Conference*.

[43] Gibbons P. (2001) Distinct sampling for highly accurate answers to distinct value queries and event reports. *VLDB Conference*.

[44] Gilbert A., Kotidis Y., Muthukrishnan S., Strauss M. (2001) Surfing Wavelets on Streams: One Pass Summaries for Approximate Aggregate Queries. *VLDB Conference*.

[45] Gilbert A., Kotidis Y., Muthukrishnan S., Strauss M. (2003) One-pass wavelet decompositions of data streams. *IEEE TKDE*, 15(3), pp. 541–554. (Extended version of [44])

[46] Gilbert A., Kotidis Y., Muthukrishnan S., Strauss M. (2002) How to summarize the universe: Dynamic Maintenance of quantiles. *VLDB Conference*.

[47] Gilbert A., Guha S., Indyk P., Kotidis Y., Muthukrishnan S., Strauss M. (2002) Fast small-space algorithms for approximate histogram maintenance. *ACM STOC Conference*.

[48] Gionis A., Datar M., Indyk P., Motwani R. (2002) Maintaining Stream Statistics over Sliding Windows. *SODA Conference.*

[49] Greenwald M., Khanna S. (2001) Space Efficient Online Computation of Quantile Summaries. *ACM SIGMOD Conference*, 2001.

[50] Greenwald M., Khanna S. (2004) Power-Conserving Computation of Order-Statistics over Sensor Networks. *ACM PODS Conference.*

[51] Guha S. (2005). Space efficiency in Synopsis construction algorithms. *VLDB Conference.*

[52] Guha S., Kim C., Shim K. (2004) XWAVE: Approximate Extended Wavelets for Streaming Data. *VLDB Conference*, 2004.

[53] Guha S., Shim K., Woo J. (2004) REHIST: Relative Error Histogram Construction algorithms. *VLDB Conference.*

[54] Guha S., Koudas N., Shim K. (2001) Data-Streams and Histograms. *ACM STOC Conference.*

[55] Guha S., Harb B. (2005) Wavelet Synopses for Data Streams: Minimizing Non-Euclidean Error. *ACM KDD Conference.*

[56] Guha S., Koudas N. (2002) Approximating a Data Stream for Querying and Estimation: Algorithms and Performance Evaluation. *ICDE Conference.*

[57] Guha S., Indyk P., Muthukrishnan S., Strauss M. (2002) Histogramming data streams with fast per-item processing. *Proceedings of ICALP.*

[58] Hellerstein J., Haas P., Wang H. (1997) Online Aggregation. *ACM SIGMOD Conference.*

[59] Ioannidis Y., Poosala V. (1999) Histogram-Based Approximation of Set-Valued Query-Answers. *VLDB Conference.*

[60] Ioannidis Y., Poosala V. (1995) Balancing Histogram Optimality and Practicality for Query Set Size Estimation. *ACM SIGMOD Conference.*

[61] Indyk P., Koudas N., Muthukrishnan S. (2000) Identifying Representative Trends in Massive Time Series Data Sets Using Sketches. *VLDB Conference.*

[62] Indyk P. (2000) Stable Distributions, Pseudorandom Generators, Embeddings, and Data Stream Computation, *IEEE FOCS.*

[63] Jagadish H., Koudas N., Muthukrishnan S., Poosala V., Sevcik K., and Suel T. (1998) Optimal Histograms with Quality Guarantees. *VLDB Conference.*

[64] Johnson W., Lindenstrauss J. (1984) Extensions of Lipshitz mapping into Hilbert space. *Contemporary Mathematics*, Vol 26, pp. 189–206.

[65] Karras P., Mamoulis N. (2005) One-pass wavelet synopses for maximum error metrics. *VLDB Conference.*

[66] Keim D. A., Heczko M. (2001) Wavelets and their Applications in Databases. *ICDE Conference.*

[67] Kempe D., Dobra A., Gehrke J. (2004) Gossip Based Computation of Aggregate Information. *ACM PODS Conference.*

[68] Kollios G., Byers J., Considine J., Hadjieleftheriou M., Li F.(2005) Robust Aggregation in Sensor Networks. *IEEE Data Engineering Bulletin.*

[69] Kooi R. (1980) The optimization of queries in relational databases. *Ph. D Thesis*, Case Western Reserve University.

[70] Manjhi A., Shkapenyuk V., Dhamdhere K., Olston C. (2005) Finding (recently) frequent items in distributed data streams. *ICDE Conference.*

[71] Manku G., Rajagopalan S, Lindsay B. (1998) Approximate medians and other quantiles in one pass and with limited memory. *ACM SIGMOD Conference.*

[72] Manku G., Rajagopalan S, Lindsay B. (1999) Random Sampling for Space Efficient Computation of order statistics in large datasets. *ACM SIGMOD Conference.*

[73] Matias Y., Vitter J. S., Wang M. (1998) Wavelet-based histograms for selectivity estimation. *ACM SIGMOD Conference.*

[74] Matias Y., Vitter J. S., Wang M. (2000) Dynamic Maintenance of Wavelet-based histograms. *VLDB Conference.*

[75] Matias Y., Urieli D. (2005) Optimal workload-based wavelet synopsis. *ICDT Conference.*

[76] Manku G., Motwani R. (2002) Approximate Frequency Counts over Data Streams. *VLDB Conference.*

[77] Muthukrishnan S. (2004) Workload Optimal Wavelet Synopses. *DIMACS Technical Report.*

[78] Muthukrishnan S., Poosala V., Suel T. (1999) On Rectangular Partitioning in Two Dimensions: Algorithms, Complexity and Applications, *ICDT Conference.*

[79] Muthukrishnan S., Strauss M., Zheng X. (2005) Workload-Optimal Histograms on Streams. *Annual European Symposium*, Proceedings in *Lecture Notes in Computer Science*, 3669, pp. 734-745

[80] Olston C., Jiang J., Widom J. (2003) Adaptive Filters for Continuous Queries over Distributed Data Streams. *ACM SIGMOD Conference.*

[81] Piatetsky-Shapiro G., Connell C. (1984) Accurate Estimation of the number of tuples satisfying a condition. *ACM SIGMOD Conference.*

[82] Polyzotis N., Garofalakis M. (2002) Structure and Value Synopsis for XML Data Graphs. *VLDB Conference.*

[83] Polyzotis N., Garofalakis M. (2006) XCluster Synopses for Structured XML Content. *IEEE ICDE Conference.*

[84] Poosala V., Ganti V., Ioannidis Y. (1999) Approximate Query Answering using Histograms. *IEEE Data Eng. Bull.*

[85] Poosala V., Ioannidis Y., Haas P., Shekita E. (1996) Improved Histograms for Selectivity Estimation of Range Predicates. *ACM SIGMOD Conference.*

[86] Poosala V., Ioannidis Y. (1997) Selectivity Estimation without the Attribute Value Independence assumption. *VLDB Conference.*

[87] Rao P., Moon B. (2006) SketchTree: Approximate Tree Pattern Counts over Streaming Labeled Trees, *ICDE Conference.*

[88] Schweller R., Gupta A., Parsons E., Chen Y. (2004) Reversible Sketches for Efficient and Accurate Change Detection over Network Data Streams. *Internet Measurement Conference Proceedings.*

[89] Stolnitz E. J., Derose T., Salesin T. (1996) *Wavelets for computer graphics: theory and applications*, Morgan Kaufmann.

[90] Thaper N., Indyk P., Guha S., Koudas N. (2002) Dynamic Multidimensional Histograms. *ACM SIGMOD Conference.*

[91] Thomas D. (2006) Personal Communication.

[92] Vitter J. S. (1985) Random Sampling with a Reservoir. *ACM Transactions on Mathematical Software*, Vol. 11(1), pp 37–57.

[93] Vitter J. S., Wang M. (1999) Approximate Computation of Multidimensional Aggregates of Sparse Data Using Wavelets. *ACM SIGMOD Conference.*

Chapter 10

A SURVEY OF JOIN PROCESSING IN DATA STREAMS

Junyi Xie and Jun Yang
Department of Computer Science
Duke University
{junyi,junyang}@cs.duke.edu

1.   Introduction

Given the fundamental role played by joins in querying relational databases, it is not surprising that *stream join* has also been the focus of much research on streams. Recall that relational (theta) join between two non-streaming relations R_1 and R_2, denoted $R_1 \bowtie_\theta R_2$, returns the set of all pairs $\langle r_1, r_2 \rangle$, where $r_1 \in R_1$, $r_2 \in R_2$, and the join condition $\theta(r_1, r_2)$ evaluates to *true*. A straightforward extension of join to streams gives the following semantics (in rough terms): At any time t, the set of output tuples generated thus far by the join between two streams S_1 and S_2 should be the same as the result of the relational (non-streaming) join between the sets of input tuples that have arrived thus far in S_1 and S_2.

Stream join is a fundamental operation for relating information from different streams. For example, given two stream of packets seen by network monitors placed at two routers, we can join the streams on packet ids to identify those packets that flowed through both routers, and compute the time it took for each such packet to reach the other router. As another example, an online auction system may generate two event streams: One signals opening of auctions and the other contains bids on the open auctions. A stream join is needed to relate bids with the corresponding open-auction events. As a third example, which involves a non-equality join, consider two data streams that arise in monitoring a cluster machine room, where one stream contains load information collected from different machines, and the other stream contains temperature readings from various sensors in the room. Using a stream join, we can look for possible correlations between loads on machines and temperatures at different locations in the machine room. In this case, we need to relate temperature readings and load data with close, but necessarily identical, spatio-temporal coordinates.

What makes stream join so special to warrant new approaches different from conventional join processing? In the stream setting, input tuples arrive continuously, and result tuples need to be produced continuously as well. We cannot assume that the input data is already stored or indexed, or that the input rate can be controlled by the query plan. Standard join algorithms that use blocking operations, e.g., sorting, no longer work. Conventional methods for cost estimation and query optimization are also inappropriate, because they assume finite input. Moreover, the long-running nature of stream queries calls for more adaptive processing strategies that can react to changes and fluctuations in data and stream characteristics. The "stateful" nature of stream joins adds another dimension to the challenge. In general, in order to compute the complete result of a stream join, we need to retain all past arrivals as part of the processing state, because a new tuple may join with an arbitrarily old tuple arrived in the past. This problem is exacerbated by unbounded input streams, limited processing resources, and high performance requirements, as it is impossible in the long run to keep all past history in fast memory.

This chapter provides an overview of research problems, recent advances, and future research directions in stream join processing. We start by elucidating the model and semantics for stream joins in Section 2. Section 3 focuses on join state management—the important problem of how to cope with large and potentially unbounded join state given limited memory. Section 4 covers fundamental algorithms for stream join processing. Section 5 discusses aspects of stream join optimization, including objectives and techniques for optimizing multi-way joins. We conclude the chapter in Section 6 by pointing out several related research areas and proposing some directions for future research.

2.   Model and Semantics

Basic Model and Semantics.   A stream is an unbounded sequence of stream tuples of the form $\langle s, t \rangle$ ordered by t, where s is a relational tuple and t is the *timestamp* of the stream tuple. Following a "reductionist" approach, we conceptually regard the *(unwindowed) stream join* between streams S_1 and S_2 to be a view defined as the (bag) relational join between two append-only bags S_1 and S_2. Whenever new tuples arrive in S_1 or S_2, the view must be updated accordingly. Since relational join is monotonic, insertions into S_1 and S_2 can result only in possible insertions into the view. The sequence of resulting insertions into the view constitutes the output stream of the stream join between S_1 and S_2. The timestamp of an output tuple is the time at which the insertion should be reflected in view, i.e., the larger of the timestamps of the two input tuples.

Alternatively, we can describe the same semantics operationally as follows: To compute the stream join between S_1 and S_2, we maintain a *join state* containing all tuples received so far from S_1 (which we call S_1's join state) and those from S_2 (which we call S_2's join state). For each new tuple s_1 arriving in S_1, we record s_1 in S_1's join state, probe S_2's join state for tuples joining with s_1, and output the join result tuples. New tuples arriving in S_2 are processed in a symmetrical fashion.

Semantics of Sliding-Window Joins.   An obvious issue with unwindowed stream joins is that the join state is unbounded and will eventually outgrow memory and storage capacity of the stream processing system. One possibility is to restrict the scope of the join to a recent window, resulting in a *sliding-window stream join*. For binary joins, we call the two input streams *partner stream* of each other. Operationally, a *time-based sliding window* of duration w on stream S restricts each new partner stream tuple to join only with S tuples that arrived within the last w time units. A *tuple-based sliding window* of size k restricts each new partner stream tuple to join only with the last k tuples arrived in S. Both types of windows "slide" forward, as time advances or new stream tuples arrive, respectively. The sliding-window semantics enables us to purge from the join state any tuple that has fallen out of the current window, because future arrivals in the partner stream cannot possibly join with them.

Continuous Query Language, or *CQL* for short [2], gives the semantics of a sliding-window stream join by regarding it as a relational join view over the sliding windows, each of which contains the bag of tuples in the current window of the respective stream. New stream tuples are treated as insertion into the windows, while old tuples that fall out of the windows are treated as deletions. The resulting sequences of updates on the join view constitutes the output stream of the stream join. Note that deletions from the windows can result in deletions from the view. Therefore, sliding-window stream joins are not monotonic. The presence of deletions in the output stream does complicate semantics considerably. Fortunately, in many situations users may not care about these deletions at all, and CQL provides an `Istream` operator for removing them from the output stream. For a time-based sliding-window join, even if we do not want to ignore deletions in the output stream, it is easy to infer when an old output tuple needs to be deleted by examining the timestamps of the input tuples that generated it. For this reason, time-based sliding-window join under the CQL semantics is classified as a *weak non-monotonic* operator by Golab and Özsu [24]. However, for a tuple-based sliding-window join, how to infer deletions in the output stream timely and efficiently without relying on explicitly generated "negative tuples" still remains an open question [24].

There is an alternative definition of sliding-window stream joins that does not introduce non-monotonicity. For a time-based sliding-window join with duration w, we simply regard the stream join between S_1 and S_2 as a relational join view over append-only bags S_1 and S_2 with an extra "window join condition": $-w \leq S_1.t - S_2.t \leq w$. As in the case of an unwindowed stream join, the output stream is simply the sequence of updates on the view resulting from the insertions into S_1 and S_2. Despite the extra window join condition, join remains monotonic; deletions never arise in the output stream because S_1 and S_2 are append-only. This definition of time-based sliding-window join has been used by some, e.g., [10, 27]. It is also possible to define a tuple-based sliding-window join as a monotonic view over append-only bags (with the help of an extra attribute that records the sequence number for each tuple in an input stream), though the definition is more convoluted. This alternative semantics yields the same sequence of insertions as the CQL semantics. In the remainder of this chapter, we shall assume this semantics and ignore the issue of deletions in the output stream.

Relaxations and Variations of the Standard Semantics.   The semantics of stream joins above requires the output sequence to reflect the complete sequence of states of the underlying view, in the exact same order. In some settings this requirement is relaxed. For example, the stream join algorithms in [27] may generate output tuples slightly out of order. The XJoin-family of algorithms (e.g., [41, 33, 38]) relaxes the single-pass stream processing model and allows some tuples to be spilled out from memory and onto disk to be processed later, which means that output tuples may be generated out of order. In any case, the correct output order can be reconstruct from the tuple timestamps. Besides relaxing the requirement on output ordering, there are also variations of sliding windows that offer explicit control over what states of the view can be ignored. For example, with the "jumping window" semantics [22], we divide the sliding window into a number of sub-windows; when the newest sub-window fills up, it is appended to the sliding window while the oldest sub-window in the sliding window is removed, and then the query is re-evaluated. This semantics induces a window that is "jumping" periodically instead of sliding gradually.

Semantics of Joins between Streams and Database Relations.   Joins between streams and time-varying database relations have also been considered [2, 24]. Golab and Özsu [24] proposed a *non-retroactive relation* semantics, where each stream tuple joins only with the state of the time-varying database relation at the time of its arrival. Consequently, an update on the database relation does not retroactively apply to previously generated output tuples. This semantics is also supported by CQL [2], where the query can be interpreted as a join between the database relation and a zero-duration sliding window over the stream containing only those tuples arriving at the current time. We shall assume this semantics in our later discussion on joining streams and database relations.

3.   State Management for Stream Joins

In this section, we turn specifically to the problem of *state management* for stream joins. As discussed earlier, join is stateful operator; without the sliding-window semantics, computing the complete result of a stream join generally requires keeping unbounded state to remember all past tuples [1]. The question is: What is the most effective use of the limited memory resource? How do we decide what part of the join state to keep and what to discard? Can we mitigate the problem by identifying and purging "useless" parts of the join state without affecting the completeness of the result? When we run out of memory and are no longer able to produce the complete result, how do we then measure the "error" in an incomplete result, and how do we manage the join state in a way to minimize this error?

Join state management is also relevant even for sliding-window joins, where the join state is bounded by the size of the sliding windows. Sometimes, sliding windows may be quite large, and any further reduction of the join state is welcome because memory is often a scarce resource in stream processing systems. Moreover, if we consider a more general stream processing model where streams are processed not just in fast main memory but instead in a memory hierarchy involving smaller, faster caches as well as larger, slower disks, join state management generalizes into the problem of deciding how to ferry data up and down the memory hierarchy to maximize processing efficiency.

One effective approach towards join state management is to exploit "hard" constraints in the input streams to reduce state. For example, we might know that for a stream, the join attribute is a key, or the value of the join attribute always increases over time. Through reasoning with these constraints and the join condition, we can sometimes infer that certain tuples in the join state cannot contribute to any future output tuples. Such tuples can then be purged from the join state without compromising result completeness. In Section 3.1, we examine two techniques that generalize constraints in the stream setting and use them for join state reduction.

Another approach is to exploit statistical properties of the input streams, which can be seen as "soft" constraints, to help make join state management decisions. For example, we might know (or have observed) that the frequency of each join attribute value is stable over time, or that the join attribute values in a stream can be modeled by some stochastic process, e.g., random walk. Such knowledge allows us to estimate the benefit of keeping a tuple in the join state (for example, as measured by how many output tuples it is expected to generate over a period of time). Because of the stochastic nature of such knowledge, we usually cannot guarantee result completeness. However, this approach can be used to minimize the expected error in the incomplete result, or to optimize the organization of the join state in a memory hierarchy to maximize performance. We discuss this approach in Section 3.2.

3.1 Exploiting Constraints

k-Constraints. Babu et al. [7] introduced k-*constraints* for join state reduction. The parameter k is an *adherence parameter* that specifies how closely a stream adheres to the constraint. As an example of k-constraints, consider first a "strict" *ordered-arrival* constraint on stream S, which requires that the join attribute values of S tuples never decrease over time. In a network monitoring application, a stream of TCP/IP packets transmitted from a source to a destination should arrive in the order of their source timestamps (denoted by t_s to distinguish them from the tuple timestamps t). However, suppose that for efficiency, we instead use UDP, a less reliable protocol with no guarantee on the delivery order. Nonetheless, if we can bound the extent of packet reordering that occur in practice, we can relax the constraint into an ordered-arrival k-constraint: For any tuple s, a tuple s' with an earlier source timestamp (i.e., $s'.t_s < s.t_s$) must arrive as or before the k-th tuple following s. A smaller k implies a tighter constraint; a constraint with $k = 0$ becomes strict.

To see how k-constraints can be used for join state reduction, suppose we join the packet stream S in the above example with another stream S' using the condition $|S.t_s - S'.t_s| \leq 10$. Without any constraint on $S.t_s$, we must remember all S' tuples in the join state, because any future S tuple could arrive with a joining t_s value. With the ordered-arrival k-constraint on $S.t_s$, however, we can purge a tuple $s' \in S'$ from the join state as soon as k tuples have arrived in S following some S tuple with $t_s > s'.t_s + 10$. The reason is that the k-constraint guarantees any subsequent S tuples will have source timestamps strictly greater than $s'.t_s + 10$ and therefore not join with s'. Other k-constraints considered by [7] include generalizations of referential integrity constraints and *clustered-arrival* constraints.

Although k-constraints provide some "slack" through the adherence parameter k, strictly speaking they are still hard constraints in that we assume the conditions must hold strictly after k arrivals. Babu et al. also developed techniques for monitoring streams for k-constraints and determining the value of k at runtime. Interestingly, k-constraints with dynamically observed k become necessarily soft in nature: They can assert that the constraints hold with high probability, but cannot guarantee them with absolute certainty.

Punctuations. In contrast to k-constraints, whose forms are known a priori, *punctuations*, introduced by Tucker et al. [40], are constraints that are dynamically inserted into a stream. Specifically, a punctuation is a tuple of patterns specifying a predicate that must evaluate to *false* for all future data tuples in the stream. For example, consider an auction system with two streams: $Auction(id, info, t)$ generates a tuple at the opening of each auction (with a unique auction id), and $Bid(auction\_id, price, t)$ contains bids for open auctions. When an auction with id a_i closes, the system inserts a punctuation $\langle a_i, * \rangle$ into the Bid stream to signal that there will be no more bids for auction a_i. Also, since auction ids are unique, following the opening of every auction a_j, the system can also insert a punctuation $\langle a_j, * \rangle$ into $Auction$ to signal that will be no other auctions with the same id.

Ding et al. [17] developed a stream join algorithm called PJoin to exploit punctuations. When a punctuation arrives in a stream, PJoin examines the join state of the partner stream and purges those tuples that cannot possibly join with future arrivals. For example, upon the arrival of a punctuation $\langle a_i, * \rangle$ in Bid, we can purge any $Auction$ tuples in the join state with id a_i (provided that they have already been processed for join with all past Bid tuples). PJoin also propagates punctuations to the output stream. For example, after receiving $\langle a_i, * \rangle$ from *both* input streams, we can propagate $\langle a_i, *, * \rangle$ to the output, because we are sure that no more output tuple with a_i can be generated. Punctuation propagation is important because propagated punctuations can be further exploited by downstream operators that receive the join output stream as their input. Ding and Rundensteiner [18] further extended their join algorithm to work with sliding windows, which allow punctuations to be propagated quicker. For example, suppose that we set the sliding window to 24 hours, and 24 hours have past after we saw punctuation $\langle a_i, * \rangle$ from $Auction$. Even if we might not have seen $\langle a_i, * \rangle$ yet from Bid, in this case we can still propagate $\langle a_i, *, * \rangle$ to the output, because future Bid tuples cannot join with an $Auction$ tuple that has already fallen outside the sliding window.

While punctuations are more flexible and generally more expressive than k-constraints, they do introduce some processing overhead. Besides the overhead of generating, processing, and propagating punctuations, we note that some past punctuations need to be retained as part of the join state, thereby consuming more memory. For stream joins, past punctuations cannot be purged until we can propagate them, so it is possible to accumulate many punctuations. Also, not all punctuations are equally effective in join state reduction, and their effectiveness may vary for different join conditions. We believe that further research on the trade-off between the cost and the benefit of punctuations is needed, and that managing the "punctuation state" poses an interesting problem parallel to join state management itself.

3.2   Exploiting Statistical Properties

Strictly speaking, both k-constraints and punctuations are hard constraints. Now, we explore how to exploit "soft" constraints, or statistical properties of input streams, in join state management. Compared with hard constraints, these soft constraints can convey more information relevant to join state management. For example, consider again the UDP packet stream discussed in Section 3.1. Depending on the characteristics of the communication network, k may need to be very large for the ordered-arrival k-constraint to hold. However, it may turn out that 99% of the time the extent of packet reordering is limited to a much smaller k', and that 80% of the time reordering is limited to an even smaller k''. Soft, statistical constraints are better at capturing these properties and enabling optimization based on common cases rather than the worst case.

Given a limited amount of memory to hold the join state, for each incoming stream tuple, we need to make a decision—not unlike a cache replacement decision—about whether to discard the new tuple (after joining it with the partner stream tuples in the join state) or to retain it in the join state; in the latter case, we also need to decide which old tuple to discard from the join state to make space for the new one. In the following, we shall use the term "cache" to refer to the memory available for keeping the join state.

Before we proceed, we need to discuss how to evaluate a join state management strategy. There are two major perspectives, depending on the purpose of join state management. The first perspective assumes the single-pass stream processing model where output tuples can be produced only from the part of the join state that we choose to retain in cache. In this case, our goal is to minimize the error in (or to maximize the quality of) the output stream compared with the complete result. A number of popular measures have been defined from this perspective:

- *Max-subset.* This measure, introduced by Das et al. [15], aims at producing as many output tuples as possible. (Note that any reasonable stream join algorithm would never produce any incorrect output tuples, so we can ignore the issue of false positives.) Because input streams are unbounded, we cannot compare two strategies simply by comparing the total numbers of output tuples they produce—both may be infinite. The approach taken by Srivastava and Widom [37] is to consider the ratio between the number of output tuples produced up to some time t and the number of tuples in the complete result up to t. Then, a reasonable goal is to maximize this ratio as t tends to infinity.

- *Sampling rate.* Like max-subset, this measure aims at producing as many output tuples as possible, but with the additional requirement that the set of output tuples constitutes a uniform random sample of the complete join result. Thus, the goal is to maximize the sampling rate. This measure is first considered in the stream join setting by [37].

- *Application-defined importance.* This measure is based on the notion of importance specific to application needs. For example, *Aurora* [39] allows applications to define *value-based quality-of-service functions* that specify the utilities of output tuples based on their attribute values. The goal in this case is to maximize the utility of the join result.

The second perspective targets expected performance rather than result completeness. This perspective relaxes the single-pass processing model by allowing tuples to be spilled out from memory and onto disk to be processed later in "mop-up" phases. Assuming that we still produce the complete answer, our goal is to minimize the total processing cost of the online and the mop-up phases. One measure defined from this perspective is the *archive metric* proposed by [15]. This measure has also been used implicitly by the XJoin-family of algorithms ([41, 33, 38], etc.). As it is usually more expensive to process tuples that have been spilled out to disk, a reasonable approximation is to try to leave as little work as possible to the mop-up phases; this goal roughly compatible with max-subset's objective of getting as much as possible done online.

In the remainder of this section, we focus first and mostly on the max-subset measure. Besides being a reasonable measure in its own right, techniques developed for max-subset are roughly in line with the archive metric, and can be generalized to certain application-defined importance measures through appropriate weighting. Next, we discuss the connection between classic caching and join state management, and state management for joins between streams and database relations. Finally, we briefly discuss the sampling-rate measure towards the end of this section.

Max-Subset Measure. Assuming perfect knowledge of the future arrivals in the input streams, the problem of finding the optimal sequence (up to a given time) of join state management decisions under max-subset can be cast as a network flow problem, and can be solved offline in time polynomial to the length of the sequence and the size of the cache [15]. In practice, however, we need an online algorithm that does not have perfect knowledge of the future. Unfortunately, without any knowledge (statistical or otherwise) of the input streams, no online algorithm—not even a randomized one—can be k-competitive (i.e., generating at least $1/k$ as many tuples as an optimal offline algorithm) for any k independent of the length of the input streams [37]. This hardness result highlights the need to exploit statistical properties of the input streams. Next, we review previous work in this area, starting with specific scenarios and ending with a general approach.

Frequency-Based Model.   In the *frequency-based model*, the join attribute value of each new tuple arriving in a stream is drawn independently from a probability distribution that is stationary (i.e., it does not change over time). This model is made explicit by [37] and discussed as special case by [45], although it has been implicitly assumed in the development of many join state management techniques. Under this model, assuming an unwindowed stream equijoin, we can calculate the *benefit* of a tuple s as the product of the partner stream arrival rate and the probability that a new partner stream tuple has the same join attribute value as s. This benefit measures how many output tuples s is expected to generate per unit time in the future. A straightforward strategy is to replace the tuple with the lowest benefit. This strategy, called **PROB**, was proposed by [15], and can be easily shown to be optimal for unwindowed joins. For sliding-window joins, an alternative strategy called **LIFE** was proposed, which weighs a tuple's benefit by its remaining lifetime in the sliding window. Unfortunately, neither **PROB** nor **LIFE** is known to be optimal for sliding-window joins. To illustrate, suppose that we are faced with the choice between two tuples s_1 and s_2, where s_1 has a higher probability of joining with an incoming tuple, but s_2 has a longer lifetime, allowing it to generate more output tuples than s_2 eventually. **PROB** would prefer s_1 while **LIFE** would prefer s_2; however, neither choice is always better, as we will see later in this section.

The frequency-based model is also implicitly assumed by [38] in developing the **RPJ** (*rate-based progressive join*) algorithm. **RPJ** stores the in-memory portion of each input stream's join state as a hash table, and maintains necessary statistics for each hash partition; statistics for individual join attribute values within each partition are computed assuming local uniformity. When **RPJ** runs out of memory, it flushes the partition with lowest benefit out to disk. This strategy is analogous to **PROB**.

Kang et al. [30] assumed a simplified version of the frequency-based model, where each join attribute value occurs with equal frequency in both input streams (though stream arrival rates may differ). With this simplification, the optimal strategy is to prefer keeping the slower stream in memory, because the tuples from the slower stream get more opportunities to join with an incoming partner stream tuple. This strategy is also consistent with **PROB**. More generally, *random load shedding* [39, 4], or **RAND** [15], which simply discards input stream tuples at random, is also justifiable under the max-subset measure by this equal-frequency assumption.

Age-Based Model.   The *age-based model* of [37] captures a scenario where the stationarity assumption of the frequency-based model breaks down because of correlated tuple arrivals in the input streams. Consider the *Auction* and *Bid* example from Section 3.1. A recent *Auction* tuple has a much better chance of joining with a new *Bid* than an old *Auction* tuple. Furthermore, we may be able to assume that the bids for each open auction follow a similar arrival pattern. The age-based model states that, for each tuple s in stream S (with partner stream S'), the expected number of output tuples that s generates at each time step during its lifetime is given by a function $p(\Delta t)$, where Δt denotes the age of the tuple (i.e., how long it has been in the join state). The age-based model further assumes that function $p(\Delta t)$ is the same for all tuples from the same stream, independent of their join attribute values. At the first glance, this assumption may appear quite strong: If we consider two tuples with the same join attribute value arriving at two different times t_1 and t_2, we should have $p(t - t_1) = p(t - t_2)$ for all t when both tuples are in the join state, which would severely limit the form of function p. However, this issue will not arise, for example, if the join attribute is a key of the input stream (e.g., *Auction*). Because foreign-key joins are so common, the age-based model may be appropriate in many settings.

An optimal state management strategy for the age-based model, called *AGE*, was developed by [37]. Given the function $p(\Delta t)$, AGE calculates an optimal age Δt_o such that the expected number of output tuples generated by a tuple per unit time is maximized when it is kept until age Δt_o. Intuitively, if every tuple in the cache is kept for exactly Δt_o time steps, then we are making the most efficient use of every slot in the cache. This optimal strategy is possible if the arrival rate is high enough to keep every cache slot occupied. If not, we can keep each tuple to an age beyond the optimal, which would still result in an optimal strategy assuming that $p(\Delta t)$ has no local minima. A heuristic strategy for the case where $p(\Delta t)$ has local minima is also provided in [37].

Towards General Stochastic Models.   There are many situations where the input stream follows neither the frequency-based model nor the age-based model. For example, consider a measurement stream S_1 generated by a network of sensors. Each stream tuple carries a timestamp t_m recording the time at which the measurement was taken by the sensor (which is different from the stream timestamp t). Because of processing delays at the senors, transmission delays in the network, and a network protocol with no in-order delivery guarantee (e.g., UDP), the t_m values do not arrive in order, but may instead follow a discretized bounded normal distribution centered at the current time minus the average latency. Figure 10.1 shows the pdf (probability density function) of this distribution, which moves right as time progresses. Suppose there is a second stream S_2 of timestamped measurements of a different type coming from another network of sensors, which is slower and less reliable. The resulting distribution has a higher variance and looser bounds, and lags slightly behind that of S_1. To correlate measurements from S_1 and S_2 by time, we use an equijoin on t_m. Intuitively, as the pdf curve for S_2 moves over the join attribute

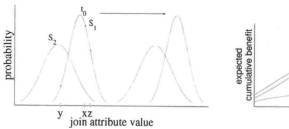

Figure 10.1.   Drifting normal distributions.   *Figure 10.2.*   Example ECBs.

value of a cached S_1 tuple, this tuple gets a chance of joining with each incoming S_2 tuple, with a probability given by S_2's pdf at that time. Clearly, these streams do not follow the frequency-based model, because the frequency of each t_m value varies over time. They do not follow the age-based model either, because each arriving tuple may have a different $p(\Delta t)$ function, which depends on the location of the partner stream's pdf. Blindly applying a specific join state management strategy without verifying its underlying assumption may lead to very poor performance. To illustrate, consider the two tuples x and y from stream S_2 in Figure 10.1 currently cached at time t_0 as part of the join state. Which tuple should we choose to discard when we are low on memory? Intuitively, it is better to discard y since it has almost already "missed" the moving pdf of S_1 and is therefore unlikely to join with future S_1 tuples. Unfortunately, if we use the past to predict future, **PROB** might make the exact opposite decision: y would be kept because it probably has joined more times with S_1 *in the past* than x.

Work by Xie et al. [45] represents a first step towards developing general techniques to exploit a broader class of statistical properties, without being tied to particular models or assumptions. A general question posed by [45] is, given the stochastic processes modeling the join attribute values of the input stream tuples, what join state management strategy has the best expected performance? In general, the stochastic processes can be non-stationary (e.g., the join attribute value follows a random walk, or its mean drifts over time) and correlated (e.g., if one stream has recently produced a certain value, then it becomes more likely for the other stream to produce the same value).

Knowing the stochastic processes governing the input streams gives us considerable predictive power, but finding the optimal join state management strategy is still challenging. A brute-force approach (called *FlowExpect* in [45]) would be the following. Conceptually, starting from the current time and the current join state, we enumerate all possible sequences of future "state management actions" (up to a given length), calculate the expected number of output tuples for each sequence, and identify the optimal sequence. This search problem can be formulated and solved as a network flow problem. The first action in the optimal sequence is taken at the current time. As soon as any new tuple arrives, we solve the problem again with the new join state, and take the first action in the new optimal sequence. The process then repeats. Interestingly, Xie et al. [45] showed that it is not enough to consider all possible sequences of *unconditional* state management actions; we must also consider strategies that make actions *conditional* upon the join attribute values of future tuples. An example of a conditional action at a future time t might be: "If the new tuple arriving at t has value 100 as its join attribute, then discard the new tuple; otherwise use it to replace the tuple currently occupying the fifth cache slot." Unfortunately, searching through the enormous space of conditional action sequences is not practical. Therefore, we need to develop simpler, more practical approaches.

It turns out that under certain conditions, the best state management action is clear. Xie et al. [45] developed an *ECB dominance test* (or *dom-test* for short) to capture these conditions. From the stochastic processes governing the input streams, we can compute a tuple s's *ECB* (*expected cumulative benefit*) with respect to the current time t_0 as a function $B_s(\Delta t)$, which returns the number of output tuples that s is expected to generate over the period $(t_0, t_0 + \Delta t]$. As a concrete example, Figure 10.2 plots the ECBs of tuples x, y, and z from stream S_2 in Figure 10.1. Intuitively, we prefer removing tuples with the "lowest" ECBs from the cache. The dom-test states that, if the ECB of tuple s_1 *dominates* that of tuple s_2 (i.e., $B_{s_1}(\Delta t) \geq B_{s_2}(\Delta t)$ for all $\Delta t > 0$), then keeping s_1 is better than or equally good as keeping s_2. For example, from Figure 10.2, we see that tuple y is clearly the least preferable among the three. However, because the ECBs of x and z cross over, the dom-test is silent on the choice between x and z. To handle "incomparable" ECBs such as these, Xie et al. proposed a heuristic measure that combines the ECB with a heuristic "survival probability" function $L_s(\Delta t)$ estimating the probability for tuple s to be still cached at time $t_0 + \Delta t$. Intuitively, if we estimate that x and z will be replaced before the time when their ECBs cross, then x is more preferable; otherwise, z is more preferable. Although the heuristic strategy cannot guarantee optimality in all cases, it always agrees with the decision of the dom-test whenever that test is applicable.

It is instructive to see how the general techniques above apply to specific scenarios. To begin, consider the simple case of unwindowed stream joins under the frequency-based model. The ECB of a tuple s is simply a linear function $B_s(\Delta t) = b(s)\Delta t$, where $b(s)$ is the number of output tuples that s is expected to generate per unit time, consistent with the definition of "benefit" discussed earlier in the context of the frequency-based model. Obviously, for two tuples s_1 and s_2, s_1's ECB dominates s_2's ECB if and only if $b(s_1) \geq b(s_2)$. Therefore, the dom-test basically yields **PROB**, and provides a proof of its optimality. The case of sliding-window joins is considerably more complex, and as discussed earlier, the optimal state management strategy is not known. As illustrated by Figure 10.3, the ECB of a tuple s consists of two connected pieces: The first piece has slope $b(s)$, while the second piece is flat and begins at the time $l(s)$ when s will drop out of the sliding window. While the dom-test does not help in comparing s_1 and s_2 in Figure 10.3, some insight can still be gained from their ECBs. Suppose we decide to cache s_1 and discard s_2. If at time $l(s_1)$ when s_1 exits the join state, a new tuple will be available to take its place and produce at least $B_{s_2}(l(s_2)) - B_{s_1}(l(s_1))$ output tuples during $(l(s_1), l(s_2)]$, then our decision is justified. Still, the exact condition that guarantees the optimality of the decision is complex, and will be an interesting problem for further investigation.

Finally, let us try applying the ECB-based analysis to the age-based model, for which we know that **AGE** [37] is optimal. Under the age-based model, every new tuple has the same ECB $B_0(\Delta t)$ at the time of its arrival. As the tuple ages in the cache, its ECB "shifts": The ECB of a tuple at age t is $B_t(\Delta t) = B_0(t + \Delta t) - B_0(t)$. For some shapes of B_0, it is possible to have ECBs that are not comparable by the dom-test. Figure 10.4 illustrates one such example; the marks on the ECB curves indicate when the respective tuples reach their optimal ages. Between two tuples *old* and *new* in Figure 10.4, the correct decision (by **AGE**) is to ignore the new tuple and keep caching the old tuple (until it reaches its optimal age). Unfortunately, however, the dom-test is unable to come to any conclusion, for the following two reasons. First, the dom-test actually provides a stronger optimality guarantee than **AGE**: The dom-test guarantees the optimality of its decisions over *any* time period; in contrast, **AGE** is optimal when the period tends to infinity. Second, the dom-test examines only the two ECBs in question and does not make use of any global information. However, in order to realize that replacing *old* tuple is not

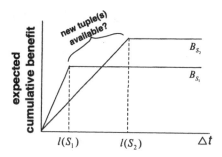

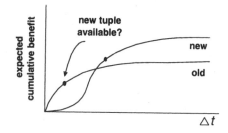

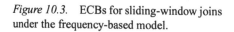

Figure 10.3. ECBs for sliding-window joins under the frequency-based model.   *Figure 10.4.* ECBs under the age-based model.

worthwhile in Figure 10.4, we need to be sure that when we do discard *old* when it reaches its optimal age, there will be a new tuple available at that time with high enough benefit to make up for the loss in discarding *new* earlier. Indeed, the age-based model allows us to make this conclusion from its assumption that every incoming tuple has the same ECB. It remains an open problem to develop better, general techniques to overcome the limitations of the dom-test without purposefully "special-casing" for specific scenarios.

An important practical problem is that we may not know in advance the parameter values of the stochastic processes modeling the input streams. One possibility is to use existing techniques to compute stream statistics online, or offline over the history of observations. Another approach proposed by [45] is to monitor certain statistics of the past behavior of the cache and input streams, and use them to estimate the expected benefit of caching. A notable feature of the proposed method is that it considers the form of the stochastic process in order to determine what statistics to monitor. This feature is crucial because, for time-dependent processes, the past is not always indicative of the future. For example, suppose that the join attribute values in a stream follow a distribution whose shape is stationary but mean is drifting over time. Simply tracking the frequency of each value is not meaningful as it changes all the time. Instead, we can subtract the current mean from each observed value, and track the frequency of these offset values, which will remain constant over time.

One direction for future research is to investigate how statistical properties of the input stream propagate to the output of stream joins. This problem is important if we want to apply the techniques in this section to more complex stream queries where the output of a stream join may be the input to another. While there has been some investigation of the non-streaming version of this problem related to its application in query optimization, there are many statistical properties unique to streams (e.g., trends, orderedness, clusteredness) whose propagation through queries is not yet fully understood.

Relationship to Classic Caching.   A natural question is how the stream join state management problem differs from the classic caching problem. Many cache replacement policies have been proposed in the past, e.g., LFD (longest-forward distance), LRU (least-recently used), LFU (least-frequently used), etc. All seem applicable to our problem. After all, our problem comes down to deciding what to retain in a cache to serve as many "reference requests" by the partner stream as possible. As pointed out by [45], there is a subtle but important difference between caching stream tuples and caching regular objects. When caching regular objects, we can recover from mistakes easily: The penalty of not caching an object is limited to a single cache miss, after which the object would be brought in and cached if needed. In contrast, in the case of stream join state management, a mistake can cost a lot more: If we discard a tuple completely from the join state, it is irrevocably gone, along with all output tuples that it could generate in the future. This difference explains why LFD, the optimal replacement policy for classic caching, turns out to be suboptimal for stream join state management, where references beyond the first one also matter.

Even if we relax the single-pass stream processing model and allow state to be spilled out to disk, join state management still differs from classic caching. The reason is that stream processing systems, e.g., those running the XJoin-family of algorithms, typically recover missing output tuples by processing flushed tuples later offline. In other words, "cache misses" are not processed online as in classic caching—random disk accesses may be too slow for stream applications, and just to be able to detect that there has indeed been a cache miss (as opposed to a new tuple that does not join with any previous arrivals) requires maintaining extra state.

Despite their differences, classic caching and stream join state management can be tackled under the same general analytical framework proposed by [45]. In fact, classic caching can be reduced to stream join state management, and can be analyzed using ECBs, in some cases yielding provably optimal results that agree with or extend classic ones. Such consistency is evidence of the strong link between the two problems, and a hint that some results on classic caching could be brought to bear on the state management problem for stream joins.

Joining with Database Relation.   Interestingly, unlike the case of joining two streams, state management for joining a stream and a database relation under the non-retroactive relation semantics (Section 2) is practically identical to classic caching [45]. First, it is easy to see that there is no benefit at all in caching any stream tuples, because under the non-retroactive relation semantics they do not join with any future updates to the relation. On the other hand, for tuples in the database relation, their current version can be cached in fast memory to satisfy reference requests by stream tuples. Upon a cache miss, the disk-resident relation can be probed. It would be interesting to investigate whether it makes sense to defer handling of misses to XJoin-style "mop-up" phases. However, care must be taken to avoid joining old stream tuples with newer versions of the database relation.

The Sampling-Rate Measure.   By design, join state management strategies optimized for max-subset favor input tuples that are more likely to join with the partner stream, causing such tuples to be overrepresented in the result. While this bias is not a problem in many contexts, it can be an issue if a statistically meaningful sample is desired, e.g., to obtain unbiased statistics of the join result. In this case, we should use the sampling-rate measure.

Getting an unbiased random sample of a join result has long been recognized as a difficult problem [34, 12]. The straightforward approach of sampling each input uniformly and then joining them does not work—the result may be arbitrarily skewed and small compared with the actual result. The hardness result from [37] states that for arbitrary input streams, if the available memory is insufficient to retain the entire sliding windows (or the entire history, if the join is unwindowed), then it is impossible to guarantee a uniform random sample for any nonzero sampling rate. The problem is that any tuple we choose to discard may turn out to be the one that will generate all subsequent output tuples.

Srivastava and Widom [37] developed a procedure for generating unbiased random samples of join results under the frequency-based and age-based models. The procedure requires knowledge of the model parameters, and uses them to determine the maximum sampling rate under the constraint that the probability of running out of memory at runtime is sufficiently small. The procedure keeps each stream tuple in the join state until the tuple will not contribute any more result tuples to the sample. Note that not all join result tuples that can be obtained from the join state will actually be output—many may need to be discarded in order to keep the sample unbiased. This inefficient use of resources is unavoidable because of the stringent requirement of a truly random sample. A statistically weaker form of sampling called *cluster sampling*, which uses resources more efficiently, was also considered by [37]. Cluster sampling is still unbiased, but is no longer independent; i.e., the inclusion of tuples is not independent of each other. Which type of sampling is appropriate depends on how the join result will be used.

4. Fundamental Algorithms for Stream Join Processing

Symmetric hash join (SHJ) is a simple hashing-based join algorithm, which has been used to support highly pipelined processing in parallel database systems [44]. It assumes that the entire join state can be kept in main memory; the join state for each input stream is stored in a hash table. For each incoming S tuple, SHJ inserts it into the hash table for S, and uses it to probe the hash table for the partner stream of S to identify joining tuples. SHJ can be extended to support the sliding-window semantics and the join statement management strategies in Section 3, though SHJ is limited to the single-pass stream processing model. Golab et al. [22] developed main-memory data structures especially suited for storing sliding windows, with efficient support for removing tuples that have fallen out of the sliding windows.

Both *XJoin* [41] and *DPHJ* (*double pipelined hash join*) of *Tukwila* [29] extend SHJ by allowing parts of the hash tables to be spilled out to disk for later processing. This extension removes the assumption that the entire join state must be kept in memory, greatly enhancing the applicability of the algorithm. Tukwila's DPHJ processes disk-resident tuples only when both inputs are exhausted; XJoin schedules joins involving disk-resident tuples whenever the inputs are blocked, and therefore is better suited for stream joins with unbounded inputs. One complication is the possibility of producing duplicate output tuples. XJoin pioneers the use of timestamp marking for detecting duplicates. Timestamps record the period when a tuple was in memory, and the times when a memory-resident hash partition was used to join with the corresponding disk-resident partition of the partner stream. From these timestamps, XJoin is able to infer which pairs of tuples have been processed before.

XJoin is the basis for many stream join algorithms developed later, e.g., [33, 38]. RPJ [38] is the latest in the series. One of the main contributions of RPJ, discussed earlier in Section 3.2, is a statistics-based flushing strategy that tries to keep in memory those tuples that are more likely to join. In contrast, XJoin flushes the largest hash partition; HMJ (*hash merge join*) of [33] always flushes corresponding partitions together, and tries to balance memory allocation between incoming streams. Neither XJoin nor HMJ takes tuple join probabilities into consideration. Unlike HMJ, which joins all previously flushed data whenever entering a disk-join phase, RPJ breaks down the work into smaller units, which offer more scheduling possibilities. In particular, RPJ also uses statistics to prioritize disk-join tasks in order to maximize output rate.

There are a number of interesting open issues. First, can we exploit statistics better by allowing flushing of individual tuples instead of entire hash partitions? This extension would allow us to apply the fine-grained join state management techniques from Section 3.2 to the XJoin-family of algorithms. However, the potential benefits must be weighed against the overhead in statistics collection and bookkeeping to avoid duplicates. Second, is it ever beneficial to reintroduce a tuple that has been previously flushed to disk back into memory? Again, what would be the bookkeeping overhead involved? Third, can we develop better statistics collection methods for RPJ? Currently, it maintains statistics on the partition level, but the hash function may map tuples with very different statistics to the same partition.

Sorting-based join algorithms, such as the sort-merge join, have been traditionally deemed inappropriate for stream joins, because sorting is a blocking operation that requires seeing the entire input before producing any output. To circumvent this problem, Dittrich et al. [20] developed an algorithm called PMJ (*progressive merge join*) that is sorting-based but non-blocking. In fact, both RPJ and HMJ use PMJ for joining disk-resident parts of the join state. The idea of PMJ is as follows. During the initial sorting phase that creates the initial runs, PMJ sorts portions of both input streams in parallel, and immediately produces join result tuples from the corresponding runs that are in memory at the same time. During the subsequent merge phases that merge shorter runs into longer ones, PMJ again processes both input streams in parallel, and joins them while their runs are in memory at the same times. To ensure that the output contains no duplicates, PMJ does not join tuples from corresponding shorter runs that have been joined in a previous phase; the duplicate avoidance logic is considerably simpler than XJoin. Of course, PMJ pays some price for its non-blocking feature—it does incur a moderate amount of overhead compared to the basic sort-merge join. On the other hand, PMJ also inherits the advantages of sorting-based algorithms over hashing-bashed algorithms, including in particular the ability to handle non-equality joins. A more thorough performance comparison between PMJ and XJoin for equijoins would be very useful.

5. Optimizing Stream Joins

Optimizing Response Time. Viglas and Naughton [42] introduced the notion of *rate-based optimization* and considered how to estimate the output rate of stream operators. An important observation is that standard cost analysis based on total processing cost is not applicable in the stream setting, because infinite costs resulted from unbounded inputs cannot be compared directly. Even if one can "hack" the analysis by assuming a large (yet bounded) input, classic analysis may produce incorrect estimate of the output rate since it ignores the rate at which inputs (or intermediate result streams) are coming. Specifically, classic analysis assumes that input is available at all times, but in practice operators could be blocked by the input. The optimization objectives considered by [42] are oriented towards response time: For a stream query, how can we produce the largest number of output tuples in a given amount of time, or produce a given number of output tuples in the shortest amount of time?

As an example of response-time optimization, Hammad et al. [28] studied shared processing of multiple sliding-window joins, focusing on developing scheduling strategies aimed at reducing response times across queries. More broadly speaking, work on non-blocking join algorithms, e.g., XJoin and PMJ discussed earlier, also incorporate response-time considerations.

Optimizing Unit-Time Processing Cost. Kang et al. [30] were among the first to focus specifically on optimization of stream joins. They made the same observation as in [42] that optimizing the total processing cost is no longer appropriate with unbounded input. However, instead of optimizing response time, they propose to optimize the processing cost per unit time, which is equivalent to the average processing cost per tuple weighted by the arrival rate. Another important observation made by [30] is that the best processing strategy may be *asymmetric*; i.e., different methods may be used for joining a new S_1 tuple with S_2's join state and for joining a new S_2 tuple with S_1's join state. For example, suppose that S_1 is very fast and S_2 is very slow. We may index S_2's join state as a hash table while leaving S_1's join state not indexed. The reason for not indexing S_1 is that its join state is frequently updated (because S_1 is fast) but rarely queried (because S_2 is slow).

Ayad and Naughton [4] provided more comprehensive discussions of optimization objectives for stream queries. An important observation is that, given enough processing resources, the steady-state output rate of a query is independent of the execution plan and therefore should be not be the objective of query optimization; a cost-based objective should be used in this case instead. Another interesting point is that load shedding considerations should be incorporated into query optimization: If we simply shed load from a plan that was originally optimized assuming sufficient resources, the resulting plan may be suboptimal.

Optimizing Multi-Way Stream Joins.   XJoin can be used to implement multi-way joins in a straightforward manner. For instance, a four-way join among S_1, S_2, S_3, and S_4 can be implemented as a series of XJoins, e.g., $((S_1 \text{ XJoin } S_2) \text{ XJoin } S_3) \text{ XJoin } S_4$. Since XJoin needs to store both of its inputs in hash tables, the example plan above in effect materializes the intermediate results $S_1 \text{ XJoin } S_2$ and $(S_1 \text{ XJoin } S_2) \text{ XJoin } S_3$. An obvious disadvantage of this plan is that these intermediate results can become quite large and costly to maintain. Another disadvantage is that this plan is static and fixes the join order. For example, a new S_3 tuple must be joined with the materialized $S_1 \text{ XJoin } S_2$ first, and then with S_4; the option of joining the new S_3 tuple first with S_4 is simply not available.

Viglas et al. [43] proposed *MJoin* to combat the above problems. MJoin maintains a hash table for each input involved in the multi-way join. When a tuple arrives, it is inserted into the corresponding hash table, and then used to probe all other hash tables in some order. This order can be different for tuples from different input streams, and can be determined based on join selectivities. Similar to XJoin, MJoin can flush join state out to disk when low on memory. Flushing is random (because of the assumption of a simple statistical model), but for the special case of star joins (where all streams join on the same attribute), flushing is "coordinated": When flushing one tuple, joining tuples from other hash tables are also flushed, because no output tuples can be produced unless joining tuples are found in all other hash tables. Note that coordinated flushing does not bring the same benefit for binary joins, because in this case output tuples are produced by joining an incoming tuple with the (only) partner stream hash table, not by joining two old tuples from difference hash tables.

Finding the optimal join order in MJoin is challenging. A simple heuristic that tracks selectivity for each hash table independently would have trouble with the following issues: (1) Selectivities can be correlated; e.g., a tuple that already joins with S_1 will be more likely to join with S_2. (2) Selectivities may vary among individual tuples; e.g., one tuple may join with many S_1 tuples but few S_2 tuples, while another tuple may behave in the exact opposite way. The first issue is tackled by [5], who provided a family of algorithms for adaptively finding the optimal order to apply a series of filters (joining a tuple with a stream can be regarded as subjecting the tuple to a filter) through runtime profiling. In particular, the *A-Greedy* algorithm is able to capture correlations among filter selectivities, and is guaranteed to converge to an ordering within a constant factor of the optimal. The theoretical guarantee extends to star joins; for general join graphs, though *A-Greedy* still can be used, the theoretical guarantee no longer holds. The second issue is recently addressed by an approach called *CBR* [9], or *content-based routing*, which makes the choice of query plan dependent on the values of the incoming tuple's "classifier attributes," whose values strongly correlate with operator selectivities. In effect, *CBR* is able to process each incoming tuple with a customized query plan.

One problem with MJoin is that it may incur a significant amount of recomputation. Consider again the four-way join among $S_1, \ldots, S_4$, now processed by a single MJoin operator. Whenever a new tuple s_3 arrives in S_3, MJoin in effect executes the query $S_1 \bowtie S_2 \bowtie \{s_3\} \bowtie S_4$; similarly, whenever a new tuple s_4 arrives in S_4, MJoin executes $S_1 \bowtie S_2 \bowtie S_3 \bowtie \{s_4\}$. The common subquery $S_1 \bowtie S_2$ is processed over and over again for these S_3 and S_4 tuples. In contrast, the XJoin plan $((S_1 \text{ XJoin } S_2) \text{ XJoin } S_3) \text{ XJoin } S_4$ materializes all its intermediate results in hash tables, including $S_1 \bowtie S_2$; new tuples from S_3 and S_4 simply have to probe this hash table, thereby avoiding recomputation. The optimal solution may well lie between these two extremes, as pointed out by [6]. They proposed an adaptive caching strategy, *A-Caching*, which starts with MJoins and adds join subresult caches adaptively. *A-Caching* profiles cache benefit and cost online, selects caches dynamically, and allocates memory to caches dynamically. With this approach, the entire spectrum of caching options from MJoins to XJoins can be explored.

A number of other papers also consider multi-way stream joins. Golab and Özsu [23] studied processing and optimization of multi-way sliding-window joins. Traditionally, we eagerly remove (expire) tuples that are no longer part of the sliding window, and eagerly generate output tuples whenever input arrives. The authors proposed algorithms supporting lazy expiration and lazy evaluation as alternatives, which achieve higher efficiency at the expense of higher memory requirements and longer response times, respectively. Hammad et al. [27] considered multi-way stream joins where a time-based window constraint can be specified for each pair (or, in general, subset) of input streams. An interesting algorithm called *FEW* is proposed, which computes a forward point in time before which all arriving tuples can join, thereby avoiding repeated checking of window constraints.

Eddies [3] are a novel approach towards stream query processing and optimization that is markedly different from the standard plan-based approaches.

Eddies eliminate query plans entirely by routing each input tuple adaptively across the operators that need to process it. Interestingly, in eddies, the behavior of *SteM* [36] mimics that of MJoin, while *STAIRS* [16] is able to emulate XJoin. Note that while eddies provide the *mechanisms* for adapting the processing strategy on an individual tuple basis, currently their *policies* typically do not result in plans that change for every incoming tuple. It would be nice to see how features of CBR can be supported in eddies.

6.   Conclusion

In this chapter, we have presented an overview of research problems and recent advances in join processing for data streams. Stream processing is a young and exciting research area, yet it also has roots in and connections to well-established areas in databases as well as computer science in general. In Section 3.2, we have already discussed the relationship between stream join state management and classic caching. Now, let us briefly re-examine parts of this chapter in light of their relationship to *materialized views* [25].

The general connection between stream processing and materialized views has long been identified [8]. This connection is reflected in the way that we specify the semantics of stream joins—by regarding them as views and defining their output as the view update stream resulting from base relation updates (Section 2). Recall that the standard semantics requires the output sequence to reflect the exact sequence of states of the underlying view, which is analogous to the notion of *complete and strong consistency* of a data warehouse view with respect to its source relations [46]. The connection does not stop at the semantics. The problem of determining what needs to be retained in the state to compute a stream join is analogous to the problem of deriving auxiliary views to make a join view *self-maintainable* [35]. Just as constraints can be used to reduce stream join state (Section 3.1), they have also been used to help expire data from data warehouses without affecting the maintainability of warehouse views [21]. For a stream join $S_1 \bowtie \cdots \bowtie S_n$, processing an incoming tuple from stream S_i is analogous to maintaining a join view incrementally by evaluating a maintenance query $S_1 \bowtie \cdots \bowtie \Delta S_i \bowtie \cdots \bowtie S_n$. Since there are n different forms of maintenance queries (one for each i), it is natural to optimize each form differently, which echoes the intuition behind the asymmetric processing strategy of [30] and MJoin [43]. In fact, we can optimize the maintenance query for each *instance* of ΔS_i, which would achieve the same goal of supporting a customized query plan for each tuple as CBR [9]. Finally, noticing that the maintenance queries run frequently and share many common subqueries, we may choose to materialize some subqueries as additional views to improve query performance, which is also what A-Caching [6] tries to accomplish.

Of course, despite high-level similarities, techniques from the two areas—data streams and materialized views—may still differ significantly in actual details. Nonetheless, it would be nice to develop a general framework that unifies both areas, or, less ambitiously, to apply ideas from one area to the other. Many such possibilities exist. For example, methods and insights from the well-studied problems of *answering query using views* [26] and *view selection* [14] could be extended and applied to data streams: Given a set of stream queries running continuously in a system, what materialized views (over join states and database relations) and/or additional stream queries can we create to improve the performance of the system? Another area is distributed stream processing. Distributed stream processing can be regarded as view maintenance in a distributed setting, which has been studied extensively in the context of data warehousing. Potentially applicable in this setting are techniques for making warehouse self-maintainable [35], optimizing view maintenance queries across distributed sources [31], ensuring consistency of multi-source warehouse views [46], etc. Conversely, stream processing techniques can be applied to materialized views as well. In particular, view maintenance could benefit from optimization techniques that exploit update stream statistics (Section 3.2). Also, selection of materialized views for performance can be improved by adaptive caching techniques (Section 5).

Besides the future work directions mentioned above and throughout the chapter, another important direction worth exploring is the connection between data stream processing and *distributed event-based systems* [19] such as publish/subscribe systems. Such systems need to scale to thousands or even millions of subscriptions, which are essentially continuous queries over event streams. While efficient techniques for handling continuous selections already exist, scalable processing of continuous joins remains a challenging problem. Hammad et al. [28] considered shared processing of stream joins with identical join conditions but different sliding-window durations. We need to consider more general query forms, e.g., joins with different join conditions as well as additional selection conditions on input streams. *NiagaraCQ* [13] and *CACQ* [32] are able to group-process selections and share processing of identical join operations. However, there is no group or shared processing of joins with different join conditions, and processing selections separately from joins limits optimization potentials. *PSoup* [11] treats queries as data, thereby allowing set-oriented processing of queries with arbitrary join and selection conditions. Still, new indexing and processing techniques must be developed for the system to be able to process each event in time sublinear in the number of subscriptions.

Acknowledgments

This work is supported by a NSF CAREER Award under grant IIS-0238386. We would also like to thank Shivnath Babu, Yuguo Chen, Kamesh Munagala, and members of the Duke Database Research Group for their discussions.

References

[1] Arasu, A., Babcock, B., Babu, S., McAlister, J., and Widom, J. (2002). Characterizing memory requirements for queries over continuous data streams. In *Proceedings of the 2002 ACM Symposium on Principles of Database Systems*, pages 221–232, Madison, Wisconsin, USA.

[2] Arasu, A., Babu, S., and Widom, J. (2003). The CQL continuous query language: Semantic foundations and query execution. Technical Report 2003–67, InfoLab, Stanford University.

[3] Avnur, R. and Hellerstein, J. M. (2000). Eddies: Continuously adaptive query processing. In *Proceedings of the 2000 ACM SIGMOD International Conference on Management of Data*, pages 261–272, Dallas, Texas, USA.

[4] Ayad, A. and Naughton, J. F. (2004). Static optimization of conjunctive queries with sliding windows over infinite streams. In *Proceedings of the 2004 ACM SIGMOD International Conference on Management of Data*, pages 419–430, Paris, France.

[5] Babu, S., Motwani, R., Munagala, K., Nishizawa, I., and Widom, J. (2004a). Adaptive ordering of pipelined stream filters. In *Proceedings of the 2004 ACM SIGMOD International Conference on Management of Data*, pages 407–418, Paris, France.

[6] Babu, S., Munagala, K., Widom, J., and Motwani, R. (2005). Adaptive caching for continuous queries. In *Proceedings of the 2005 International Conference on Data Engineering*, Tokyo, Japan.

[7] Babu, S., Srivastava, U., and Widom, J. (2004b). Exploiting k-constraints to reduce memory overhead in continuous queries over data streams. *ACM Transactions on Database Systems*, 29(3):545–580.

[8] Babu, S. and Widom, J. (2001). Continuous queries over data streams. *ACM SIGMOD Record*.

[9] Bizarro, P., Babu, S., DeWitt, D., and Widom, J. (2005). Content-based routing: Different plans for different data. In *Proceedings of the 2005 International Conference on Very Large Data Bases*, Trondheim, Norway.

[10] Carney, D., Çetintemel, U., Cherniack, M., Convey, C., Lee, S., Seidman, G., Stonebraker, M., Tatbul, N., and Zdonik, S. B. (2002). Monitoring streams - a new class of data management applications. In *Proceedings of the 2002*

International Conference on Very Large Data Bases, pages 215–226, Hong Kong, China.

[11] Chandrasekaran, S. and Franklin, M. J. (2003). PSoup: a system for streaming queries over streaming data. *The VLDB Journal*, 12(2):140–156.

[12] Chaudhuri, S., Motwani, R., and Narasayya, V. R. (1999). On random sampling over joins. In *Proceedings of the 1999 ACM SIGMOD International Conference on Management of Data*, pages 263–274, Philadelphia, Pennsylvania, USA.

[13] Chen, J., DeWitt, D. J., Tian, F., and Wang, Y. (2000). NiagaraCQ: A scalable continuous query system for internet databases. In *Proceedings of the 2000 ACM SIGMOD International Conference on Management of Data*, pages 379–390, Dallas, Texas, USA.

[14] Chirkova, R., Halevy, A. Y., and Suciu, D. (2001). A formal perspective on the view selection problem. In *Proceedings of the 2001 International Conference on Very Large Data Bases*, pages 59–68, Roma, Italy.

[15] Das, A., Gehrke, J., and Riedewald, M. (2003). Approximate join processing over data streams. In *Proceedings of the 2003 ACM SIGMOD International Conference on Management of Data*, pages 40–51, San Diego, California, USA.

[16] Deshpande, A. and Hellerstein, J. M. (2004). Lifting the burden of history from adaptive query processing. In *Proceedings of the 2004 International Conference on Very Large Data Bases*, pages 948–959, Toronto, Canada.

[17] Ding, L., Mehta, N., Rundensteiner, E., and Heineman, G. (2004). Joining punctuated streams. In *Proceedings of the 2004 International Conference on Extending Database Technology*, Heraklion, Crete, Greece.

[18] Ding, L. and Rundensteiner, E. A. (2004). Evaluating window joins over punctuated streams. In *Proceedings of the 2004 International Conference on Information and Knowledge Management*, pages 98–107, Washington DC, USA.

[19] Dingel, J. and Strom, R., editors (2005). *Proceedings of the 2005 International Workshop on Distributed Event Based Systems*, Columbus, Ohio, USA.

[20] Dittrich, J.-P., Seeger, B., Taylor, D. S., and Widmayer, P. (2002). Progressive merge join: A generic and non-blocking sort-based join algorithm. In *Proceedings of the 2002 International Conference on Very Large Data Bases*, pages 299–310, Hong Kong, China.

[21] Garcia-Molina, H., Labio, W., and Yang, J. (1998). Expiring data in a warehouse. In *Proceedings of the 1998 International Conference on Very Large Data Bases*, pages 500–511, New York City, New York, USA.

[22] Golab, L., Garg, S., and Özsu, T. (2004). On indexing sliding windows over on-line data streams. In *Proceedings of the 2004 International Conference on Extending Database Technology*, Heraklion, Crete, Greece.

[23] Golab, L. and Özsu, M. T. (2003). Processing sliding window multi-joins in continuous queries over data streams. In *Proceedings of the 2003 International Conference on Very Large Data Bases*, pages 500–511, Berlin, Germany.

[24] Golab, L. and Özsu, M. T. (2005). Update-pattern-aware modeling and processing of continuous queries. In *Proceedings of the 2005 ACM SIG-MOD International Conference on Management of Data*, pages 658–669, Baltimore, Maryland, USA.

[25] Gupta, A. and Mumick, I. S., editors (1999). *Materialized Views: Techniques, Implementations, and Applications*. MIT Press.

[26] Halevy, A. Y. (2001). Answering queries using views: A survey. *The VLDB Journal*, 10(4):270–294.

[27] Hammad, M. A., Aref, W. G., and Elmagarmid, A. K. (2003a). Stream window join: Tracking moving objects in sensor-network databases. In *Proceedings of the 2003 International Conference on Scientific and Statistical Database Management*, pages 75–84, Cambridge, Massachusetts, USA.

[28] Hammad, M. A., Franklin, M. J., Aref, W. G., and Elmagarmid, A. K. (2003b). Scheduling for shared window joins over data streams. In *Proceedings of the 2003 International Conference on Very Large Data Bases*, pages 297–308, Berlin, Germany.

[29] Ives, Z. G., Florescu, D., Friedman, M., Levy, A. Y., and Weld, D. S. (1999). An adaptive query execution system for data integration. In *Proceedings of the 1999 ACM SIGMOD International Conference on Management of Data*, pages 299–310, Philadelphia, Pennsylvania, USA.

[30] Kang, J., Naughton, J. F., and Viglas, S. (2003). Evaluating window joins over unbounded streams. In *Proceedings of the 2003 International Conference on Data Engineering*, pages 341–352, Bangalore, India.

[31] Liu, B. and Rundensteiner, E. A. (2005). Cost-driven general join view maintenance over distributed data sources. In *Proceedings of the 2005 International Conference on Data Engineering*, pages 578–579, Tokyo, Japan.

[32] Madden, S., Shah, M. A., Hellerstein, J. M., and Raman, V. (2002). Continuously adaptive continuous queries over streams. In *Proceedings of the 2002 ACM SIGMOD International Conference on Management of Data*, Madison, Wisconsin, USA.

[33] Mokbel, M. F., Lu, M., and Aref, W. G. (2004). Hash-merge join: A non-blocking join algorithm for producing fast and early join results. In

Proceedings of the 2004 International Conference on Data Engineering, pages 251–263, Boston, Massachusetts, USA.

[34] Olken, F. (1993). *Random Sampling from Databases.* PhD thesis, University of California at Berkeley.

[35] Quass, D., Gupta, A., Mumick, I. S., and Widom, J. (1996). Making views self-maintainable for data warehousing. In *Proceedings of the 1996 International Conference on Parallel and Distributed Information Systems,* pages 158–169, Miami Beach, Florida, USA.

[36] Raman, V., Deshpande, A., and Hellerstein, J. M. (2003). Using state modules for adaptive query processing. In *Proceedings of the 2003 International Conference on Data Engineering,* pages 353–364, Bangalore, India.

[37] Srivastava, U. and Widom, J. (2004). Memory-limited execution of windowed stream joins. In *Proceedings of the 2004 International Conference on Very Large Data Bases,* pages 324–335, Toronto, Canada.

[38] Tao, Y., Yiu, M. L., Papadias, D., Hadjieleftheriou, M., and Mamoulis, N. (2005). RPJ: Producing fast join results on streams through rate-based optimization. In *Proceedings of the 2005 ACM SIGMOD International Conference on Management of Data,* pages 371–382, Baltimore, Maryland, USA.

[39] Tatbul, N., Cetintemel, U., Zdonik, S. B., Cherniack, M., and Stonebraker, M. (2003). Load shedding in a data stream manager. In *Proceedings of the 2003 International Conference on Very Large Data Bases,* pages 309–320, Berlin, Germany.

[40] Tucker, P. A., Maier, D., Sheard, T., and Fegaras, L. (2003). Exploiting punctuation semantics in continuous data streams. *IEEE Transactions on Knowledge and Data Engineering,* 15(3):555–568.

[41] Urhan, T. and Franklin, M. J. (2001). Dynamic pipeline scheduling for improving interactive query performance. In *Proceedings of the 2001 International Conference on Very Large Data Bases,* pages 501–510, Roma, Italy.

[42] Viglas, S. D. and Naughton, J. F. (2002). Rate-based query optimization for streaming information sources. In *Proceedings of the 2002 ACM SIGMOD International Conference on Management of Data,* pages 37–48, Madison, Wisconsin, USA.

[43] Viglas, S. D., Naughton, J. F., and Burger, J. (2003). Maximizing the output rate of multi-way join queries over streaming information sources. In *Proceedings of the 2003 International Conference on Very Large Data Bases,* pages 285–296, Berlin, Germany.

[44] Wilschut, A. N. and Apers, P. M. G. (1991). Dataflow query execution in a parallel main-memory environment. In *Proceedings of the 1991 Interna-*

tional Conference on Parallel and Distributed Information Systems, pages 68–77, Miami Beach, Florida, USA.

[45] Xie, J., Yang, J., and Chen, Y. (2005). On joining and caching stochastic streams. In *Proceedings of the 2005 ACM SIGMOD International Conference on Management of Data*, pages 359–370, Baltimore, Maryland, USA.

[46] Zhuge, Y., Garcia-Molina, H., and Wiener, J. L. (1998). Consistency algorithms for multi-source warehouse view maintenance. *Distributed and Parallel Databases*, 6(1):7–40.

Chapter 11

INDEXING AND QUERYING DATA STREAMS

Ahmet Bulut

Citrix Online Division
5385 Hollister Avenue
Santa Barbara, CA 93111
ahmet.bulut@citrix.com

Ambuj K. Singh

Department of Computer Science
University of California Santa Barbara
Santa Barbara, CA 93106
ambuj@cs.ucsb.edu

Abstract   Online monitoring of data streams poses a challenge in many data-centric applications including network traffic management, trend analysis, web-click streams, intrusion detection, and sensor networks. Indexing techniques used in these applications have to be time and space efficient while providing a high quality of answers to user queries: (1) queries that monitor aggregates, such as finding surprising levels ("volatility" of a data stream), and detecting bursts, and (2) queries that monitor trends, such as detecting correlations and finding similar patterns. Data stream indexing becomes an even more challenging task, when we take into account the dynamic nature of underlying raw data. For example, bursts of events can occur at variable temporal modalities from hours to days to weeks. We focus on a multi-resolution indexing architecture. The architecture enables the discovery of "interesting" behavior online, provides flexibility in user query definitions, and interconnects registered queries for real-time and in-depth analysis.

Keywords: stream indexing, monitoring real-time systems, mining continuous data flows, multi-resolution index, synopsis maintenance, trend analysis, network traffic analysis.

1.   Introduction

Raw stream data, such as faults and alarms generated by network traffic monitors and log records generated by web servers, are almost always at low level and too large to maintain in main memory. One can instead summarize the data and compute synopsis structures at meaningful abstraction levels on the fly. The synopsis is a small space data structure, and can be updated incrementally as new stream values arrive. Later in operational cycle, it can be used to discover interesting behavior, which prompts in-depth analysis at lower levels of abstraction [10].

Consider the following application in astrophysics: the sky is constantly observed for high-energy particles. When a particular astrophysical event happens, a shower of high-energy particles arrives in addition to the background noise. This yields an unusually high number of detectable events (high-energy photons) over a certain time period, which indicates the existence of a *Gamma Ray Burst*. If we know the duration of the shower, we can maintain a count on the total number of events over sliding windows of the known window size and raise an alarm when the moving sum is above a threshold. Unfortunately, in many cases, we cannot predict the duration of the burst period. The burst of high-energy photons might last for a few milliseconds, a few hours, or even a few days [31].

Finding similar patterns in a time series database is a well studied problem [1, 13]. The features of a time series sequence are extracted using a sliding window, and inserted into an index structure for query efficiency. However, such an approach is not adequate for data stream applications, since it requires a time consuming feature extraction step with each incoming data item. For this purpose, incremental feature extraction techniques that use the previous feature in computing the new feature have been proposed to accelerate per-item processing [30]. A batch technique can further decrease the per-item processing cost by computing a new feature periodically instead of every time unit [22]. A majority of these techniques assume *a priori* knowledge on query patterns. However in a real world situation, a user might want to know all time periods during which the movement of a particular stock follows a certain interesting trend, which itself can be generated automatically by a particular application [26]. In order to address this issue, a multi-resolution indexing scheme has been proposed [16]. This work addresses off-line time series databases, and does not consider how well the proposed scheme extends to a real-time streaming algorithm.

Continuous queries that run indefinitely, unless a query lifetime has been specified, fit naturally into the mold of data stream applications. Examples of these queries include monitoring a set of conditions or events to occur, detecting a certain trend in the underlying raw data, or in general discovering relations between various components of a large real time system. The kinds of queries that are of interest from an application point of view can be listed as follows: (1) monitoring aggregates, (2) monitoring or finding patterns, and (3) detecting correlations. Each of these queries requires data management over some history of values, and not just over the most recently reported values [9]. For example in case of aggregate queries, the system monitors whether the current window aggregate deviates significantly from that aggregate in most time periods of the same size. In case of correlation queries, the self-similar nature of sensor measurements may be reflected as temporal correlations at some resolution over the course of the stream [24]. Therefore, the system has to maintain historical data along with the current data in order to be able to answer these queries.

A key vision in developing stream management systems of practical value is to interconnect queries in a monitoring infrastructure. For example, an unusual volatility of a stream may trigger an in-depth trend analysis. Unified system solutions can lay ground for tomorrow's information infrastructures by providing users with a rich set of interconnected querying capabilities [8].

2.   Indexing Streams

In this section, we introduce a multi-resolution indexing architecture, and then later in Section 3, show how it can be utilized to monitor user queries efficiently. Multi-resolution approach imposes an inherent restriction on what constitutes a meaningful query. The core part of the scheme is the *feature extraction* at multiple resolutions. A *dynamic* index structure is used to index features for query efficiency. The system architecture is shown in Figure 11.1. The key architecture aspects are:

- The features at higher resolutions are computed using the features at lower resolutions; therefore, all features are computed in a single pass.

- The system guarantees the accuracy provided to user queries by provable error bounds.

- The index structure has tunable parameters to trade accuracy for speed and space. The per-item processing cost and the space overhead can be tuned according to the application requirements by varying the update rate and the number of coefficients maintained in the index structure.

2.1   Preliminaries and definitions

We adapt the use of $x[i]$ to refer to the i-th entry of stream x, and $x[i_1 : i_2]$ to refer to the subsequence of entries at positions i_1 through i_2.

DEFINITION 2.1 *A feature is the result of applying a characteristic function over a possibly normalized set of stream values in order to acquire a higher level information or concept.*

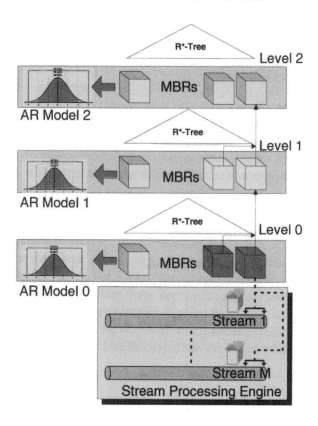

Figure 11.1. The system architecture for a multi-resolution index structure consisting of 3 levels and stream-specific auto-regressive (AR) models for capturing multi-resolution trends in the data.

The widely used characteristic functions are (1) aggregate functions, such as summation, maximum, minimum, and average, (2) orthogonal transformations, such as discrete wavelet transform (DWT) and discrete fourier transform (DFT), and (3) piecewise linear approximations. Normalization is performed in case of DWT, DFT, and linear approximations. The interested reader can refer to the Sections 3.2 and 3.3 for more details.

2.2   Feature extraction

The features at a specific resolution are obtained with a sliding window of a fixed length w. The sliding window size doubles as we go up a resolution, i.e., a level. In the rest of the paper, we will use the terms "level" and "resolution" interchangeably. We denote a newly computed feature at resolution i as $\mathcal{F}_i$. Figure 11.2 shows an example where we have three resolutions with corresponding sliding window sizes of 2, 4 and 8. With each arrival of a new stream value, features $\mathcal{F}_0$, $\mathcal{F}_1$, and $\mathcal{F}_2$, i.e., one for each resolution, can be computed. However, this requires maintaining all the stream values within a time window equal to the size of the largest sliding window, i.e., 8 in our running example. The per-item processing cost and the space required is linear in the size of the largest window [16].

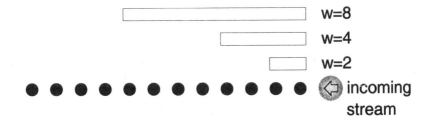

Figure 11.2. Exact feature extraction, update rate $T = 1$.

For a given window w of values $y = x[t - w + 1], \ldots, x[t]$, an *incremental* transformation $F(y)$ is used to compute features. The type of transformation F depends on the monitoring query. For example, F is SUM for burst detection, MAX−MIN for volatility detection, and DWT for detecting correlations and finding surprising patterns. For most real time series, the first f ($f << w$) DWT coefficients retain most of the energy of the signal. Therefore, we can safely disregard all but the very first few coefficients to retain the salient features (e.g., the overall trend) of the original signal.

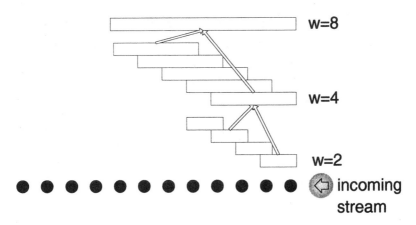

Figure 11.3. Incremental feature extraction, update rate $T = 1$.

Using an incremental transformation leads to a more efficient way of computing features at all resolutions. Level-1 features are computed using level-0 features, and level-2 features are computed using level-1 features. In general, we can use lower level features to compute higher level features [3]. Figure 11.3 depicts this new way of computation. This new algorithm has a lower per-item processing cost, since we can compute $\mathcal{F}_1$ and $\mathcal{F}_2$ in constant time. The following lemma establishes this result.

LEMMA 11.1 *The new feature $\mathcal{F}_j$ at level j for the subsequence $x[t-w+1:t]$ can be computed "exactly" using the features $\mathcal{F}'_{j-1}$ and $\mathcal{F}_{j-1}$ at level $j-1$ for the subsequences $x[t-w+1:t-w/2]$ and $x[t-w/2+1:t]$ respectively.*

Proof $\mathcal{F}_j$ is $max(\mathcal{F}'_{j-1}, \mathcal{F}_{j-1})$, $min(\mathcal{F}'_{j-1}, \mathcal{F}_{j-1})$, $\mathcal{F}'_{j-1} + \mathcal{F}_{j-1}$ for MAX, MIN, and SUM respectively. For DWT, see Lemma 11.4 in Section 2.4. ∎

However, the space required for this scheme is also linear in the size of the largest window. The reason is that we need to maintain half of the features at the lower level to compute the feature at the upper level incrementally. If we can trade accuracy for space, then we can decrease the space overhead by computing features approximately. At each resolution level, every c of the feature vectors are combined in a box, or in other words, a minimum bounding rectangle (MBR). Figure 11.4 depicts this scheme for $c = 2$. Since each MBR B contains c features, it has an extent along each dimension. In case of SUM, $B[1]$ corresponds to the smallest sum, and $B[2]$ corresponds to the largest sum among all c sums. In general, $B[2i]$ denotes the low coordinate and $B[2i + 1]$ denotes the high coordinate along the i-th dimension. Note that for SUM, MAX and MIN, B has a single dimension. However, for DWT the number of dimensions f is application dependent.

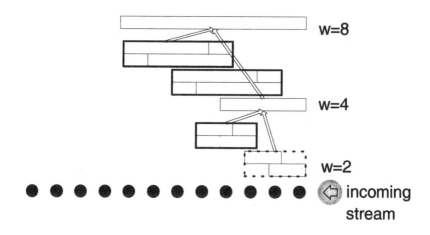

Figure 11.4. Approximate feature extraction, update rate $T = 1$.

This new approach decreases the space overhead by a factor of c. Since the extent information of the MBRs is used in the computation, the newly computed feature will also be an extent. The following lemma proves this result.

LEMMA 11.2 *The new feature $\mathcal{F}_j$ at level j can be computed "approximately" using the MBRs B_1 and B_2 that contain the features $\mathcal{F}'_{j-1}$ and $\mathcal{F}_{j-1}$ at level $j - 1$ respectively.*

Proof

$$max(B_1[1], B_2[1]) \leq \mathcal{F}_j \leq max(B_1[2], B_2[2])$$
$$min(B_1[1], B_2[1]) \leq \mathcal{F}_j \leq min(B_1[2], B_2[2])$$
$$B_1[1] + B_2[1] \leq \mathcal{F}_j \leq B_1[2] + B_2[2]$$
See Lemma 11.5 in Section 2.4

for MAX, MIN, SUM and DWT respectively. ∎

Using MBRs instead of individual features exploits the fact that there is a strong spatio-temporal correlation between the consecutive features. Therefore, it is natural to extend the computation scheme to eliminate this redundancy. Instead of computing a new feature at each data arrival, one can employ a batch computation such that a new feature is computed periodically, at every T time unit. This allows us to maintain features instead of MBRs. Figure 11.5 shows this scheme with $T = 2$. The new scheme has a clear advantage in terms of accuracy; however it can dismiss potentially interesting events that may occur between the periods.

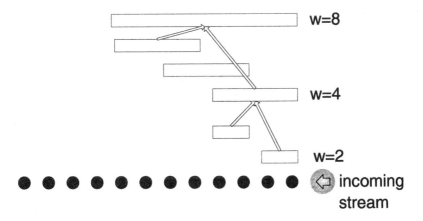

Figure 11.5. Incremental feature extraction, update rate $T = 2$.

Depending on the box capacity and the update rate T_j at a given level j (the rate at which we compute a new feature), there are two general feature computation algorithms:

- **Online algorithm:** Update rate T_j is equal to 1. The box capacity c is variable. It is used for aggregate monitoring queries.

- **Batch algorithm:** Update rate T_j is greater than 1, i.e., $T_j > 1$. The box capacity is set to $c = 1$. The setting $T_j=W$ can be used for finding surprising patterns and detecting correlations. SWAT is a batch algorithm with $T_j = 2^j$[7].

We establish the time and space complexity of a given algorithm in terms of c and T_j in the following theorem. Assume that W denotes the sliding window size at the lowest resolution $j = 0$.

THEOREM 11.3 $\mathcal{F}_j$ *at level j for a stream can be computed incrementally in constant time and in space* $\Theta(2^{j-1}W/cT_{j-1})$.

Proof $\mathcal{F}_j$ at level j is computed using the features at level $j - 1$ in constant time as shown in Lemmas 11.1 and 11.2. The number of features that need to be maintained at level $j - 1$ for incremental computation at level j is $2^{j-1}W$. Therefore, depending on the box capacity and update rate, the space complexity at level $j - 1$ is $\Theta(2^{j-1}W/cT_{j-1})$. ∎

2.3   Index maintenance

As new values stream in, new features are computed and inserted into the corresponding index structures while features that are out of history of interest are deleted to save space. Coefficients are computed at multiple resolutions starting from level 0 up to a configurable level J: at each level a sliding window is used to extract the appropriate features. Computation of features at higher levels is accelerated using the MBRs at lower levels. The MBRs belonging to a specific stream are threaded together in order to provide a sequential access to the summary information about the stream. This approach results in a constant retrieval time of the MBRs. The complete algorithm is shown in Algorithm 1.

Features at a given level are maintained in a high dimensional index structure. The index combines information from all the streams, and provides a scalable access medium for answering queries over multiple data streams. However, each MBR inserted into the index is specific to a single stream. The R*-Tree family of index structures are used for indexing MBRs at each level [5]. An R*-Tree, a variant of R-Tree [15], is a spatial access method, which splits feature space in hierarchically nested, possibly overlapping MBRs. In order to support frequent updates, the techniques for predicting MBR boundaries outlined in [20] can be used to decrease the cost of index maintenance.

Algorithm 1 Compute_Coefficients(Stream S)

Require: $B^S_{j,i}$ denotes the i-th MBR at level j for stream S.

1: **begin procedure**
2:   $w := W$ (the window size at the lowest resolution);
3:   t_{now} denotes the current discrete time;
4:   **for** $j := 0$ to J **do**
5:     $B^S_{j,i} :=$ the current MBR at level j for stream S;
6:     **if** $j = 0$ **then**
7:       $y := S[t_{\text{now}} - w + 1 : t_{\text{now}}]$;
8:       normalize y if $F = DWT$;
9:       $\mathcal{F}_j := F(y)$;
10:     **else**
11:       find MBR B^S_{j-1,i_1} that contains the feature
12:         for the subsequence $S[t_{\text{now}} - w + 1 : t_{\text{now}} - \frac{w}{2}]$;
13:       find MBR B^S_{j-1,i_2} that contains the feature
14:         for the subsequence $S[t_{\text{now}} - \frac{w}{2} + 1 : t_{\text{now}}]$;
15:       $\mathcal{F}_j := F(B^S_{j-1,i_1}, B^S_{j-1,i_2})$;
16:     **end if**
17:     **if** number of features in $B^S_{j,i} < c$ (box capacity) **then**
18:       insert $\mathcal{F}_j$ into $B^S_{j,i}$;
19:     **else**
20:       insert $B^S_{j,i}$ into index at level j;
21:       start a new MBR $B^S_{j,i+1}$;
22:       insert $\mathcal{F}_j$ into $B^S_{j,i+1}$;
23:     **end if**
24:     adjust the sliding window size to $w := w * 2$;
25:   **end for**
26: **end procedure**

2.4   Discrete Wavelet Transform

The approximation coefficients are defined through the inner product of the input signal with $\phi_{j,k}$, the shifted and dilated versions a low-pass scaling function ϕ_0. In the same vein, the detail coefficients are defined through the inner product of the input signal with $\psi_{j,k}$, the shifted and dilated versions the wavelet basis function ψ_0.

$$\phi_{j,k}(t) = 2^{-j/2}\phi_0(2^{-j}t - k), \; j,k \in Z \tag{11.1}$$

$$\psi_{j,k}(t) = 2^{-j/2}\psi_0(2^{-j}t - k), \; j,k \in Z \tag{11.2}$$

We show how to compute approximation coefficients. Detail coefficients at level j are computed using approximation coefficients at level $j-1$. Using Equation 11.1, the approximation signal at level j for the signal x is obtained by

$$A_j^{(x)} = \sum_k \langle x, \phi_{j,k}\rangle \phi_{j,k}$$

In the same manner, the approximation signal at level $j+1$ for x is

$$A_{j+1}^{(x)} = \sum_k \langle x, \phi_{j+1,k}\rangle \phi_{j+1,k}$$

To compute $A_{j+1}^{(x)}$, we need to compute coefficients $\langle x, \phi_{j+1,n}\rangle$. Using the twin-scale relation for ϕ, we can compute $\langle x, \phi_{j+1,n}\rangle$ from $\langle x, \phi_{j,k}\rangle$ [21]. This can mathematically be expressed as

$$\langle x, \phi_{j+1,n}\rangle \;=\; \sum_k h_{k-2n}\langle x, \phi_{j,k}\rangle \tag{11.3}$$

$$C_{j,n} \;=\; \sum_k \tilde{h}_{n-k}\langle x, \phi_{j,k}\rangle \tag{11.4}$$

$$\langle x, \phi_{j+1,n}\rangle \;=\; C_{j,2n} \tag{11.5}$$

where h_k and $\tilde{h}$ are low-pass reconstruction and decomposition filters respectively. Note that the terms "approximation signal" and "approximation coefficients" are used interchangeably.

LEMMA 11.4 *The approximation coefficients at level j, $1 \le j \le J$, for a signal $x[t - w + 1 : t]$ can be computed exactly using the approximation coefficients at level $j-1$ for the signals $x[t-w+1 : t-w/2]$ and $x[t-w/2+1 : t]$.*

Proof Let x, x_1 and x_2 denote signals $x[t - w + 1 : t]$, $x[t - w + 1 : t - w/2]$, and $x[t - w/2 + 1 : t]$ respectively. At a particular scale $j - 1$, the shape of the wavelet scaling function $\phi_{j-1,0}$ is kept the same, while it is translated to

$$x_2' = \left\{ \begin{array}{ll} 0, & 1 \leq n \leq w/2 \\ x_2[n - w/2], & w/2 + 1 \leq n \leq w \end{array} \right\}$$

$$\langle x, \phi_{j-1,k} \rangle = \langle x_1' + x_2', \phi_{j-1,k} \rangle = \langle x_1', \phi_{j-1,k} \rangle + \langle x_2', \phi_{j-1,k} \rangle$$

This result is due to the linearity of the wavelet transformation. Using Equations 11.4, 11.5, and the coefficients $\langle x, \phi_{j-1,k} \rangle$, we can obtain the approximation signal $A_j^{(x)}$ at level j for x. ∎

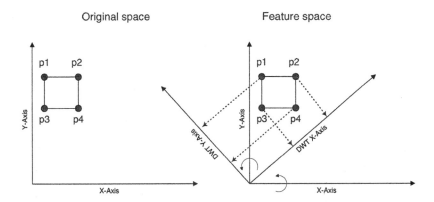

Figure 11.6. Transforming an MBR using discrete wavelet transform. Transformation corresponds to rotating the axes (the rotation angle = 45° for Haar wavelets)

LEMMA 11.5 *One can compute approximation coefficients on a hyper-rectangle* $B \in \Re^{f'}$ *with low coordinates* $[x_{l_1}, \ldots, x_{l_{f'}}]$ *and high coordinates* $[x_{h_1}, \ldots, x_{h_{f'}}]$.

Proof The most recent approximation coefficients for a sliding window of values x at a given resolution j can be computed on the extent information of the MBRs B_{j-1,i_1} and B_{j-1,i_2} in $\Re^f$ at level $j - 1$ that contain the coefficients of the corresponding two halves of x. These MBRs are merged together using Lemma 11.4 to get an MBR B in $\Re^{f'}$, where f' is larger than f (e.g., f' is $2f$ for Haar wavelets). The MBR B approximates the coefficients at level $j - 1$ for x. First compute the coefficients for each one of the $2^{f'}$ corners of B, and find the tightest MBR $A^{(B)}$ in $\Re^f$ that encloses the resulting $2^{f'}$ coefficients in $\Re^f$. The coefficients at level j for x, i.e., the feature $A_j^{(x)}$, lies inside the MBR $A^{(B)}$. This is true for any such unitary transformation as wavelet transformation that rotates the axes as shown in Figure 11.6. This algorithm has a processing time of $\Theta(2^{f'} f)$, where f and f' are constant for a specific application. ∎

The coefficients at level j for x, i.e., the feature $A_j^{(x)}$, lies inside the MBR $A^{(B)}$. This is true for any such unitary transformation as wavelet transformation that rotates the axes as shown in Figure 11.6. This algorithm has a processing time of $\Theta(2^{f'} f)$, where f and f' are constant for a specific application. ∎

Error bound.   Wavelet transformation corresponds to the rotation of the axes in the original space. An input MBR B in the original space is transformed to a new shape S in the feature space (see Figure 11.6). The resulting shape S is projected on each dimension in the feature space, and the tightest MBR $A^{(B)}$ that encloses S is identified. The MBR $A^{(B)}$ contains the feature $A^{(x)}$. The volume of $A^{(B)}$ is a function of the projection along each dimension. Since the wavelet transformation is a distance preserving transformation, the length along each dimension can be at most two times the original length.

3.   Querying Streams

In this section, monitoring queries that are important from an application point of view, such as deviant aggregates, interesting patterns, and trend correlations, are presented.

3.1   Monitoring an aggregate query

In this class of queries, aggregates of data streams are monitored over a set of time intervals [31]: "Report all occurrences of Gamma Ray bursts from a timescale of minutes to a timescale of days". Formally, given a *bounded* window size w, an aggregate function F, and a threshold τ associated with the window, the goal is to report all those time instances such that the aggregate applied to the subsequence $x[t - w + 1 : t]$ exceeds the corresponding window threshold, i.e., check if

$$F(x[t - w + 1 : t]) \geq \tau \qquad (11.6)$$

where t denotes the current time. The threshold values τ can either be specified as part of the input, or they can be determined using historical data.

The algorithm.   Assume that the query window size is a multiple of W. An aggregate query with window size w and threshold τ is answered by first partitioning the window into multiple sub-windows, $w_1, w_2, \ldots, w_n$ such that $0 \leq j_1 < \ldots < j_i < j_{i+1} < \ldots < j_n \leq J$, and $w_i = W 2^{j_i}$. For a given window of length bW, the partitioning corresponds to the ones in the binary representation of b such that $\sum_{i=1}^{n} 2^{j_i} = b$. The current aggregate over a window of size w is computed using the sub-aggregates for sub-windows in the partitioning. Assume that $W = 2$ and $c = 2$. Consider a query window $w = 26$. The binary representation of $b = 13$ is 1101, and therefore the query is partitioned into three sub-windows $w_0 = 2$, $w_2 = 8$, and $w_3 = 16$. Figure 11.7 shows the decomposition of the query and the composition of the aggregate together. The current aggregate over a window of size $w = 26$ is approximated using the extents of MBRs that contain the corresponding sub-aggregates. The computation is approximate in the sense that the algorithm returns an interval $\mathcal{F}$ such that the upper coordinate $\mathcal{F}[2]$ is always greater than or equal to the true aggregate. If $\mathcal{F}[2]$ is larger than the threshold τ, the most recent subsequence of length w is retrieved, and the true aggregate is computed. If this value exceeds the query threshold, an alarm is raised. The complete algorithm is shown in Algorithm 2.

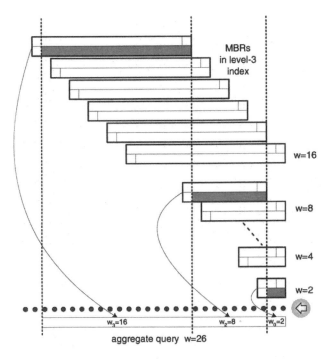

Figure 11.7. Aggregate query decomposition and approximation composition for a query window of size $w = 26$.

The accuracy. The false alarm rate of this approximation is quantified as follows: assume that bursts of events are monitored, i.e., F is SUM. Let X denote the sum within sliding window $w = bW$. If the threshold τ is set to $\mu_X(1 - \Phi(p))$, the inequation

$$Pr\left(\frac{X - \mu_X}{\mu_X} \geq \frac{\tau - \mu_X}{\mu_X}\right) \leq p \tag{11.7}$$

Algorithm 2 Aggregate_Query(Stream S, Window w, Threshold τ)

1: **begin procedure**
2: initialize t to t_{now}, the current discrete time;
3: partition w into n parts as $w_1, w_2, \ldots, w_n$;
4: initialize aggregate $\mathcal{F}$;
5: **for** $i := 1$ to $i := n$ **do**
6:     find the resolution level j such that $w_i = W2^j$;
7:     MBR B contains the feature on $S[t - w_i + 1 : t]$;
8:     merge sub-aggregate B to $\mathcal{F} := F(B, \mathcal{F})$;
9:     adjust offset to $t := t - w_i$ for next sub-window;
10: **end for**
11: **if** $\tau \le \mathcal{F}[2]$ **then**
12:     retrieve $S[t_{now} - w + 1 : t_{now}]$;
13:     **if** $\tau \le F(S[t_{now} - w + 1 : t_{now}])$ **then**
14:         raise an alarm;
15:     **end if**
16: **end if**
17: **end procedure**

---

holds for a given sufficiently small p, where Φ denotes the normal cumulative distribution function. Monitor the burst based on windows with size Tw such that $1 \le T < 2$, where $2^{j-1}W < w \le 2^jW$. This approach corresponds to monitoring the burst via one of the levels in the index structure [31]. Let Z denote the sum within sliding window Tw. We assume that

$$\frac{Z - \mu_Z}{\mu_Z} \sim \text{Norm}(0, 1) \tag{11.8}$$

Assuming $\mu_Z = T\mu(X)$, the false alarm rate is equal to $Pr(Z > \tau)$, which implies

$$Pr\left(\frac{Z - T\mu(X)}{T\mu(X)} \ge \frac{\tau - T\mu(X)}{T\mu(X)}\right) = \Phi\left(1 - \frac{1 - \Phi^{-1}(p)}{T}\right) \tag{11.9}$$

According to Equation 11.9, for a fixed value of p, the smaller T is, the smaller is the false alarm rate. If sub-aggregates for sub-windows $w_1, w_2, \ldots, w_n$ are used for computing the final aggregate on a given query window of size w and threshold τ, a smaller T can be achieved. The sub-aggregate for sub-window w_i is stored in an MBR at level j_i. An MBR at level j_i corresponds to a monitoring window of size $2^{j_i}W + c - 1$. Then, effectively a burst is monitored using a window of size $bW + \log b * (c - 1)$ such that:

$$T' = \frac{bW + \log b * (c - 1)}{bW} = 1 + \frac{\log b * (c - 1)}{bW} \tag{11.10}$$

where T' decreases with increasing b. For example, for $c = W = 64$ and $b = 12$, we have $T' = 1.2987$ and $T = 1.3333$. This implies that the sub-window sum approximation reduces the false alarm rate to a minimal amount with the optimal being at $T' = 1$. In fact, the optimal is reached with $c = 1$. However the space consumption in this case is much larger.

3.2 Monitoring a pattern query

In this class of queries, a pattern database is continuously monitored over dynamic data streams: "Identify all temperature sensors in a weather monitoring sensornet that currently exhibit an interesting trend". Formally, given a query sequence Q and a threshold value r, find the set of streams that are within distance r to the query sequence Q. The distance measure we adopt is the Euclidean distance (L_2) between the corresponding normalized sequences. We normalize a window of values $x[1], \ldots, x[w]$ as follows:

$$\hat{x}[i] = \frac{x[i]}{\sqrt{w} * R_{max}} \qquad i = 1, \ldots, w \qquad (11.11)$$

thereby mapping it to the unit hyper-sphere. We establish the notion of similarity between sequences as follows: a stream sequence x and a query sequence Q are considered to be *r-similar* if

$$L_2(\hat{x}, \hat{Q}) = \sqrt{\sum_{i=0}^{N-1} (\hat{x}_i - \hat{Q}_i)^2} \leq r \qquad (11.12)$$

The online algorithm. Given a query sequence Q and a threshold value r, partition Q into multiple sub-queries, $Q_1, Q_2, \ldots, Q_n$ such that $0 \leq j_1 < \ldots < j_i < j_{i+1} < \ldots < j_n \leq J$, and $|Q_i| = W2^{j_i}$. Assume that the first sub-query Q_1 has resolution j_1. A range query with radius r is performed on the index constructed at resolution j_1. The initial candidate box set R is refined using the hierarchical radius optimization proposed in [16]. Briefly, for each MBR $B \in R$, this technique is used to refine the original radius r to $r' = \sqrt{r^2 - d_{min}(Q_1, B)^2}$ for the next sub-query Q_2, where $d_{min}(p, B)$ for a point p and an MBR B is defined as the minimum Euclidean distance of the query point p to the MBR B [25]. The same procedure is applied recursively until the last sub-query Q_n is processed, resulting in a final set of MBRs C to check for true matches. The complete algorithm is shown in Algorithm 3.

The batch algorithm. Let the update rate for each index level j be $T_j = W$. The stream is divided into W-step sliding windows of size w. Let $|S|$ denote the size of the stream. Then, there are $\lfloor (|S| - w + 1)/W \rfloor$ many such windows. Given a query sequence Q, W-many prefixes of size w are extracted as $Q[0 :$

Algorithm 3 Pattern_Query_Online(Query Q)

1:**begin procedure**
2: partition Q into n parts as $Q_1, Q_2, \ldots, Q_n$;
3: find the resolution level j_1 such that $|Q_1| = W2^{j_1}$;
4: $R :=$ Range_Query(Index$_{j_1}$, DWT($\hat{Q}_1$),Q.r);
5: $C :=$ Hierarchical_Radius_Refinement(R,Q);
6: post-process C to discard false alarms;
7:**end procedure**

$w-1], Q[1 : w], \ldots, Q[W-1 : w+W-1]$. Each prefix query is used to identify potential candidates. In order to clarify the ensuing development, we note that a single prefix query would suffice in case an online algorithm with $T_j = 1$ was used for index construction. This approach is similar to the technique proposed in [22], where a single resolution index is constructed using a sliding window of maximum allowable size w that satisfies $1 \leq \lfloor(\min(Q) - W + 1)/w\rfloor$. Note that min(Q) is the *a priori* information regarding the minimum query length. However, in a multi-resolution index, a given query can be answered using any index at resolution j that satisfies $1 \leq \lfloor(|Q| - W + 1)/(2^j W)\rfloor$. The accuracy of this multi-resolution search algorithm can be improved by extracting disjoint windows along with each prefix in order to refine the original query radius using a multi-piece search technique [13]. The number of such disjoint windows is at most $p = \lfloor(|Q| - W + 1)/w\rfloor$. We illustrate these concepts on a query window of size $|Q| = 9$ as shown in Figure 11.8, where $J = 1$ and $W = 2$. The prefixes are shown as $i = 0$ and $i = 1$ along with the corresponding disjoint windows. Each and every feature extracted over Q is inserted into a query MBR B. The MBR B is extended in each dimension by a fraction of the query radius, i.e., $r/\sqrt{p}$. A range query is performed on the index at level j using B, and a set R of candidate features is retrieved. The set R is post-processed to discard false alarms. The complete algorithm is shown in Algorithm 4.

3.3   Monitoring a correlation query

In this class of queries, all stream pairs that are correlated within a user specified threshold r at some level of abstraction are reported continuously. The correlation between two sequences x and y can be reduced to the Euclidean distance between their z-norms [30]. The z-norm of a sequence $x[1], \ldots, x[w]$ is defined as follows:

$$\hat{x}[i] = \frac{x[i] - \mu_x}{\sqrt{\sum_{i=1}^{w}(x[i] - \mu_x)^2}} \qquad i = 1, \ldots, w \qquad (11.13)$$

where μ_x is the arithmetic mean. The correlation coefficient between sequences x and y is computed using the L_2 distance between $\hat{x}$ and $\hat{y}$ as $1 - L_2^2(\hat{x}, \hat{y})/2$.

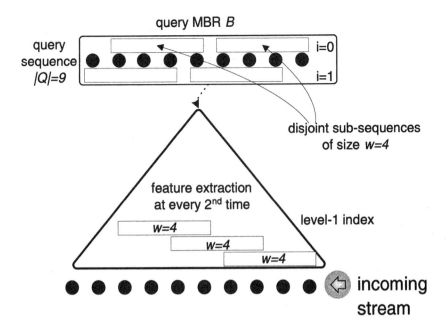

Figure 11.8. Subsequence query decomposition for a query window of size $|Q| = 9$.

Algorithm 4 Pattern_Query_Batch(Query Q)

1: **begin procedure**
2: find the largest level j such that $2^j W + W - 1 \le |Q|$;
3: initialize query MBR B to empty;
4: let w be equal to $2^j W$, level-j sliding window size;
5: **for** $i := 0$ to $i := W - 1$ **do**
6:    **for** $k := 0$ to $k := \lfloor (|Q| - i)/w \rfloor$ **do**
7:       extract k^{th} disjoint subsequence of the query
8:          sequence into $y := Q[i + kw : i + (k + 1)w - 1]$;
9:       insert DWT($\hat{y}$) into MBR B;
10:    **end for**
11: **end for**
12: compute radius refinement factor $p := \lfloor (|Q| - W + 1)/w \rfloor$;
13: enlarge query MBR B by $Q.r/\sqrt{p}$;
14: $R := $ Range_Query(Index$_j$, B);
15: post-process R to discard false alarms;
16: **end procedure**

The algorithm. Whenever a new feature $\mathcal{F}_j$ of a stream S is computed at level j, a range query on the index at level j is performed with $\mathcal{F}_j$ as the query center and the radius set to the correlation threshold r. In a system with M

synchronized streams, this involves execution of $O(M)$ range queries at every data arrival.

4.   Related Work

Shifted-Wavelet Tree (SWT) is a wavelet-based summary structure proposed for detecting bursts of events on a stream of data [31]. For a given set of query windows $w_1, w_2, \ldots, w_m$ such that $2^L W \leq w_1 \leq w_2 \leq \ldots \leq w_m \leq 2^U W$, SWT maintains $U - L$ moving aggregates using a wavelet tree for incremental computation. A window w_i is monitored by the lowest level j, $L \leq j \leq U$, that satisfies $w_i \leq 2^j W$. Therefore, associated with each level j, $L \leq j \leq U$, there is a threshold τ_j equal to the smallest of the thresholds of windows $w_{i_1}, \ldots, w_{i_j}$ monitored by that level. Whenever the moving sum at some level j exceeds the level threshold τ_j, all query windows associated with this level are checked using a brute force approach.

MR-Index addresses variable length queries over time series data [16]. Wavelets are used to extract features from a time series at multiple resolutions. At each resolution, a set of feature vectors are combined into an MBR and stored sequentially in the order they are computed. A given query is decomposed into multiple sub-queries such that each sub-query has resolution corresponding to a resolution at the index. A given set of candidate MBRs are refined using each query as a filter to prune out non-potential candidates.

Versions of piecewise constant approximation are proposed for time series similarity matching. Specifically, adaptive piecewise constant approximation (APCA) represents data regions of great fluctuations with several short segments, while data regions of less fluctuations are represented with fewer, long segments [17]. An extension of this approximation allows error specification for each point in time [23]. The resulting approach can approximate data with fidelity proportional to its age. GeneralMatch, a refreshingly new idea in similarity matching, divides the data sequences into disjoint windows, and the query sequence into sliding windows [22]. This approach is the dual of the conventional approaches, i.e., dividing the data sequence into sliding windows, and the query sequence into disjoint windows. The overall framework is based on answering pattern queries using a single-resolution index built on a specific choice of window size. The allowed window size depends on the minimum query length, which has to be provided a-priori before the index construction.

Methods based on multi-variate linear regression are considered for analyzing co-evolving time sequences [28]. For a given stream S, its current value (dependent variable) is expressed as a linear combination of values of the same and other streams (independent variables) under sliding window model. Given v independent variables and a dependent variable y with N samples each, the model identifies the best b independent variables in order to compute the current value of the dependent variable in $O(Nbv^2)$ time.

StatStream is a state of the art system proposed for monitoring a large number of streams in real time [30]. It subdivides the history of a stream into a fixed number of basic windows and maintains DFT coefficients for each basic window. This allows a batch update of DFT coefficients over the entire history. It superimposes an orthogonal regular grid on the feature space, and partitions the space into cells of diameter r, the correlation threshold. Each stream is mapped to a number of cells (exactly how many depends on the "lag time") in the feature space based on a subset of its DFT coefficients. It uses proximity in this feature space to report correlations [6].

5. Future Directions

We are witnessing the blurring of the traditional boundaries between Networks and Databases, especially in the emerging areas of sensor and peer-to-peer networks. Data stream processing in these application domains requires networked data management, solutions of which borrow ideas from both disciplines. We believe that researchers from these two communities should share their expertise, results, terminologies, and contributions. This exchange can promote ideas that will influence and foster continued research in the areas of sensor and peer-to-peer networks. The following three research avenues are promising future directions to pursue further: (1) Distributed monitoring systems, (2) Probabilistic modeling of sensor networks, and (3) Publish-subscribe systems.

5.1 Distributed monitoring systems

In today's rapidly growing networks, data streams arrive at widely dispersed locations. Assume that a system administrator wants to analyze the local network traffic and requires access to data collections that are maintained at different locations. The rapid growth of such collections fed by data streams makes it virtually impossible to simply store a collection at every location where it is possibly queried. This prompts a need to design more scalable approaches for disseminating the information of a data stream. Each peer monitoring station characterizes its stream of data in terms of a model (signature) and transmits this information to a central site using an adaptive communication protocol. The abstraction levels of signatures collected at the server can be quite different. A higher level corresponds to coarser statistics. Therefore, it contains less representative information, and incurs smaller transmission cost. A lower level corresponds to finer statistics. Therefore, it has more characteristics information; however it incurs larger transmission cost. Naturally, there is an interplay of opposing factors, i.e., accuracy vs. overhead. At the server, tasks that involve information from multiple clients are executed. The question is to find an optimal data acquisition strategy that maximizes the conservation of limited system resources [11].

A more specific scenario arises in a sensor-net: anomalous event detection is a collaborative task, which involves aggregation of measurements from a number of sensors. Only if a certain number of these sensors signify an alarm and a consensus is reached, a drill-down analysis is performed to collect more information and take affirmative steps. After computing a local fingerprint on its stream of data, each sensor needs to diffuse this fingerprint into the net in order to reach a consensus on the alarming incidents [2]. This work can introduce "reactive monitoring" into sensor networks.

5.2   Probabilistic modeling of sensor networks

Embedded low-power sensing devices revolutionize the way we collect and process information for building emergency response systems. Miniature sensors are deployed to monitor ever-changing conditions in their surroundings. Statistical models such as stochastic models and multivariate regression enable capturing intra-sensor and inter-sensor dependencies in order to model sensor network data accurately. Such models can be used in backcasting missing sensor values, forecasting future data values, and guiding efficient data acquisition. Current mathematical models allow decomposing the main research problem into subproblems [14]. This in turn leads to a natural way of computing model components incrementally and in a distributed manner.

Regressive models are proposed for computing inter-scale and intra-scale correlations among wavelet coefficients [24]. The magnitude of these coefficients is used for detecting interesting events such as seasonal components and bursts of events. The wavelet coefficients at a given level j in the multiresolution index are expressed as a function of the k previous coefficients at the same level plus noise ϵ (optionally including coefficients from upper levels) i.e.,

$$A_j^{(x_0)} = \beta_1 A_j^{(x_1)} + \ldots + \beta_k A_j^{(x_k)} + \epsilon_{j,t} \qquad (11.14)$$

where the symbol $x_i, 0 \le i \le k$, denotes $x[t - (i+1) * 2^j W + 1 : t - i * 2^j W]$, and the term $\epsilon_{j,t}$ denotes the noise added. Recursive Least Squares is used to update these regressive models incrementally [29]. Further research efforts are encouraged on exploring how to use these models for compressing and querying stream information regarding the past, and more importantly in a feedback loop for setting the query window parameters automatically and in a semantically meaningful manner.

5.3   Content distribution networks

Publish-and-subscribe services provide the ability to create persistent queries or subscriptions to new content. In a typical content based pub-sub system, content providers send structured content to instances of *pub-sub service*, which are responsible for sending messages to the subscribers of each particular content [27]. The pub-sub system forms a semantic layer on the top of a monitoring infrastructure by providing a query interface: events of interest are specified using an appropriate continuous query language [19]. Furthermore, it realizes the reactive part of the whole infrastructure by sending notifications about events of interest to users. Recent advances in application layer multicast for content delivery address the scalability issues that usually arise in data stream applications with large receiver sets [4]. However, the problem of providing real-time guarantees for time-critical user tasks under stringent constraints still needs exploration.

6. Chapter Summary

In this chapter, we presented a space and time efficient architecture to extract features over streams and index these features for improving query performance. The maintenance cost in the index structure is leveraged by computing transformation coefficients online: the coefficients at higher levels are computed over the index that stores the coefficients at lower levels. This approach decreases per-item processing time considerably, and minimizes the space required for incremental computation. The index structure has an adaptive time-space complexity depending on the update rate and the number of coefficients maintained, and guarantees the approximation quality by provable error bounds.

References

[1] R. Agrawal, C. Faloutsos, and A. Swami. Efficient similarity search in sequence databases. In *FODO*, pages 69–84, 1993.

[2] A. Akella, A. Bharambe, M. Reiter, and S. Seshan. Detecting DDoS attacks on ISP networks. In *MPDS*, 2003.

[3] A. Arasu and J. Widom. Resource sharing in continuous sliding-window aggregates. In *VLDB*, pages 336–347, 2004.

[4] S. Banerjee, B. Bhattacharjee, and C. Kommareddy. Scalable Application Layer Multicast. In *SIGCOMM*, pages 205–217, 2002.

[5] N. Beckmann, H. Kriegel, R. Schneider, and B. Seeger. The R*-tree: An efficient and robust access method for points and rectangles. In *SIGMOD*, pages 322–331, 1990.

[6] J. Bentley, B. Weide, and A. Yao. Optimal expected time algorithms for closest point problems. In *ACM Trans. on Math. Software*, volume 6, pages 563–580, 1980.

[7] A. Bulut and A. Singh. SWAT: Hierarchical stream summarization in large networks. In *ICDE*, pages 303–314, 2003.

[8] A. Bulut and A. Singh. A unified framework for monitoring data streams in real time. In *ICDE*, pages 44–55, 2005.

[9] D. Carney, U. Cetintemel, M. Cherniack, C. Convey, S. Lee, G. Seidman, M. Stonebraker, N. Tatbul, and S. Zdonik. Monitoring streams - a new class of data management applications. In *VLDB*, pages 215–226, 2002.

[10] Y. Chen, G. Dong, J. Han, J. Pei, B. Wah, and J. Wang. Online analytical processing stream data: Is it feasible? In *DMKD*, 2002.

[11] A. Deshpande, C. Guestrin, S. Madden, J. Hellerstein, and W. Hong. Model-driven data acquisition in sensor networks. In *VLDB*, pages 588–599, 2004.

[12] P. Dinda. CMU, Aug 97 Load Trace. In Host Load Data Archive *http://www.cs.northwestern.edu/~pdinda/LoadTraces/*.

[13] C. Faloutsos, M. Ranganathan, and Y. Manolopoulos. Fast subsequence matching in time-series databases. In *SIGMOD*, pages 419–429, 1994.

[14] C. Guestrin, P. Bodi, R. Thibau, M. Paski, and S. Madden. Distributed regression: an efficient framework for modeling sensor network data. In *IPSN*, pages 1–10, 2004.

[15] A. Guttman. R-trees: A dynamic index structure for spatial searching. In *SIGMOD*, pages 47–57, 1984.

[16] T. Kahveci and A. Singh. Variable length queries for time series data. In *ICDE*, pages 273–282, 2001.

[17] E. Keogh, K. Chakrabarti, S. Mehrotra, and M. Pazzani. Locally adaptive dimensionality reduction for indexing large time series databases. In *SIGMOD*, pages 151 – 162, 2001.

[18] E. Keogh and T. Folias. Time Series Data Mining Archive. In *http://www.cs.ucr.edu/~eamonn/TSDMA*, 2002.

[19] Y. Law, H. Wang, and C. Zaniolo. Query languages and data models for database sequences and data streams. In *VLDB*, pages 492–503, 2004.

[20] M. Lee, W. Hsu, C. Jensen, B. Cui, and K. Teo. Supporting frequent updates in R-Trees: A bottom-up approach. In *VLDB*, pages 608–619, 2003.

[21] S. Mallat. *A Wavelet Tour of Signal Processing*. Academic Press, 2 edition, 1999.

[22] Y. Moon, K. Whang, and W. Han. General match: a subsequence matching method in time-series databases based on generalized windows. In *SIGMOD*, pages 382–393, 2002.

[23] T. Palpanas, M. Vlachos, E. Keogh, D. Gunopulos, and W. Truppel. Online amnesic approximation of streaming time series. In *ICDE*, pages 338–349, 2004.

[24] S. Papadimitriou, A. Brockwell, and C. Faloutsos. AWSOM: Adaptive, hands-off stream mining. In *VLDB*, pages 560–571, 2003.

[25] N. Roussopoulos, S. Kelley, and F. Vincent. Nearest neighbor queries. pages 71–79, 1995.

[26] H. Wu, B. Salzberg, and D. Zhang. Online event-driven subsequence matching over financial data streams. In *SIGMOD*, pages 23–34, 2004.

[27] B. Wyman and D. Werner. Content-based Publish-Subscribe over APEX. In *Internet-Draft*, April, 2002.

[28] B. Yi, N. Sidiropoulos, T. Johnson, H. Jagadish, C. Faloutsos, and A. Biliris. Online data mining for co-evolving time sequences. In *ICDE*, 2000.

[29] P. Young. *Recursive Estimation and Time-Series Analysis: An Introduction*. Springer-Verlag, 1984.

[30] Y. Zhu and D. Shasha. Statstream: Statistical monitoring of thousands of data streams in real time. In *VLDB*, pages 358–369, 2002.

[31] Y. Zhu and D. Shasha. Efficient elastic burst detection in data streams. In *SIGKDD*, pages 336 – 345, 2003.

Chapter 12

DIMENSIONALITY REDUCTION AND FORECASTING ON STREAMS

Spiros Papadimitriou,[1] Jimeng Sun,[2] and Christos Faloutsos[2]

[1]*IBM Watson Research Center,*
Hawthorne, NY, USA
spapadim@us.ibm.com

[2]*Carnegie Mellon University*
Pittsburgh, PA, USA
jimeng@cs.cmu.edu, christos@cs.cmu.edu

Abstract   We consider the problem of capturing correlations and finding hidden variables corresponding to trends on collections of time series streams. Our proposed method, SPIRIT, can incrementally find correlations and hidden variables, which summarise the key trends in the entire stream collection. It can do this quickly, with no buffering of stream values and without comparing pairs of streams. Moreover, it is any-time, single pass, and it dynamically detects changes. The discovered trends can also be used to immediately spot potential anomalies, to do efficient forecasting and, more generally, to dramatically simplify further data processing.

Introduction

In this chapter, we consider the problem of capturing correlations and finding hidden variables corresponding to trends on collections of semi-infinite, time series data streams, where the data consist of tuples with n numbers, one for each time tick t.

Streams often are inherently correlated (e.g., temperatures in the same building, traffic in the same network, prices in the same market, etc.) and it is possible to reduce hundreds of numerical streams into just a handful of *hidden variables* that compactly describe the key trends and dramatically reduce the complexity of further data processing. We propose an approach to do this incrementally.

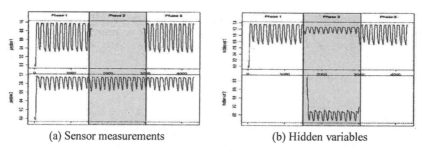

(a) Sensor measurements   (b) Hidden variables

Figure 12.1. Illustration of problem. Sensors measure chlorine in drinking water and show a daily, near sinusoidal periodicity during phases 1 and 3. During phase 2, some of the sensors are "stuck" due to a major leak. The extra hidden variable introduced during phase 2 captures the presence of a new trend. SPIRIT can also tell us which sensors participate in the new, "abnormal" trend (e.g., close to a construction site). In phase 3, everything returns to normal.

We describe a motivating scenario, to illustrate the problem we want to solve. Consider a large number of sensors measuring chlorine concentration in a drinkable water distribution network (see Figure 12.1, showing 15 days worth of data). Every five minutes, each sensor sends its measurement to a central node, which monitors and analyses the streams in real time.

The patterns in chlorine concentration levels normally arise from water demand. If water is not refreshed in the pipes, existing chlorine reacts with pipe walls and micro-organisms and its concentration drops. However, if fresh water flows in at a particular location due to demand, chlorine concentration rises again. The rise depends primarily on how much chlorine is originally mixed at the reservoirs (and also, to a small extent, on the distance to the closest reservoir—as the distance increases, the peak concentration drops slightly, due to reactions along the way). Thus, since demand typically follows a periodic pattern, chlorine concentration reflects that (see Figure 12.1a, bottom): it is high when demand is high and vice versa.

Assume that at some point in time, there is a major leak at some pipe in the network. Since fresh water flows in constantly (possibly mixed with debris from the leak), chlorine concentration at the nodes near the leak will be close to peak at all times.

Figure 12.1a shows measurements collected from two nodes, one away from the leak (bottom) and one close to the leak (top). At any time, a human operator would like to know how many trends (or *hidden variables*) are in the data and ask queries about them. Each hidden variable essentially corresponds to a group of correlated streams.

In this simple example, SPIRIT discovers the correct number of hidden variables. Under normal operation, only one hidden variable is needed, which corresponds to the periodic pattern (Figure 12.1b, top). Both observed variables follow this hidden variable (multiplied by a constant factor, which is the *participation weight* of each observed variable into the particular hidden variable). Mathematically, the hidden variables are the *principal components* of the observed variables and the participation weights are the entries of the *principal direction* vectors (more precisely, this is true under certain assumptions, which will be explained later).

However, during the leak, a second trend is detected and a new hidden variable is introduced (Figure 12.1b, bottom). As soon as the leak is fixed, the number of hidden variables returns to one. If we examine the hidden variables, the interpretation is straightforward: The first one still reflects the periodic demand pattern in the sections of the network under normal operation. All nodes in this section of the network have a participation weight of ≈ 1 to the "periodic trend" hidden variable and ≈ 0 to the new one. The second hidden variable represents the additive effect of the catastrophic event, which is to cancel out the normal pattern. The nodes close to the leak have participation weights ≈ 0.5 to both hidden variables.

Summarising, SPIRIT can tell us the following (Figure 12.1): (i) Under normal operation (phases 1 and 3), there is one trend. The corresponding hidden variable follows a periodic pattern and all nodes participate in this trend. All is well. (ii) During the leak (phase 2), there is a *second* trend, trying to cancel the normal trend. The nodes with non-zero participation to the corresponding hidden variable can be immediately identified (e.g., they are close to a construction site). An abnormal event may have occurred in the vicinity of those nodes, which should be investigated.

Matters are further complicated when there are hundreds or thousands of nodes and more than one demand pattern. However, as we show later, SPIRIT is still able to extract the key trends from the stream collection, follow trend drifts and immediately detect outliers and abnormal events. Besides providing a concise summary of key trends/correlations among streams, SPIRIT can successfully deal with missing values and its discovered hidden variables can be used to do very efficient, resource-economic forecasting.

There are several other applications and domains to which SPIRIT can be applied. For example, (i) given more than 50,000 securities trading in US, on a second-by-second basis, detect patterns and correlations [27], (ii) given traffic measurements [24], find routers that tend to go down together.

Contributions

The problem of pattern discovery in a large number of co-evolving streams has attracted much attention in many domains. We introduce *SPIRIT (Streaming Pattern dIscoveRy in multIple Time-series)*, a comprehensive approach to discover correlations that effectively and efficiently summarise large collections of streams. SPIRIT satisfies the following requirements:

(i) It is *streaming*, i.e., it is incremental, scalable, *any-time*. It requires very memory and processing time per time tick. In fact, both are independent of the stream length t.

(ii) It scales *linearly* with the number of streams n, not quadratically. This may seem counter-intuitive, because the naïve method to spot correlations across n streams examines all $O(n^2)$ pairs.

(iii) It is *adaptive*, and fully *automatic*. It dynamically detects changes (both gradual, as well as sudden) in the input streams, and automatically determines the number k of hidden variables.

The correlations and hidden variables we discover have multiple uses. They provide a succinct summary to the user, they can help to do fast forecasting and detect outliers, and they facilitate interpolations and handling of missing values, as we discuss later.

The rest of the chapter is organized as follows: Section 1 discusses related work, on data streams and stream mining. Section 2 and 3 overview some of the background. Section 5 describes our method and Section 6 shows how its output can be interpreted and immediately utilized, both by humans, as well as for further data analysis. Section 7 discusses experimental case studies that demonstrate the effectiveness of our approach. In Section 8 we elaborate on the efficiency and accuracy of SPIRIT. Finally, in Section 9 we conclude.

1.   Related work

Much of the work on stream mining has focused on finding interesting patterns in a single stream, but multiple streams have also attracted significant interest. Ganti et al. [8] propose a generic framework for stream mining. 10 propose a one-pass k-median clustering algorithm. 6 construct a decision tree online, by passing over the data only once. Recently, 12 and 22 address the problem of finding patterns over concept drifting streams. 19 proposed a method to find patterns in a single stream, using wavelets. More recently, 18 consider approximation of time-series with *amnesic* functions. They propose novel techniques suitable for streaming, and applicable to a wide range of user-specified approximating functions.

15 propose parameter-free methods for classic data mining tasks (i.e., clustering, anomaly detection, classification), based on compression. 16 perform clustering on different levels of wavelet coefficients of multiple time series. Both approaches require having all the data in advance. Recently, 2 propose a framework for *Phenomena Detection and Tracking (PDT)* in sensor networks. They define a phenomenon on discrete-valued streams and develop query execution techniques based on multi-way hash join with PDT-specific optimizations.

CluStream (1) is a flexible clustering framework with online and offline components. The online component extends micro-cluster information (26) by incorporating exponentially-sized sliding windows while coalescing micro-cluster summaries. Actual clusters are found by the offline component. Stat-Stream (27) uses the DFT to summarise streams within a finite window and then compute the highest pairwise correlations among all pairs of streams, at each timestamp. BRAID (20) addresses the problem of discovering lag correlations among multiple streams. The focus is on time and space efficient methods for finding the earliest and highest peak in the cross-correlation functions between all pairs of streams. Neither CluStream, StatStream or BRAID explicitly focus on discovering hidden variables.

9 improve on discovering correlations, by first doing dimensionality reduction with random projections, and then periodically computing the SVD. However, the method incurs high overhead because of the SVD re-computation and it can not easily handle missing values. Also related to these is the work of 4, which uses a different formulation of linear correlations and focuses on compressing historical data, mainly for power conservation in sensor networks. MUSCLES (24) is exactly designed to do forecasting (thus it could handle missing values). However, it can not find hidden variables and it scales poorly for a large number of streams n, since it requires at least quadratic space and time, or expensive reorganisation (*selective MUSCLES*).

Finally, a number of the above methods usually require choosing a sliding window size, which typically translates to buffer space requirements. Our approach does not require any sliding windows and does not need to buffer *any* of the stream data.

In conclusion, none of the above methods simultaneously satisfy the requirements in the introduction: "any-time" streaming operation, scalability on the number of streams, adaptivity, and full automation.

2. Principal component analysis (PCA)

Here we give a brief overview of PCA (13) and explain the intuition behind our approach. We use standard matrix algebra notation: vectors are lower-case bold, matrices are upper-case bold, and scalars are in plain font. The transpose of matrix $\mathbf{X}$ is denoted by $\mathbf{X}^T$. In the following, $\mathbf{x}_t \equiv [x_{t,1}\, x_{t,2}\, \cdots\, x_{t,n}]^T \in \mathbb{R}^n$ is the column-vector. of stream values at time t. We adhere to the common convention of using column vectors and writing them out in transposed form. The stream data can be viewed as a continuously growing $t \times n$ matrix $\mathbf{X}_t :=$ $[\mathbf{x}_1\, \mathbf{x}_2\, \cdots\, \mathbf{x}_t]^T \in \mathbb{R}^{t \times n}$, where one new row is added at each time tick t. In the chlorine example, $\mathbf{x}_t$ is the measurements column-vector at t over all the sensors, where n is the number of chlorine sensors and t is the measurement timestamp.

Typically, in collections of n-dimensional points $\mathbf{x}_t \equiv [x_{t,1} \ldots, x_{t,n}]^T$, $t = 1, 2, \ldots$, there exist correlations between the n dimensions (which correspond

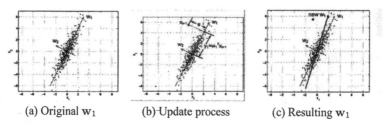

| (a) Original w_1 | (b) Update process | (c) Resulting w_1 |

Figure 12.2.   Illustration of updating w_1 when a new point x_{t+1} arrives.

to streams in our setting). These can be captured by principal components analysis (PCA). Consider for example the setting in Figure 12.2. There is a visible linear correlation. Thus, if we represent every point with its projection on the direction of w_1, the error of this approximation is very small. In fact, the first principal direction w_1, is the *optimal* in the following sense.

DEFINITION 12.1 (FIRST PRINCIPAL COMPONENT) *Given a collection of n-dimensional vectors* $x_\tau \in \mathbb{R}^n$, $\tau = 1, 2, \ldots, t$, *the first principal direction* $w_1 \in \mathbb{R}^n$ *is the vector minimizing the sum of squared residuals, i.e.,*

$$w_1 := \arg\min_{\|w\|=1} \sum_{\tau=1}^{t} \|x_\tau - (ww^T)x_\tau\|^2.$$

The projection of x_τ *on* w_1 *is the* first principal component *(PC)* $y_{\tau,1} := w_1^T x_\tau$, $\tau = 1, \ldots, t$.

Note that, since $\|w_1\| = 1$, we have $(w_1 w_1^T)x_\tau = (w_1^T x_\tau)w_1 = y_{\tau,1} w_1 =: \tilde{x}_\tau$, where $\tilde{x}_\tau$ is the projection of $y_{\tau,1}$ back into the original n-D space. That is, $\tilde{x}_\tau$ is the *reconstruction* of the original measurements from the first PC $y_{\tau,1}$. More generally, PCA will produce k vectors $w_1, w_2, \ldots, w_k$ such that, if we represent each n-D data point $x_t := [x_{t,1} \cdots x_{t,n}]$ with its k-D projection $y_t := [w_1^T x_t \cdots w_k^T x_t]^T$, then this representation minimises the squared error $\sum_\tau \|x_t - \tilde{x}_t\|^2$. Furthermore, the principal directions are orthogonal, so the principal components $y_{\tau,i}, 1 \le i \le k$ are *by construction uncorrelated*, i.e., if $y^{(i)} := [y_{1,i}, \ldots, y_{t,i}, \ldots]^T$ is the stream of the i-th principal component, then $\left(y^{(i)}\right)^T y^{(j)} = 0$ if $i \ne j$.

OBSERVATION 2.1 (DIMENSIONALITY REDUCTION) *If we represent each n-dimensional point* $x_\tau \in \mathbb{R}^n$ *using all n principal components, then the error* $\|x_\tau - \tilde{x}_\tau\| = 0$. *However, in typical datasets, we can achieve a very small error using only k principal components, where* $k \ll n$.

In the context of the chlorine example, each point in Figure 12.2 would correspond to the 2-D projection of x_τ (where $1 \le \tau \le t$) onto the first two principal directions, w_1 and w_2, which are the most important according to the

Table 12.1. Description of notation.

Symbol	Description
$\mathbf{x}, \ldots$	Column vectors (lowercase boldface).
$\mathbf{A}, \ldots$	Matrices (uppercase boldface).
$\mathbf{x}_t$	The n stream values $\mathbf{x}_t := [x_{t,1} \cdots x_{t,n}]^T$ at time t.
n	Number of streams.
$\mathbf{w}_i$	The i-th participation weight vector (i.e., principal direction).
k	Number of hidden variables.
$\mathbf{y}_t$	Vector of hidden variables (i.e., principal components) for $\mathbf{x}_t$, i.e., $\mathbf{y}_t \equiv [y_{t,1} \cdots y_{t,k}]^T := [\mathbf{w}_1^T \mathbf{x}_t \cdots \mathbf{w}_k^T \mathbf{x}_t]^T$.
$\tilde{\mathbf{x}}_t$	Reconstruction of $\mathbf{x}_t$ from the k hidden variable values, i.e., $\tilde{\mathbf{x}}_t := y_{t,1}\mathbf{w}_1 + \cdots + y_{t,k}\mathbf{w}_k$.
E_t	Total energy up to time t.
$\tilde{E}_{t,i}$	Total energy captured by the i-th hidden variable, up to time t.
f_E, F_E	Lower and upper bounds on the fraction of energy we wish to maintain via SPIRIT's approximation.

distribution of $\{\mathbf{x}_\tau \mid 1 \leq \tau \leq t\}$. The principal components $y_{\tau,1}$ and $y_{\tau,2}$ are the coordinates of these projections in the orthogonal coordinate system defined by $\mathbf{w}_1$ and $\mathbf{w}_2$.

However, batch methods for estimating the principal components require time that depends on the duration t, which grows to infinity. In fact, the principal directions are the eigenvectors of $\mathbf{X}_t^T \mathbf{X}_t$, which are best computed through the singular value decomposition (SVD) of $\mathbf{X}_t$. Space requirements also depend on t. Clearly, in a stream setting, it is impossible to perform this computation at every step, aside from the fact that we don't have the space to store all past values. In short, we want a method that does not need to store *any* past values.

3. Auto-regressive models and recursive least squares

In this section we review some of the background on forecasting.

Auto-regressive (AR) modeling

Auto-regressive models are the most widely known and used—more information can be found in, e.g., (3). The main idea is to express x_t as a function of its previous values, plus (filtered) noise ϵ_t:

$$x_t = \phi_1 x_{t-1} + \ldots + \phi_W x_{t-W} + \epsilon_t, \qquad (12.1)$$

where W is a the forecasting window size. Seasonal variants (SAR, SAR(I)MA) also use window offsets that are multiples of a single, fixed period (i.e., besides terms of the form y_{t-i}, the equation contains terms of the form y_{t-Si} where S is a constant).

If we have a collection of n time series $x_{t,i}$, $1 \leq i \leq n$ then multivariate AR simply expresses $x_{t,i}$ as a linear combination of previous values of all streams (plus noise), i.e.,

$$x_{t,i} = \phi_{1,1}x_{t-1,1} + \dots + \phi_{1,W}x_{t-W,1} +$$
$$\dots +$$
$$\phi_{n,1}x_{t-1,n} + \dots + \phi_{n,W}x_{t-W,n} + \epsilon_t. \qquad (12.2)$$

Recursive Least Squares (RLS)

Recursive Least Squares (RLS) is a method that allows dynamic update of a least-squares fit. The least squares solution to an overdetermined system of equations $\mathbf{Xb} = \mathbf{y}$ where $\mathbf{X} \in \mathbb{R}^{m \times k}$ (measurements), $\mathbf{y} \in \mathbb{R}^m$ (output variables) and $\mathbf{b} \in \mathbb{R}^k$ (regression coefficients to be estimated) is given by the solution of $\mathbf{X}^T\mathbf{Xb} = \mathbf{X}^T\mathbf{y}$. Thus, all we need for the solution are the projections

$$\mathbf{P} \equiv \mathbf{X}^T\mathbf{X} \qquad \text{and} \qquad \mathbf{q} \equiv \mathbf{X}^T\mathbf{y}$$

We need only space $O(k^2 + k) = O(k^2)$ to keep the model up to date. When a new row $\mathbf{x}_{m+1} \in \mathbb{R}^k$ and output y_{m+1} arrive, we can update

$$\mathbf{P} \leftarrow \mathbf{P} + \mathbf{x}_{m+1}\mathbf{x}_{m+1}^T \qquad \text{and}$$
$$\mathbf{q} \leftarrow \mathbf{q} + y_{m+1}\mathbf{x}_{m+1}.$$

In fact, it is possible to update the regression coefficient vector $\mathbf{b}$ without explicitly inverting $\mathbf{P}$ to solve $P\mathbf{b} = \mathbf{P}^{-1}\mathbf{q}$. In particular (see, e.g., (25)) the update equations are

$$\mathbf{G} \leftarrow \mathbf{G} - (1 + \mathbf{x}_{m+1}^T\mathbf{Gx}_{m+1})^{-1}\mathbf{Gx}_{m+1}\mathbf{x}_{m+1}^T\mathbf{G} \qquad (12.3)$$
$$\mathbf{b} \leftarrow \mathbf{b} - \mathbf{Gx}_{m+1}(\mathbf{x}_{m+1}^T\mathbf{b} - y_{m+1}), \qquad (12.4)$$

where the matrix $\mathbf{G}$ can be initialized to $\mathbf{G} \leftarrow \epsilon\mathbf{I}$, with ϵ a small positive number and $\mathbf{I}$ the $k \times k$ identity matrix.

RLS and AR   In the context of auto-regressive modeling (Eq. 12.1), we have one equation for each stream value $x_{w+1}, \dots, x_t, \dots$, i.e., the m-th row of the $\mathbf{X}$ matrix above is

$$\mathbf{X}_m = [x_{m-1} \ x_{m-2} \ \cdots \ x_{m-w}]^T \in \mathbb{R}^w$$

and $z_m = x_m$, for $t - w = m = 1, 2, \dots$ ($t > w$). In this case, the solution vector $\mathbf{b}$ consists precisely of the auto-regression coefficients in Eq. 12.1, i.e.,

$$\mathbf{b} = [\phi_1 \ \phi_2 \ \cdots \ \phi_w]^T \in \mathbb{R}^w.$$

RLS can be similarly used for multivariate AR model estimation.

4. MUSCLES

MUSCLES (MUlti-SequenCe LEast Squares) (24) tries to predict the value of one stream, $x_{t,i}$ based on the previous values from all streams, $x_{t-l,j}, l > 1, 1 \leq j \leq n$ and current values from other streams, $x_{t,j}, j \neq i$. It uses multivariate autoregression, thus the prediction $\hat{x}_{t,i}$ for a given stream i is, similar to Eq. 12.2

$$\hat{x}_{t,i} = \phi_{1,0}x_{t,1} + \phi_{1,1}x_{t-1,1} + \cdots + \phi_{1,W}x_{t-W,1} +$$
$$\cdots +$$
$$\phi_{i-1,0}x_{t-1,i-1} + \phi_{i-1,1}x_{t-1,i-1} + \cdots + \phi_{i-1,w}x_{t-W,i-1} +$$
$$\phi_{i,1}x_{t-1,i} + \cdots + \phi_{i,w}x_{t-W,i} +$$
$$\phi{i+1,0}x_{t,i+1} + \phi_{i+1,1}x_{t-1,i+1} + \cdots + \phi_{i+1,w}x_{t-W,i+1} +$$
$$\cdots +$$
$$\phi_{n,0}x_{t,n} + \phi_{n,1}x_{t-1,n} + \cdots + \phi_{n,W}x_{t-W,n} + \epsilon_t.$$

and employs RLS to continuously update the coefficients $\phi_{i,j}$ such that the prediction error

$$\sum_{\tau=1}^{t}(\hat{x}_{\tau,i} - x_{\tau,i})^2$$

is minimized. Note that the above equation has one dependent variable (the estimate $\hat{x}_{t,i}$) and $v = W * n + n - 1$ independent variables (the past values of all streams plus the current values of all other streams except i).

Exponentially forgetting MUSCLES employs a forgetting factor $0 < \lambda \leq 1$ and minimizes instead

$$\sum_{\tau=1}^{t}\lambda^{t-\tau}(\hat{x}_{\tau,i} - x_{\tau,i})^2.$$

For $\lambda < 1$, errors for old values are downplayed by a geometric factor, and hence it permits the estimate to adapt as sequence characteristics change.

Selective MUSCLES

In case we have too many time sequences (e.g., $n = 100,000$ nodes in a network, producing information about their load every minute), even the incremental version of MUSCLES will suffer. The solution we propose is based on the conjecture that we do not really need information from every sequence to make a good estimation of a missing value much of the benefit of using multiple sequences may be captured by using only a small number of carefully selected other sequences. Thus, we propose to do some preprocessing of a training set, to find a promising subset of sequences, and to apply MUSCLES only to those promising ones (hence the name Selective MUSCLES).

Assume that sequence i is the one notoriously delayed and we need to estimate its "delayed" values $x_{t,i}$. For a given tracking window span W, among the $v = W * n + n - 1$ independent variables, we have to choose the ones that are most useful in estimating the delayed value of $x_{t,i}$. More generally, we want to solve the following

PROBLEM 4.1 (SUBSET SELECTION) *Given v independent variables $x_1, x_2, \ldots, x_v$ and a dependent variable y with N samples each, find the best b ($< v$) independent variables to minimize the mean-square error for $\hat{y}$ for the given samples.*

We need a measure of goodness to decide which subset of b variables is the best we can choose. Ideally, we should choose the best subset that yields the smallest estimation error in the future. Since, however, we don't have future samples, we can only infer the *expected estimation error* (EEE for short) from the available samples as follows:

$$\text{EEE}(\mathcal{S}) = \sum_{t=1}^{N} (y[t] - \hat{y}_{\mathcal{S}}[t])^2$$

where $\mathcal{S}$ is the selected subset of variables and $\hat{y}_{\mathcal{S}}[t]$ is the estimation based on $\mathcal{S}$ for the t-th sample. Note that, thanks to Eq. 12.3, $\text{EEE}(\mathcal{S})$ can be computed in $O(N \cdot \|\mathcal{S}\|^2)$ time. Let's say that we are allowed to keep only $b = 1$ independent variable. Which one should we choose? Intuitively, we could try the one that has the highest (in absolute value) correlation coefficient with y. It turns out that this is indeed optimal: (to satisfy the unit variance assumption, we will normalize samples by the sample variance within the window.)

LEMMA 12.2 *Given a dependent variable y, and v independent variables with unit variance, the best single variable to keep to minimize $\text{EEE}(\mathcal{S})$ is the one with the highest absolute correlation coefficient with y.*

Proof For a single variable, if a is the least squares solution, we can express the error in matrix form as

$$\text{EEE}(\{x_i\}) = \|\mathbf{y}\|^2 - 2a(\mathbf{y}^T \mathbf{x}_i) + a^2 \|\mathbf{x}_i\|^2.$$

Let d and p denote $\|\mathbf{x}_i\|^2$ and $(\mathbf{x}^T\mathbf{y})$, respectively. Since $a = d^{-1}p$, $\text{EEE}(\{x_i\}) = \|y\|^2 - p^2 d^{-1}$. To minimize the error, we must choose x_i which maximize p^2 and minimize d. Assuming unit-variance ($d = 1$), such x_i is the one with the biggest correlation coefficient to y. This concludes the proof.

The question is how we should handle the case when $b > 1$. Normally, we should consider all the possible groups of b independent variables, and try to pick the best. This approach explodes combinatorially; thus we propose to use a greedy algorithm. At each step s, we select the independent variable x_s that minimizes the EEE for the dependent variable y, in light of the $s-1$ independent variables that we have already chosen in the previous steps.

Bottleneck of the algorithm is clearly the computation of EEE. Since it computes EEE approximately $O(v \cdot b)$ times and each computation of EEE requires $O(N \cdot b^2)$ in average, the overall complexity mounts to $O(N \cdot v \cdot b^3)$. To reduce the overhead, we observe that intermediate results produced for $\text{EEE}(S)$ can be re-used for $\text{EEE}(S \cup \{x\})$.

LEMMA 12.3 *The complexity of the greedy selection algorithm is $O(N \cdot v \cdot b^2)$.*

Proof Let S^+ be $S \cup \{x\}$. The core in computing $\text{EEE}(S^+)$ is the inverse of $\mathbf{D}_{S^+} = (\mathbf{X}_{S^+}^T \mathbf{X}_{S^+})$. Thanks to block matrix inversion formula (14) (p. 656) and the availability of D_S^{-1} from the previous iteration step, it can be computed in $O(N \cdot |S| + |S|^2)$. Hence, summing it up over $v - |S|$ remaining variables for each b iteration, we have $O(N \cdot v \cdot b^2 + v \cdot b^3)$ complexity. Since $N \gg b$, it reduces to $O(N \cdot v \cdot b^2)$.

We envision that the subset-selection will be done infrequently and off-line, say every $N = W$ time-ticks. The optimal choice of the reorganization window W is beyond the scope of this paper. Potential solutionsinclude (a) doing reorganization during off-peak hours, (b) triggering a reorganization whenever the estimation error for by increases above an application-dependent threshold etc. Also, by normalizing the training set, the unit-variance assumption in Theorem 1 can be easily satisfied.

5. Tracking correlations and hidden variables: SPIRIT

In this section we present our framework for discovering patterns in multiple streams. In the next section, we show how these can be used to perform effective, low-cost forecasting. We use auto-regression for its simplicity, but our framework allows any forecasting algorithm to take advantage of the compact representation of the stream collection.

Problem definition Given a collection of n co-evolving, semi-infinite streams, producing a value $x_{t,j}$, for every stream $1 \leq j \leq n$ and for every time-tick $t = 1, 2, \ldots$, SPIRIT does the following: (i) Adapts the number k of *hidden variables* necessary to explain/summarise the main trends in the collection. (ii) Adapts the *participation weights* $w_{i,j}$ of the j-th stream on the i-th hidden variable ($1 \leq j \leq n$ and $1 \leq i \leq k$), so as to produce an accurate summary of the stream collection. (iii) Monitors the hidden variables $y_{t,i}$, for $1 \leq i \leq k$. (iv) Keeps updating all the above efficiently.

More precisely, SPIRIT operates on the column-vectors of observed stream values $x_t \equiv [x_{t,1}, \ldots, x_{t,n}]^T$ and continually updates the participation weights $w_{i,j}$. The *participation weight vector* $\mathbf{w}_i$ for the i-th principal direction is $\mathbf{w}_i := [w_{i,1} \cdots w_{i,n}]^T$. The hidden variables $y_t \equiv [y_{t,1}, \ldots, y_{t,k}]^T$ are the projections of x_t onto each $\mathbf{w}_i$, over time (see Table 12.1), i.e.,

$$y_{t,i} := w_{i,1}x_{t,1} + w_{i,2}x_{t,2} + \cdots + w_{i,n}x_{t,n},$$

SPIRIT also adapts the number k of hidden variables necessary to capture most of the information. The adaptation is performed so that the approximation achieves a desired mean-square error. In particular, let $\tilde{x}_t = [\tilde{x}_{t,1} \cdots \tilde{x}_{t,n}]^T$ be the *reconstruction* of x_t, based on the weights and hidden variables, defined by

$$\tilde{x}_{t,j} := w_{1,j}y_{t,1} + w_{2,j}y_{t,2} + \cdots + w_{k,j}y_{t,k},$$

or more succinctly, $\tilde{\mathbf{x}}_t = \sum_{i=1}^{k} y_{i,t}\mathbf{w}_i$.

In the chlorine example, x_t is the n-dimensional column-vector of the original sensor measurements and y_t is the hidden variable column-vector, both at time t. The dimension of y_t is 1 before/after the leak ($t < 1500$ or $t > 3000$) and 2 during the leak ($1500 \le t \le 3000$), as shown in Figure 12.1.

DEFINITION 12.4 (SPIRIT TRACKING) *SPIRIT updates the participation weights $w_{i,j}$ so as to guarantee that the reconstruction error $\|\tilde{\mathbf{x}}_t - x_t\|^2$ over time is predictably small.*

This informal definition describes what SPIRIT does. The precise criteria regarding the reconstruction error will be explained later. If we assume that the x_t are drawn according to some distribution that does not change over time (i.e., under *stationarity* assumptions), then the weight vectors $\mathbf{w}_i$ converge to the principal directions. However, even if there are non-stationarities in the data (i.e., gradual drift), in practice we can deal with these very effectively, as we explain later.

An additional complication is that we often have missing values, for several reasons: either failure of the system, or delayed arrival of some measurements. For example, the sensor network may get overloaded and fail to report some of the chlorine measurements in time or some sensor may temporarily black-out. At the very least, we want to continue processing the rest of the measurements.

Tracking the hidden variables

The first step is, for a given k, to incrementally update the k participation weight vectors $\mathbf{w}_i$, $1 \le i \le k$, so as to summarise the original streams with only a few numbers (the hidden variables). In Section 5.0, we describe the complete method, which also adapts k.

For the moment, assume that the number of hidden variables k is given. Furthermore, our goal is to minimise the average reconstruction error $\sum_t \|\tilde{\mathbf{x}}_t - \mathbf{x}_t\|^2$. In this case, the desired weight vectors $\mathbf{w}_i, 1 \leq i \leq k$ are the principal directions and it turns out that we can estimate them incrementally.

We use an algorithm based on adaptive filtering techniques (23, 11), which have been tried and tested in practice, performing well in a variety of settings and applications (e.g., image compression and signal tracking for antenna arrays). We experimented with several alternatives (17, 5) and found this particular method to have the best properties for our setting: it is very efficient in terms of computational and memory requirements, while converging quickly, with no special parameters to tune. The main idea behind the algorithm is to read in the new values $\mathbf{x}_{t+1} \equiv [x_{(t+1),1}, \ldots, x_{(t+1),n}]^T$ from the n streams at time $t + 1$, and perform three steps:

1. Compute the hidden variables $y'_{t+1,i}, 1 \leq i \leq k$, based on the *current* weights $\mathbf{w}_i, 1 \leq i \leq k$, by projecting $\mathbf{x}_{t+1}$ onto these.

2. Estimate the reconstruction error (e_i below) and the energy, based on the $y'_{t+1,i}$ values.

3. Update the estimates of $\mathbf{w}_i, 1 \leq i \leq k$ and output the *actual* hidden variables $y_{t+1,i}$ for time $t + 1$.

To illustrate this, Figure 12.2b shows the e_1 and y_1 when the new data $\mathbf{x}_{t+1}$ enter the system. Intuitively, the goal is to adaptively update $\mathbf{w}_i$ so that it quickly converges to the "truth." In particular, we want to update $\mathbf{w}_i$ more when e_i is large. However, the magnitude of the update should also take into account the past data currently "captured" by $\mathbf{w}_i$. For this reason, the update is inversely proportional to the current *energy* $E_{t,i}$ of the i-th hidden variable, which is $E_{t,i} := \frac{1}{t}\sum_{\tau=1}^{t} y_{\tau,i}^2$. Figure 12.2c shows $\mathbf{w}_1$ after the update for $\mathbf{x}_{t+1}$.

Algorithm TRACKW _____

0. Initialise the k hidden variables $\mathbf{w}_i$ to unit vectors $\mathbf{w}_1 = [10 \cdots 0]^T$, $\mathbf{w}_2 = [010 \cdots 0]^T$, etc. Initialise d_i ($i = 1, \ldots k$) to a small positive value. Then:
1. As each point $\mathbf{x}_{t+1}$ arrives, initialise $\acute{\mathbf{x}}_1 := \mathbf{x}_{t+1}$.
2. For $1 \leq i \leq k$, we perform the following assignments and updates, in order:

$$y_i := \mathbf{w}_i^T \acute{\mathbf{x}}_i \qquad (y_{t+1,i} = \text{projection onto } \mathbf{w}_i)$$

$$d_i \leftarrow \lambda d_i + y_i^2 \qquad (\text{energy} \propto i\text{-th eigenval. of } \mathbf{X}_t^T \mathbf{X}_t)$$

$$\mathbf{e}_i := \acute{\mathbf{x}}_i - y_i \mathbf{w}_i \qquad (\text{error, } \mathbf{e}_i \perp \mathbf{w}_i)$$

$$\mathbf{w}_i \leftarrow \mathbf{w}_i + \frac{1}{d_i} y_i \mathbf{e}_i \qquad (\text{update PC estimate})$$

$$\acute{\mathbf{x}}_{i+1} := \acute{\mathbf{x}}_i - y_i \mathbf{w}_i \qquad (\text{repeat with remainder of } \mathbf{x}_t).$$

The *forgetting factor* λ will be discussed in Section 5.0 (for now, assume $\lambda = 1$). For each i, $d_i = tE_{t,i}$ and $\acute{x}_i$ is the component of x_{t+1} in the orthogonal complement of the space spanned by the updated estimates $w_{i'}, 1 \leq i' < i$ of the participation weights. The vectors $w_i, 1 \leq i \leq k$ are in order of importance (more precisely, in order of decreasing eigenvalue or energy). It can be shown that, under stationarity assumptions, these w_i in these equations converge to the true principal directions.

Complexity   We only need to keep the k weight vectors w_i ($1 \leq i \leq k$), each n-dimensional. Thus the total cost is $O(nk)$, both in time and in space. The update cost does not depend on t. This is a tremendous gain, compared to the usual PCA computation cost of $O(tn^2)$.

Detecting the number of hidden variables

In practice, we do not know the number k of hidden variables. We propose to estimate k on the fly, so that we maintain a high percentage f_E of the *energy* E_t. Energy thresholding is a common method to determine how many principal components are needed (13). Formally, the energy E_t (at time t) of the sequence of x_t is defined as

$$E_t := \tfrac{1}{t} \sum_{\tau=1}^{t} \|x_\tau\|^2 = \tfrac{1}{t} \sum_{\tau=1}^{t} \sum_{i=1}^{n} x_{\tau,i}^2.$$

Similarly, the energy $\tilde{E}_t$ of the reconstruction $\tilde{x}$ is defined as

$$\tilde{E}_t := \tfrac{1}{t} \sum_{\tau=1}^{t} \|\tilde{x}_\tau\|^2.$$

LEMMA 12.5 *Assuming the* $w_i, 1 \leq i \leq k$ *are orthonormal, we have*

$$\tilde{E}_t = \tfrac{1}{t} \sum_{\tau=1}^{t} \|y_\tau\|^2 = \tfrac{t-1}{t} \tilde{E}_{t-1} + \tfrac{1}{t} \|y_t\|.$$

Proof If the $w_i, 1 \leq i \leq k$ are orthonormal, then it follows easily that $\|\tilde{x}_\tau\|^2 = \|y_{\tau,1} w_1 + \cdots + y_{\tau,k} w_k\|^2 = y_{\tau,1}^2 \|w_1\|^2 + \cdots + y_{\tau,k}^2 \|w_k\|^2 = y_{\tau,1}^2 + \cdots + y_{\tau,k}^2 = \|y_\tau\|^2$ (Pythagorean theorem and normality). The result follows by summing over τ.

It can be shown that algorithm TRACKW maintains orthonormality without the need for any extra steps (otherwise, a simple re-orthonormalisation step at the end would suffice).

From the user's perspective, we have a low-energy and a high-energy threshold, f_E and F_E, respectively. We keep enough hidden variables k, so the retained energy is within the range $[f_E \cdot E_t, F_E \cdot E_t]$. Whenever we get outside these bounds, we increase or decrease k. In more detail, the steps are:

1 Estimate the full energy E_{t+1}, incrementally, from the sum of squares of $x_{\tau,i}$.

2 Estimate the energy $\tilde{E}_{(k)}$ of the k hidden variables.

3 Possibly, adjust k. We introduce a new hidden variable (update $k \leftarrow k+1$) if the current hidden variables maintain too little energy, i.e., $\tilde{E}_{(k)} < f_E E$. We drop a hidden variable (update $k \leftarrow k-1$), if the maintained energy is too high, i.e., $\tilde{E}_{(k)} > F_E E$.

The energy thresholds f_E and F_E are chosen according to recommendations in the literature (13, 7). We use a lower energy threshold $f_E = 0.95$ and an upper threshold $F_E = 0.98$. Thus, the reconstruction $\tilde{x}_t$ retains between 95% and 98% of the energy of x_t.

Algorithm SPIRIT

0. Initialise $k \leftarrow 1$ and the total energy estimates of x_t and $\tilde{x}_t$ per time tick to $E \leftarrow 0$ and $\tilde{E}_1 \leftarrow 0$. Then,
1. As each new point arrives, update w_i, for $1 \le i \le k$ (step 1, TRACKW).
2. Update the estimates (for $1 \le i \le k$)

$$E \leftarrow \frac{(t-1)E + \|x_t\|^2}{t} \quad \text{and} \quad \tilde{E}_i \leftarrow \frac{(t-1)\tilde{E}_i + y_{t,i}^2}{t}.$$

3. Let the estimate of retained energy be

$$\tilde{E}_{(k)} := \sum_{i=1}^{k} \tilde{E}_i.$$

If $\tilde{E}_{(k)} < f_E E$, then we start estimating w_{k+1} (initialising as in step 0 of TRACKW), initialise $\tilde{E}_{k+1} \leftarrow 0$ and increase $k \leftarrow k+1$. If $\tilde{E}_{(k)} > F_E E$, then we discard w_k and $\tilde{E}_k$ and decrease $k \leftarrow k-1$.

The following lemma proves that the above algorithm guarantees the relative reconstruction error is within the specified interval $[f_E, F_E]$.

LEMMA 12.6 *The relative squared error of the reconstruction satisfies*

$$1 - F_E \le \frac{\sum_{\tau=1}^{t} \|\tilde{x}_\tau - x_\tau\|^2}{\sum_t \|x_\tau\|^2} \le 1 - f_E.$$

Proof From the orthogonality of x_τ and $\tilde{x}_\tau - x_\tau$ we have $\|\tilde{x}_\tau - x_\tau\|^2 = \|x_\tau\|^2 - \|\tilde{x}_\tau\|^2 = \|x_\tau\|^2 - \|y_\tau\|^2$ (by Lemma 12.5). The result follows by summing over τ and from the definitions of E and $\tilde{E}$.

In Section 8.0 we demonstrate that the incremental weight estimates are extremely close to the principal directions computed with offline PCA.

Exponential forgetting

We can adapt to more recent behaviour by using an exponential forgetting factor, $0 < \lambda < 1$. This allows us to follow trend drifts over time. We use the same λ for the estimation of both $\mathbf{w}_i$ and of the AR models (see Section 6.0). However, we also have to properly keep track of the energy, discounting it with the same rate, i.e., the update at each step is:

$$E \leftarrow \frac{\lambda(t-1)E + \|\mathbf{x}_t\|^2}{t} \quad \text{and} \quad \tilde{E}_i \leftarrow \frac{\lambda(t-1)\tilde{E}_i + y_{t,i}^2}{t}.$$

Typical choices are $0.96 \le \lambda \le 0.98$ (11). As long as the values of $\mathbf{x}_t$ do not vary wildly, the exact value of λ is not crucial. We use $\lambda = 0.96$ throughout. A value of $\lambda = 1$ makes sense when we know that the sequence is stationary (rarely true in practice, as most sequences gradually drift). Note that the value of λ does not affect the computation cost of our method. In this sense, an exponential forgetting factor is more appealing than a sliding window, as the latter has explicit buffering requirements.

6.   Putting SPIRIT to work

We show how we can exploit the correlations and hidden variables discovered by SPIRIT to do (a) forecasting, (b) missing value estimation, (c) summarisation of the large number of streams into a small, manageable number of hidden variables, and (d) outlier detection.

Forecasting and missing values

The hidden variables $\mathbf{y}_t$ give us a much more compact representation of the "raw" variables $\mathbf{x}_t$, with guarantees of high reconstruction accuracy (in terms of relative squared error, which is less than $1 - f_E$). When our streams exhibit correlations, as we often expect to be the case, the number k of the hidden variables is much smaller than the number n of streams. Therefore, we can apply *any* forecasting algorithm to the vector of hidden variables $\mathbf{y}_t$, instead of the raw data vector $\mathbf{x}_t$. This reduces the time and space complexity by orders of magnitude, because typical forecasting methods are quadratic or worse on the number of variables.

In particular, we fit the forecasting model on the $\mathbf{y}_t$ instead of $\mathbf{x}_t$. The model provides an estimate $\hat{\mathbf{y}}_{t+1} = f(\mathbf{y}_t)$ and we can use this to get an estimate for

$$\hat{\mathbf{x}}_{t+1} := \hat{y}_{t+1,1}\mathbf{w}_1[t] + \cdots + \hat{y}_{t+1,1}\mathbf{w}_k[t],$$

using the weight estimates $\mathbf{w}_i[t]$ from the previous time tick t. We chose autoregression for its intuitiveness and simplicity, but any online method can be used.

Correlations Since the principal directions are orthogonal ($\mathbf{w}_i \perp \mathbf{w}_j, i \neq j$), the components of $\mathbf{y}_t$ are *by construction uncorrelated*—the correlations have already been captured by the $\mathbf{w}_i, 1 \leq i \leq k$. We can take advantage of this de-correlation reduce forecasting complexity. In particular for auto-regression, we found that one AR model per hidden variable provides results comparable to multivariate AR.

Auto-regression Space complexity for multivariate AR (e.g., MUSCLES (24)) is $O(n^3 \ell^2)$, where ℓ is the auto-regression window length. For AR per stream (ignoring correlations), it is $O(n\ell^2)$. However, for SPIRIT, we need $O(kn)$ space for the $\mathbf{w}_i$ and, with one AR model per y_i, the total space complexity is $O(kn + k\ell^2)$. As published, MUSCLES requires space that grows cubically with respect to the number of streams n. We believe it can be made to work with quadratic space, but this is still prohibitive. Both AR per stream and SPIRIT require space that grows linearly with respect to n, but in SPIRIT k is typically very small ($k \ll n$) and, in practice, SPIRIT requires less memory and time per update than AR per stream. More importantly, a single, independent AR model per stream cannot capture *any* correlations, whereas SPIRIT indirectly exploits the correlations present *within* a time tick.

Missing values When we have a forecasting model, we can use the forecast based on $\mathbf{x}_{t-1}$ to estimate missing values in $\mathbf{x}_t$. We then use these estimated missing values to update the weight estimates, as well as the forecasting models. Forecast-based estimation of missing values is the most time-efficient choice and gives very good results.

Interpretation

At *any* given time t, SPIRIT readily provides two key pieces of information (aside from the forecasts, etc.): (i) The number of hidden variables k. (ii) The weights $w_{i,j}, 1 \leq i \leq k, 1 \leq j \leq n$. Intuitively, the magnitude $|w_{i,j}|$ of each weight tells us how much the i-th hidden variable contributes to the reconstruction of the j-th stream.

In the chlorine example during phase 1 (see Figure 12.1), the dataset has only one hidden variable, because one sinusoidal-like pattern can reconstruct both streams (albeit with different weights for each). Thus, SPIRIT correctly identifies correlated streams. When the correlation was broken, SPIRIT introduces enough hidden variables to capture that. Finally, it also spots that, in phase 3, normal operation is reestablished and thus disposes of the unnecessary hidden variable. Section 7 has additional examples of how we can intuitively interpret this information.

Table 12.2.   Description of datasets.

Dataset	n	k	Description
Chlorine	166	2	Chlorine concentrations from EPANET.
Critter	8	1–2	Temperature sensor measurements.
River	3	1	River gauge data from USACE.
Motes	54	2–4	Light sensor measurements.

7.   Experimental case studies

In this section we present case studies on real and realistic datasets to demonstrate the effectiveness of our approach in discovering the underlying correlations among streams. In particular, we show that: (i) We capture the appropriate number of hidden variables. As the streams evolve, we capture these changes in real-time (21) and adapt the number of hidden variables k and the weights $\mathbf{w}_i$. (ii) We capture the essential behaviour with very few hidden variables and small reconstruction error. (iii) We successfully deal with missing values. (iv) We can use the discovered correlations to perform good forecasting, with *much* fewer resources. (v) We can easily spot outliers. (vi) Processing time per stream is constant. Section 8 elaborates on performance and accuracy.

Chlorine concentrations

Description   The Chlorine dataset was generated by EPANET 2.0[1] that accurately simulates the hydraulic and chemical phenomena within drinking water distribution systems. Given a network as the input, EPANET tracks the flow of water in each pipe, the pressure at each node, the height of water in each tank, and the concentration of a chemical species throughout the network, during a simulation period comprised of multiple timestamps. We monitor the chlorine concentration level at all the 166 junctions of a water distribution network, for 4310 timestamps during 15 days (one time tick every five minutes). The data was generated by using the input network with the demand patterns, pressures, flows specified at each node.

Data characteristics   The two key features are: (i) A clear global periodic pattern (daily cycle, dominating residential demand pattern). Chlorine concentrations reflect this, with few exceptions. (ii) A slight time shift across different junctions, which is due to the time it takes for fresh water to flow down the pipes from the reservoirs. Thus, most streams exhibit the same sinusoidal-

---

[1]http://www.epa.gov/ORD/NRMRL/wswrd/epanet.html

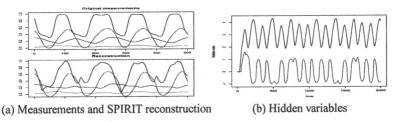

(a) Measurements and SPIRIT reconstruction   (b) Hidden variables

Figure 12.3. Chlorine dataset: (a) actual measurements and reconstruction at four junctions. We plot only 500 consecutive timestamps (the patterns repeat after that). (b) shows SPIRIT's hidden variables.

like pattern, except with gradual phase shifts as we go further away from the reservoir.

Results of SPIRIT SPIRIT can successfully summarise the data using just two numbers (hidden variables) per time tick, as opposed to the original 166 numbers. Figure 12.3a shows the reconstruction for four of the sensors (out of 166). Only two hidden variables give very good reconstruction.

Interpretation The two hidden variables (Figure 12.3b) reflect the two key dataset characteristics. The first hidden variable captures the global, periodic pattern. The second one also follows a very similar periodic pattern, but with a slight "phase shift." It turns out that the two hidden variables together are sufficient to express (via a linear combination) any other time series with an arbitrary "phase shift."

Light measurements

Description The Motes dataset consists of light intensity measurements collected using Berkeley Mote sensors, at several different locations in a lab, over a period of a month.

Data characteristics The main characteristics are: (i) A clear global periodic pattern (daily cycle). (ii) Occasional big spikes from some sensors (outliers).

Results of SPIRIT SPIRIT detects four hidden variables (see Figure 12.4b). Two of these are intermittent and correspond to outliers, or changes in the correlated trends. We show the reconstructions for some of the observed variables in Figure 12.4a.

Interpretation In summary, the first two hidden variables (see Figure 12.4b) correspond to the global trend and the last two, which are intermittently present, correspond to outliers. In particular, the first hidden variable captures the global

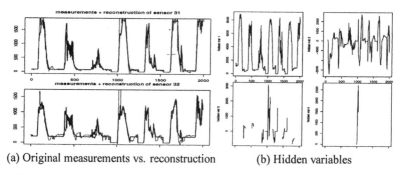

(a) Original measurements vs. reconstruction   (b) Hidden variables

Figure 12.4. Mote dataset: (a) shows the measurements (bold) and reconstruction (thin) on nodes 31 and 32. (b) the third and fourth hidden variables are intermittent and indicate anomalous behaviour (axes limits are different in each plot).

periodic pattern. The interpretation of the second one is again similar to the Chlorine dataset. The first two hidden variables together are sufficient to express arbitrary phase shifts. The third and fourth hidden variables indicate some of the potential outliers in the data. For example, there is a big spike in the 4th hidden variable at time $t = 1033$, as shown in Figure 12.4b. Examining the participation weights w_4 at that timestamp, we can find the corresponding sensors "responsible" for this anomaly, i.e., those sensors whose participation weights have very high magnitude. Among these, the most prominent are sensors 31 and 32. Looking at the actual measurements from these sensors, we see that before time $t = 1033$ they are almost 0. Then, very large increases occur around $t = 1033$, which bring an additional hidden variable into the system.

Room temperatures

Description The Critter dataset consists of 8 streams (see Figure 12.5a). Each stream comes from a small sensor[2] (aka. Critter) that connects to the joystick port and measures temperature. The sensors were placed in 5 neighbouring rooms. Each time tick represents the average temperature during one minute.

Furthermore, to demonstrate how the correlations capture information about missing values, we repeated the experiment after blanking 1.5% of the values (five blocks of *consecutive* timestamps; see Figure 12.6).

Data characteristics Overall, the dataset does not seem to exhibit a clear trend. Upon closer examination, all sensors fluctuate slightly around a constant temperature (which ranges from 22–27°C, or 72–81°F, depending on the

---

[2]http://www.ices.cmu.edu/sensornets/

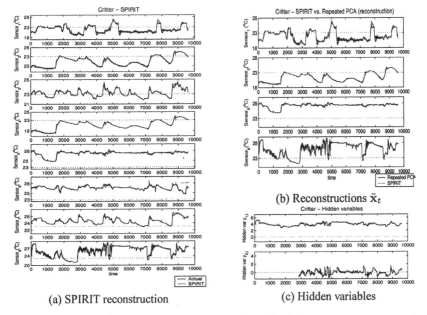

(a) SPIRIT reconstruction   (b) Reconstructions $\tilde{x}_t$

(c) Hidden variables

Figure 12.5. Critter data and SPIRIT output, for each of the temperature sensors, in (a). SPIRIT can track the overall behaviour of the entire stream collection with only two hidden variables, shown in (c). For the comparison in (b), wall clock times are 1.5 minutes (repeated PCA) versus 7 seconds (SPIRIT).

sensor). Approximately half of the sensors exhibit a more similar "fluctuation pattern."

Results of SPIRIT   SPIRIT discovers one hidden variable, which is sufficient to capture the general behaviour. However, if we utilise prior knowledge (such as, e.g., that the pre-set temperature was $23^{o}C$), we can ask SPIRIT to detect trends with respect to that. In that case, SPIRIT comes up with two hidden variables, which we explain later.

SPIRIT is also able to deal successfully with missing values in the streams. Figure 12.6 shows the results on the blanked version (1.5% of the total values in five blocks of *consecutive* timestamps, starting at a different position for each stream) of Critter. The correlations captured by SPIRIT's hidden variable often provide useful information about the missing values. In particular, on sensor 8 (second row, Figure 12.6), the correlations picked by the *single* hidden variable successfully capture the missing values in that region (consisting of 270 ticks). On sensor 7, (first row, Figure 12.6; 300 blanked values), the upward trend in the blanked region is also picked up by the correlations. Even though the trend is slightly mis-estimated, as soon as the values are observed again, SPIRIT very quickly gets back to near-perfect tracking.

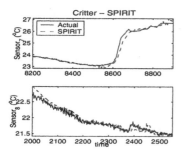
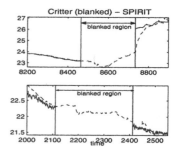

Figure 12.6. Detail of the forecasts on Critter with blanked values. The second row shows that the correlations picked by the *single* hidden variable successfully capture the missing values in that region (consisting of 270 *consecutive* ticks). In the first row (300 consecutive blanked values), the upward trend in the blanked region is also picked up by the correlations to other streams. Even though the trend is slightly mis-estimated, as soon as the values are observed again SPIRIT quickly gets back to near-perfect tracking.

Interpretation If we examine the participation weights in w_1, the largest correspond primarily to streams 5 and 6, and then to stream 8. If we examine the data, sensors 5 and 6 consistently have the highest temperatures. Sensor 8 also has a similar temperature most of the time.

However, if the sensors are calibrated based on the fact that these are building temperature measurements, where we have set the thermostat to 23°C (73°F), then SPIRIT discovers two hidden variables (see Figure 12.5c). More specifically, if we reasonably assume that we have the prior knowledge of what the temperature *should be* (note that this has nothing to do with the average temperature in the observed data) and want to discover what happens around that temperature, we can subtract it from each observation and SPIRIT will discover patterns and anomalies based on this information. Actually, this is what a human operator would be interested in discovering: "Does the system work as I expect it to?" (based on my knowledge of how it should behave) and "If not, what is wrong?" and we indeed discover this kind of information.

The interpretation of the first hidden variable is similar to that of the original signal: sensors 5 and 6 (and, to a lesser extent, 8) deviate from that temperature the most, for most of the time. Maybe the thermostats are broken or set wrong?

For w_2, the largest weights correspond to sensors 1 and 3, then to 2 and 4. If we examine the data, we notice that these streams follow a similar, fluctuating trend (close to the pre-set temperature), the first two varying more violently. The second hidden variable is added at time $t = 2016$. If we examine the plots, we see that, at the beginning, most streams exhibit a slow dip and then ascent (e.g., see 2, 4 and 5 and, to a lesser extent, 3, 7 and 8). However, a number of them start fluctuating more quickly and violently when the second hidden variable is added.

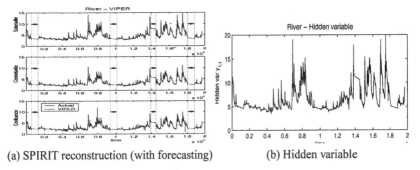

| (a) SPIRIT reconstruction (with forecasting) | (b) Hidden variable |

Figure 12.7. Actual River data (river gauges, in feet) and SPIRIT output, for each of the streams (no pun intended). The large portions with missing values across all streams are marked with dotted lines (there are also other missing values in some of the streams); about 26% of all values are missing, but this does not affect SPIRIT's tracking abilities.

River gauges

Description The dataset was collected from the USACE current river conditions website[3]. It consists of river stage (or, water level) data from three different measuring stations in the same river system (see Figure 12.7).

Data characteristics The data exhibit one common trend and has plenty of missing values (26% of all values, for all three streams).

Results and interpretation Examining the three hidden variable weights found by SPIRIT, these have ratios 1.5 : 1.1 : 1. Indeed, if we look at all 20,000 time ticks, this is what we see; all streams are very similar (since they are from the same river), with the "amplitude" of the fluctuations having roughly these proportions. Hence, one hidden variable is sufficient, the three weights compactly describe the key information and the interpretation is intuitive.

Besides recovering missing values from underlying correlations captured by the few hidden variables, SPIRIT's tracking abilities are not affected even in extreme cases.

8. Performance and accuracy

In this section we discuss performance issues. First, we show that SPIRIT requires very limited space and time. Next, we elaborate on the accuracy of SPIRIT's incremental estimates.

---

[3]http://wmw.lrp.usace.army.mil/current/

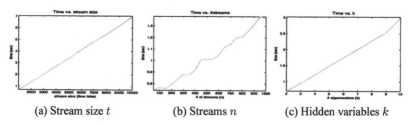

(a) Stream size t (b) Streams n (c) Hidden variables k

Figure 12.8. Wall-clock times (including time to update forecasting models). Times for AR and MUSCLES are not shown, since they are off the charts from the start (13.2 seconds in (a) and 209 in (b)). The starting values are: (a) 1000 time ticks, (b) 50 streams, and (c) 2 hidden variables (the other two held constant for each graph). It is clear that SPIRIT scales linearly.

Time and space requirements

Figure 12.8 shows that SPIRIT scales linearly with respect to number of streams n and number of hidden variables k. AR per stream and MUSCLES are essentially off the charts from the very beginning. Furthermore, SPIRIT scales linearly with stream size (i.e., requires constant processing time per tuple).

The plots were generated using a synthetic dataset that allows us to First, we choose the number k of trends and generate sine waves with different frequencies, say $y_{t,i} = \sin(2\pi i/kt), 1 \le i \le k$. Thus, all trends are pairwise linearly independent. Next, wenerate each of the n streams as random linear combinations of these k trend signals. This scheme allows us to vary k, n and the length of the streams at will. For each experiment shown, one of these parameters is varied and the other two are held fixed. The numbers in Figure 12.8 are wall-clock times of our Matlab implementation. Both AR-per-stream as well as MUSCLES (also in Matlab) are several orders of magnitude slower and thus omitted.

We have also implemented the SPIRIT algorithms in a real system (21), which can obtain measurements from sensor devices and display hidden variables and trends in real-time.

Accuracy

In terms of accuracy, everything boils down to the quality of the summary provided by the hidden variables. To this end, we show the reconstruction $\tilde{x}_t$ of x_t, from the hidden variables y_t in Figure 12.5(b). One line uses the true principal directions, the other the SPIRIT estimates (i.e., weight vectors). SPIRIT comes very close to repeated PCA.

We should note that this is an unfair comparison for SPIRIT, since repeated PCA requires (i) storing *all* stream values, and (ii) performing a very expensive SVD computation for *each* time tick. However, the tracking is still very good.

Table 12.3.   Reconstruction accuracy (mean squared error rate).

Dataset	Chlorine	Critter	Motes
MSE rate (SPIRIT)	0.0359	0.0827	0.0669
MSE rate (repeated PCA)	0.0401	0.0822	0.0448

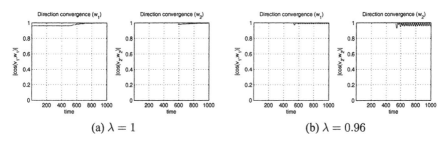

(a) $\lambda = 1$ (b) $\lambda = 0.96$

Figure 12.9.   Hidden variable tracking accuracy.

This is always the case, provided the corresponding eigenvalue is large enough and fairly well-separated from the others. If the eigenvalue is small, then the corresponding hidden variable is of no importance and we do not track it anyway.

Reconstruction error   Figure 12.3 shows the reconstruction error, $\sum \|\tilde{\mathbf{x}}_t - \mathbf{x}_t\|^2 / \sum \|\mathbf{x}_t\|^2$, achieved by SPIRIT. In every experiment, we set the energy thresholds to $[f_E, F_E] = [0.95, 0.98]$. Also, as pointed out before, we set $\lambda = 0.96$ as a reasonable default value to deal with non-stationarities that may be present in the data, according to recommendations in the literature (11). Since we want a metric of overall quality, the MSE rate weighs each observation equally and does not take into account the forgetting factor λ.

Still, the MSE rate is very close to the bounds we set. In Figure 12.3 we also show the MSE rate achieved by repeated PCA. As pointed out before, this is already an unfair comparison. In this case, we set the number of principal components k to the maximum that SPIRIT uses at any point in time. This choice favours repeated PCA even further. Despite this, the reconstruction errors of SPIRIT are close to the ideal, while using orders of magnitude less time and space.

Finally, Figure 12.9 illustrates the convergence to the "true" principal component directions on a synthetic dataset. First, we compare against the PCA of the entire data $\mathbf{X}$, with $\lambda = 1$. We see convergence is almost immediate. This is always the case, provided the corresponding eigenvalue is large enough and fairly well-separated from the others. However, if the eigenvalue is small, then the corresponding hidden variable is of no importance and we do not track it anyway. When $\lambda < 1$, the problem is harder, because the $\mathbf{w}_i$ gradually shift over time. For $\lambda < 1$, we compare against repeated PCA (using the first t rows of $\mathbf{X}$, appropriately weighted). We see that we can still track the shifting principal component directions well.

9.   Conclusion

We focus on finding patterns, correlations and hidden variables, in a large number of streams. SPIRIT has the following desirable characteristics:

(i) It discovers underlying correlations among multiple streams, incrementally and in real-time (21) and provides a very compact representation of the stream collection, via a few *hidden variables*.

(ii) It automatically estimates the number k of hidden variables to track, and it can automatically adapt, if k changes (e.g., an air-conditioner switching on, in a temperature sensor scenario).

(iii) It scales up extremely well, both on database size (i.e., number of time ticks t), and on the number n of streams. Therefore it is suitable for a large number of sensors / data sources.

(iv) Its computation demands are low: it only needs $O(nk)$ floating point operations—no matrix inversions nor SVD (both infeasible in online, any-time settings). Its space demands are similarly limited.

(v) It can naturally hook up with any forecasting method, and thus easily do prediction, as well as handle missing values.

We showed that the output of SPIRIT has a natural interpretation. We evaluated our method on several datasets, where indeed it discovered the hidden variables. Moreover, SPIRIT-based forecasting was several times faster than other methods.

Acknowledgments

We wish to thank Michael Bigrigg for the temperature sensor data and Orna Raz for collecting the river gauge data.

References

[1] Aggarwal, Charu C. Han, Jiawei, and Yu, Philip S. (2003). A framework for clustering evolving data streams. In *VLDB*.

[2] Ali, M.Ĥ. Mokbel, Mohamed F. Aref, Walid, and Kamel, Ibrahim (2005). Detection and tracking of discrete phenomena in sensor network databases. In *SSDBM*.

[3] Brockwell, Peter J. and Davis, Richard A. (1991). *Time Series: Theory and Methods*. Springer Series in Statistics. Springer-Verlag, 2nd edition.

[4] Deligiannakis, Antonios, Kotidis, Yiannis, and Roussopoulos, Nick (2004). Compressing historical information in sensor networks. In *SIGMOD*.

[5] Diamantaras, Kostas I. and Kung, Sun-Yuan (1996). *Principal Component Neural Networks: Theory and Applications*. John Wiley.

[6] Domingos, Pedro and Hulten, Geoff (2000). Mining high-speed data streams. In *KDD*.

[7] Fukunaga, Keinosuke (1990). *Introduction to Statistical Pattern Recognition*. Academic Press.

[8] Ganti, Venkatesh, Gehrke, Johannes, and Ramakrishnan, Raghu (2002). Mining data streams under block evolution. *SIGKDD Explorations*, 3(2):1–10.

[9] Guha, Sudipto, Gunopulos, Dimitrios, and Koudas, Nick (2003a). Correlating synchronous and asynchronous data streams. In *KDD*.

[10] Guha, Sudipto, Meyerson, Adam, Mishra, Nina, Motwani, Rajeev, and O'Callaghan, Liadan (2003b). Clustering data streams: Theory and practice. *IEEE TKDE*, 15(3):515–528.

[11] Haykin, Simon (1992). *Adaptive Filter Theory*. Prentice Hall.

[12] Hulten, Geoff, Spencer, Laurie, and Domingos, Pedro (2001). Mining time-changing data streams. In *KDD*.

[13] Jolliffe, I.T̃. (2002). *Principal Component Analysis*. Springer.

[14] Kailath, Thomas (1980). *Linear Systems*. Prentice Hall.

[15] Keogh, Eamonn, Lonardi, Stefano, and Ratanamahatana, Chotirat Ann (2004). Towards parameter-free data mining. In *KDD*.

[16] Lin, Jessica, Vlachos, Michail, Keogh, Eamonn, and Gunopulos, Dimitrios (2004). Iterative incremental clustering of time series. In *EDBT*.

[17] Oja, Erkki (1989). Neural networks, principal components, and subspaces. *Intl. J. Neural Syst.*, 1:61–68.

[18] Palpanas, Themistoklis, Vlachos, Michail, Keogh, Eamonn, Gunopulos, Dimitrios, and Truppel, Wagner (2004). Online amnesic approximation of streaming time series. In *ICDE*.

[19] Papadimitriou, Spiros, Brockwell, Anthony, and Faloutsos, Christos (2003). Adaptive, hands-off stream mining. In *VLDB*.

[20] Sakurai, Yasushi, Papadimitriou, Spiros, and Faloutsos, Christos (2005). BRAID: Stream mining through group lag correlations. In *SIGMOD*.

[21] Sun, Jimeng, Papadimitriou, Spiros, and Faloutsos, Christos (2005). Online latent variable detection in sensor networks. In *ICDE*. (demo).

[22] Wang, Haixun, Fan, Wei, Yu, Philip S., and Han, Jiawei (2003). Mining concept-drifting data streams using ensemble classifiers. In *KDD*.

[23] Yang, Bin (1995). Projection approximation subspace tracking. *IEEE Trans. Sig. Proc.*, 43(1):95–107.

[24] Yi, Byoung-Kee, Sidiropoulos, N.D. Johnson, Theodore, Jagadish, H.V. Faloutsos, Christos, and Biliris, Alexandros (2000). Online data mining for co-evolving time sequences. In *ICDE*.

[25] Young, Peter (1984). *Recursive Estimation and Time-Series Analysis: An Introduction*. Springer-Verlag.

[26] Zhang, Tian, Ramakrishnan, Raghu, and Livny, Miron (1996). BIRCH: An efficient data clustering method for very large databases. In *SIGMOD*.

[27] Zhu, Yunyue and Shasha, Dennis (2002). StatStream: Statistical monitoring of thousands of data streams in real time. In *VLDB*.

Chapter 13

A SURVEY OF DISTRIBUTED MINING OF DATA STREAMS

Srinivasan Parthasarathy
Department of Computer Science and Engineering
The Ohio State University
srini@cse.ohio-state.edu

Amol Ghoting
Department of Computer Science and Engineering
The Ohio State University
ghoting@cse.ohio-state.edu

Matthew Eric Otey
Department of Computer Science and Engineering
The Ohio State University
otey@cse.ohio-state.edu

Abstract   With advances in data collection and generation technologies, organizations and researchers are faced with the ever growing problem of how to manage and analyze large dynamic datasets. Environments that produce streaming sources of data are becoming common place. Examples include stock market, sensor, web click stream, and network data. In many instances, these environments are also equipped with multiple distributed computing nodes that are often located near the data sources. Analyzing and monitoring data in such environments requires data mining technology that is cognizant of the mining task, the distributed nature of the data, and the data influx rate. In this chapter, we survey the current state of the field and identify potential directions of future research.

1.   Introduction

Advances in technology have enabled us to collect vast amounts of data from various sources, whether they be from experimental observations, simulations, sensors, credit card transactions, or from networked systems. To benefit from these enhanced data collecting capabilities, it is clear that semi-automated interactive techniques such as data mining should be employed to process and analyze the data. It is also desirable to have interactive response times to client queries, as the process is often iterative in nature (with a human in the loop). The challenges to meet these criteria are often daunting as detailed next.

Although inexpensive storage space makes it possible to maintain vast volumes of data, accessing and managing the data becomes a performance issue. Often one finds that a single node is incapable of housing such large datasets. Efficient and adaptive techniques for data access, data storage and communication (if the data sources are distributed) are thus necessary. Moreover, data mining becomes more complicated in the context of dynamic databases, where there is a constant influx of data. Changes in the data can invalidate existing patterns or introduce new ones. Re-executing the algorithms from scratch leads to large computational and I/O overheads. These two factors have led to the development of distributed algorithms for analyzing streaming data which is the focus of this survey article.

Many systems use a centralized model for mining multiple data streams [2]. Under this model the distributed data streams are directed to one central location before they are mined. A schematic diagram of a centralized data stream mining system is presented in Figure 13.1. Such a model of computation is limited in several respects. First, centralized mining of data streams can result in long response time. While distributed computing resources may be available, they are not fully utilized. Second, central collection of data can result in heavy traffic over critical communication links. If these communication links have limited network bandwidth, network I/O may become a performance bottleneck. Furthermore, in power constrained domains such as sensor networks, this can result in excessive power consumption due to excessive data communication.

To alleviate the aforementioned problems, several researchers have proposed a model that is aware of the distributed sources of data, computational resources, and communication links. A schematic diagram of such a distributed stream mining system is presented in Figure 13.1 and can be contrasted with the centralized model. In the model of distributed stream mining, instead of offloading the data to one central location, the distributed computing nodes perform parts of the computation close to the data, while communicating the local models to a central site as and when needed. Such an architecture provides several benefits. First, by using distributed computing nodes, it allows the derivation of a *greater degree of parallelism*, thus reducing response time. Second, as only local models need to be communicated, communication can potentially be reduced, *improving scalability*, and reducing power consumption in power constrained domains.

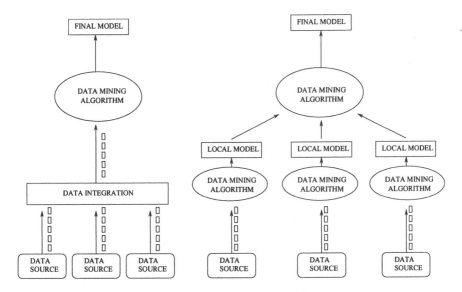

Figure 13.1. Centralized Stream Processing Architecture (left) Distributed Stream Processing Architecture (right)

This chapter presents a brief overview of distributed stream mining algorithms, systems support, and applications, together with emerging research directions. We attempt to characterize and classify these approaches as to whether they belong in the centralized model or the distributed model. The rest of this chapter is organized as follows. First, we present distributed stream mining algorithms for various mining tasks such as outlier detection, clustering, frequent itemset mining, classification, and summarization. Second, we present an overview of distributed stream mining in resource constrained domains. Third, we summarize research efforts on building systems support for facilitating distributed stream mining. Finally, we conclude with emerging research directions in distributed stream mining.

2. Outlier and Anomaly Detection

The goal in outlier or anomaly detection is to find data points that are most different from the remaining points in the data set [4]. Most outlier detection algorithms are schemes in which the distance between every pair of points is calculated, and the points most distant from all other points are marked as outliers [29]. This is an $O(n^2)$ algorithm that assumes a static data set. Such approaches are difficult to extend to distributed streaming data sets. Points in these data sets arrive at multiple distributed end-points, which may or may not be compute nodes, and must be processed incrementally. Such constraints lead us away from purely distance-based approaches, and towards more heuristic techniques. Note that the central issue in many anomaly detection systems, is to identify anomalies in *real-time* or as close to real time as possible thus making it a natural candidate for many streaming applications. Moreover, often times the data is produced at disparate sites making distributed stream mining a natural fit for this domain. In this section we review the work in outlier or anomaly detection most germane to distributed stream mining.

Various application-specific approaches to outlier/anomaly detection have been proposed in the literature. An approach [39] has been presented for distributed deviation detection in sensor networks. This approach is tailored to the sensor network domain and targets misbehaving sensors. The approach maintains density estimates of values seen by a sensor, and flags a sensor to be a misbehaving sensor if its value deviates significantly from the previously observed values. This computation is handled close to the sensors in a distributed fashion, with only results being reported to the central server as and when needed.

One of the most popular applications of distributed outlier detection is that of network intrusion detection. Recent trends have demanded a distributed approach to intrusion detection on the Internet. The first of these trends is a move towards distributed intrusions and attacks, that is to say, intrusions and attacks originating from a diverse set of hosts on the internet. Another trend is the increasing heterogeneous nature of the Internet, where different hosts, perhaps residing in the same subnetwork have differing security requirements. For example, there have been proposals for distributed firewalls [20] for fulfilling diverse security requirements. Also, the appearance of mobile and wireless computing has created dynamic network topologies that are difficult, if not impossible, to protect from a centralized location. Efficient detection and prevention of these attacks requires distributed nodes to collaborate. By itself, a node can only collect information about the state of the network immediately surrounding it, which may be insufficient to detect distributed attacks. If the nodes collaborate by sharing network audit data, host watch lists, and models of known network attacks, each can construct a better global model of the network.

Otey *et al* [36], present a distributed outlier detection algorithm targeted at distributed online streams, specifically to process network data collected at distributed sites. Their approach finds outliers based on the number of attribute dependencies violated by a data point in continuous, categorical, and mixed attribute spaces. They maintain an in-memory structure that succinctly summarizes the required dependency information. In order to find exact outliers in a distributed streaming setting, the in-memory summaries would need to be exchanged frequently. These summaries can be large, and consequently, in a distributed setting, each distributed computing node only exchanges local outliers with the other computing nodes. A point is deemed to be a global outlier if every distributed node believes it to be an outlier based on its local model of normalcy. While such an approach will only find approximate outliers, the authors show that this heuristic works well in practice. While the authors report that to find exact outliers they need to exchange a large summary which leads to excessive communication, it could be possible to exchange only decisive parts of the summary, instead of the entire summary, in order to more accurately detect the true outliers. Furthermore, their in-memory summaries are large, as they summarize a large amount of dependency information. Reducing this memory requirement could potentially allow the use of this algorithm in resource-constrained domains.

EMERALD is an approach for collaborative intrusion detection for large networks within an enterprise [42]. This approach allows for distributed protection of the network through a hierarchy of surveillance systems that analyze network data at the service, domain, and enterprise-wide levels. However, EMERALD does not provide mechanisms for allowing different organizations to collaborate. Locasto *et al* [33] examine techniques that allow different organizations to do such collaboration for enhanced network intrusion detection. If organizations can collaborate, then each can build a better model of global network activity, and more precise models of attacks (since they have more data from which to estimate the model parameters). This allows for better characterization and prediction of attacks. Collaboration is achieved through the exchange of Bloom filters, each of which encodes a list of IP addresses of suspicious hosts that a particular organization's Intrusion Detection System (IDS) has detected, as well as the ports which these suspicious hosts have accessed. The use of Bloom filters helps both to keep each collaborating organization's information confidential and to reduce the amount of data that must be exchanged.

A major limitation of this approach is that information exchanged may not be sufficient to identify distributed attacks. For example, it is possible that an attack may originate from a number of hosts, none of which are suspicious enough to be included on any organization's watch list. However, the combined audit data collected by each organization's IDS may be sufficient to detect that attack. To implement such a system, two problems must be addressed. The first is that each organization may collect disjoint sets of features. Collaborating organizations must agree beforehand on a set of common features to use. Some ideas for common standards for intrusion detection have been realized with the Common Intrusion Detection Framework (CIDF) [31]. The second problem is that of the privacy of each organization's data. It may not be practical to use Bloom filters to encode a large set of features. However, techniques do exist for privacy-preserving data mining [28, 23, 32] that will allow organizations to collaborate without compromising the privacy of their data.

There have been other approaches for detecting distributed denial-of-service attacks. Lee *et al* have proposed a technique for detecting novel and distributed intrusions based on the aforementioned CIDF [31]. The approach not only allows nodes to share information with which they can detect distributed attacks, but also allows them to distribute models of novel attacks. Yu *et al* propose a middleware-based approach to prevent distributed denial of service attacks [45]. Their approach makes use of Virtual Private Operation Environments (VPOE) to allow devices running the middleware to collaborate. These devices can act as firewalls or network monitors, and their roles can change as is necessary. Each device contains several modules, including an attack detection module, a signaling module for cooperating with other devices, and policy processing modules.

Some work in network intrusion detection has been done in the domain of mobile ad hoc networks (MANETs) [47, 18], where nodes communicate over a wireless medium. In MANETs, the topology is dynamic, and nodes must cooperate in order to route messages to their proper destinations. Because of the open communication medium, dynamic topology, and cooperative nature, MANETs are especially prone to network intrusions, and present difficulties for distributed intrusion detection.

To protect against intrusions, Zhang *at al* have proposed several intrusion detection techniques [46, 47]. In their proposed architecture, each node in the network participates in detection and response, and each is equipped with a local detection engine and a cooperative detection engine. The local detection engine is responsible for detecting intrusions from the local audit data. If a node has strong evidence that an intrusion is taking place, it can initiate a response to the intrusion. However, if the evidence is not sufficiently strong, it can initiate a global intrusion detection procedure through the cooperative detection engine. The nodes only cooperate by sharing their detection states, not their audit data, and so it is difficult for each node to build an accurate global model of the network with which to detect intrusions. In this case, intrusions detectable only at the global level (e.g. ip sweeps) will be missed. However, the authors do point out that they only use local data since the remote nodes may be compromised and their data may not be trustworthy.

In another paper [18], Huang and Lee present an alternative approach to intrusion detection in MANETs. In this work, the intrusions to be detected are attacks against the structure of the network itself. Such intrusions are those that corrupt routing tables and protocols, intercept packets, or launch network-level denial-of-service attacks. Since MANETs typically operate on battery power, it may not be cost effective for each node to constantly run its own intrusion detection system, especially when there is a low threat level. The authors propose that a more effective approach would be for a cluster of nodes in a MANET to elect one node as a monitor (the *clusterhead*) for the entire cluster. Using the assumption that each node can overhear network traffic in its transmission range, and that the other cluster members can provide (some of) the features (since the transmission ranges of the clusterhead and the other cluster members may not overlap, the other cluster members may have statistics on portions of the cluster not accessible to the clusterhead), the clusterhead is responsible for analyzing the flow of packets in its cluster in order to detect intrusions and initiate a response. In order for this intrusion detection approach to be effective, the election of the clusterhead must be fair, and each clusterhead must serve an equal amount of time. The first requirement ensures that the election of the clusterhead is unbiased (i.e. a compromised node cannot tilt the election in its favor), and the second requirement ensures that a compromised node cannot force out the current clusterhead nor remain as clusterhead for an unlimited period of time. There is a good division of labor, as the clusterhead is the only member of the cluster that must run the intrusion detection system; the other nodes need only collect data and send it to the clusterhead. However, a limitation of this approach is that not all intrusions are visible at the global level, especially given the feature set the detection system uses (statistics on the network topology, routes, and traffic). Such local intrusions include exploits of services running on a node, which may only be discernible using the content of the traffic.

3. Clustering

The goal in clustering is to partition a set of points into groups such that points within a group are similar in some sense and points in different groups are dissimilar in the same sense. In the context of distributed streams, one would want to process the data streams in a distributed fashion, while communicating the summaries, and to arrive at global clustering of the data points. Guha et al [17], present an approach for clustering data streams. Their approach produces a clustering of the points seen using small amounts of memory and time. The summarized data consists of the cluster centers together with the number of points assigned to that cluster. The k-median algorithm is used as the underlying clustering mechanism. The resulting clustering is a constant factor approximation of the true clustering. As has been shown in [16], this algorithm can be easily extended to operate in a distributed setting. Essentially, clusterings from each distributed site can be combined and clustered to find the global clustering with the same approximation factor. From a qualitative stand point, in many situations, k-median clusters are known to be less desirable than those formed by other clustering techniques. It would be interesting to see if other clustering algorithms that produce more desirable clusterings can be extended with the above methodology to operate over distributed streams.

Januzaj et al [21], present a distributed version of the density-based clustering algorithm, DBSCAN. Essentially, each site builds a local density-based clustering, and then communicates a summary of the clustering to a central site. The central site performs a density-based clustering on the summaries obtained from all sites to find a global clustering. This clustering is relayed back to the distributed sites that update their local clusterings based on the discovered global clustering. While this approach is not capable of processing dynamic data, in [13], the authors have shown that density based clustering can be performed incrementally. Therefore, a distributed and incremental version of DBSCAN can potentially be devised. However, like the distributed version presented by Januzaj *et al*, we cannot provide a guarantee on the quality of the result.

Beringer and Hullermeir consider the problem of clustering parallel data streams [5]. Their goal is to find correlated streams as they arrive synchronously. The authors represent the data streams using exponentially weighted sliding windows. The discrete Fourier transform is computed incrementally, and k-Means clustering is performed in this transformed space at regular intervals of time. Data streams belonging to the same cluster are considered to be correlated. While the processing is centralized, the approach can be tailored to correlate distributed data streams. Furthermore, the approach is suitable for online streams. It is possible that this approach can be extended to a distributed computing environment. The Fourier coefficients can be exchanged incrementally and aggregated locally to summarize remote information. Furthermore, one can potentially produce approximate results by only exchanging the significant coefficients.

4.   Frequent itemset mining

The goal in frequent itemset mining is to find groups of items or values that co-occur frequently in a transactional data set. For instance, in the context of market data analysis, a frequent two itemset could be $\{beer, chips\}$, which means that people frequently buy beer and chips together. The goal in frequent itemset mining is to find all itemsets in a data set that occur at least x number of times, where x is the minimum support parameter provided by the user.

Frequent itemset mining is both CPU and I/O intensive, making it very costly to completely re-mine a dynamic data set any time one or more transactions are added or deleted. To address the problem of mining frequent itemsets from dynamic data sets, several researchers have proposed incremental techniques [10, 11, 14, 30, 43, 44]. Incremental algorithms essentially re-use previously mined information and try to combine this information with the fresh data to efficiently compute the new set of frequent itemsets. However, it can be the case that the database may be distributed over multiple sites, and is being updated at different rates at each site, which requires the use of distributed asynchronous frequent itemset mining techniques.

Otey *et al* [38], present a distributed incremental algorithm for frequent itemset mining. The approach is capable of incrementally finding maximal frequent itemsets in dynamic data. Maximal frequent itemsets are those that do not have any frequent supersets, and the set of maximal frequent itemsets determines the complete set of frequent itemsets. Furthermore, it is capable of mining frequent itemsets in a distributed setting. Distributed sites can exchange their local maximal frequent itemsets to obtain a superset of the global maximal frequent itemsets. This superset is then exchanged between all nodes so that their local counts may be obtained. In the final round of communication, a reduction operation is performed to find the exact set of global maximal frequent itemsets.

Manku and Motwani [35], present an algorithm for mining frequent itemsets over data streams. In order to mine all frequent itemsets in constant space, they employ a down counting approach. Essentially, they update the support counts for the discovered itemsets as the data set is processed. Furthermore, for all the discovered itemsets, they decrement the support count by a specific value. As a result, itemsets that occur rarely will have their count set to zero and will be eventually eliminated from list. If they reappear later, their count is approximated. While this approach is tailored to data streams, it is not distributed. The methodology proposed in [38] can potentially be applied to this algorithm to process distributed data streams.

Manjhi *et al* [34], extend Manku and Motwani's approach to find frequent items in the union of multiple distributed streams. The central issue is how to best manage the degree of approximation performed as partial synopses from multiple nodes are combined. They characterize this process for hierarchical communication topologies in terms of a precision gradient followed by synopses as they are passed from leaves to the root and combined incrementally. They studied the problem of finding the optimal precision gradient under two alternative and incompatible optimization objectives: (1) minimizing load on the central node to which answers are delivered, and (2) minimizing worst-case load on any communication link. While this approach targets frequent items only, it would be interesting to see if it can be extended to find frequent itemsets.

5.   Classification

Hulten and Domingos [19], present a one-pass decision tree construction algorithm for streaming data. They build a tree incrementally by observing data as it streams in and splitting a node in the tree when a sufficient number of samples have been seen. Their approach uses the Hoeffding inequality to converge to a sample size. Jin and Agrawal revisit this problem and present solutions that speed up split point calculation as well as reduce the desired sample size to achieve the same level of accuracy [22]. Both these approaches are not capable of processing distributed streams.

Kargupta and Park present an approach for aggregating decision trees constructed at distributed sites [26]. As each decision tree can be represented as a numeric function, the authors propose to transmit and aggregate these trees by using their Fourier representations. They also show that the Fourier-based representation is suitable for approximating a decision tree, and thus, suitable for transmission in bandwidth-limited mobile environments. Coupled with a streaming decision tree construction algorithm, this approach should be capable of processing distributed data streams.

Chen *et al* [8], present a collective approach to mine Bayesian networks from distributed heterogeneous web-log data streams. In their approach, they learn a local Bayesian network at each site using the local data. Then each site identifies the observations that are most likely to be evidence of coupling between local and non-local variables and transmits a subset of these observations to a central site. Another Bayesian network is learned at the central site using the data transmitted from the local sites. The local and central Bayesian networks are combined to obtain a collective Bayesian network, that models the entire data. This technique is then suitably adapted to an online Bayesian learning technique, where the network parameters are updated sequentially based on new data from multiple streams. This approach is particularly suitable for mining applications with distributed sources of data streams in an environment with non-zero communication cost (e.g. wireless networks).

6.   Summarization

Bulut and Singh [6], propose a novel technique to summarize a data stream incrementally. The summaries over the stream are computed at multiple resolutions, and together they induce a unique Wavelet-based approximation tree. The resolution of approximations increases as we move from the root of the approximation tree down to its leaf nodes. The tree has space complexity $O(logN)$, where N denotes the current size of the stream. The amortized processing cost for each new data value is $O(1)$. These bounds are currently the best known for the algorithms that work under a biased query model where the most recent values are of a greater interest. They also consider the scenario in which a central source site summarizes a data stream at multiple resolutions. The clients are distributed across the network and pose queries. The summaries computed at the central site are cached adaptively at the clients. The access pattern, i.e. reads and writes, over the stream results in multiple replication schemes at different resolutions. Each replication scheme expands as the corresponding read rate increases, and contracts as the corresponding write rate increases. This adaptive scheme minimizes the total communication cost and the number of inter-site messages. While the summarization process is centralized, it can potentially be used to summarize distributed streams at distributed sites by aggregating wavelet coefficients.

The problem of pattern discovery in a large number of co-evolving streams has attracted much attention in many domains. Papadimitriou *et al* introduce SPIRIT (Streaming Pattern dIscoveRy in multIple Time-series) [40], a comprehensive approach to discover correlations that effectively and efficiently summarize large collections of streams. The approach uses very less memory and both its memory requirements and processing time are independent of the stream length. It scales linearly with the number of streams and is adaptive and fully automatic. It dynamically detects changes (both gradual and sudden) in the input streams, and automatically determines the number of hidden variables. The correlations and hidden variables discovered have multiple uses. They provide a succinct summary to the user, they can help to do fast forecasting and detect outliers, and they facilitate interpolations and handling of missing values. While the algorithm is centralized, it targets multiple distributed streams. The approach can potentially be used to summarize streams arriving at distributed sites.

Babcock and Olston [3], study a useful class of queries that continuously report the k largest values obtained from distributed data streams ("top-k monitoring queries"), which are of particular interest because they can be used to reduce the overhead incurred while running other types of monitoring queries. They show that transmitting entire data streams is unnecessary to support these queries. They present an alternative approach that significantly reduces communication. In their approach, arithmetic constraints are maintained at remote stream sources to ensure that the most recently provided top-k answer remains valid to within a user-specified error tolerance. Distributed communication is only necessary on the occasion when constraints are violated.

7. Mining Distributed Data Streams in Resource Constrained Environments

Recently, there has been a lot of interest in environments that demand distributed stream mining where resources are constrained. For instance, in the sensor network domain, due to energy consumption constraints, excessive communication is undesirable. One can potentially perform more computation and less communication to perform the same task with reduced energy consumption. Consequently, in such scenarios, data mining algorithms (specifically clustering and classification) with tunable computation and communication requirements are needed [24, 39].

A similar set of problems have recently been looked at in the network intrusion detection community. Here, researchers have proposed to offload computation related to monitoring and intrusion detection on to the network interface card (NIC) [37] with the idea of enhancing reliability and reducing the constraints imposed on the host processing environment. Initial results in this domain convey the promise of this area but there are several limiting criteria in current generation NICs (e.g. programming model, lack of floating point operations) that may be alleviated in next generation NICs.

Kargupta *et al* present Mobimine [27], a system for intelligent analysis of time-critical data using a Personal Data Assistant (PDA). The system monitors stock market data and signals interesting stock behavior to the user. Stocks are interesting if they may positively or negatively affect the stock portfolio of the user. Furthermore, to assist in the user's analysis, they transmit classification trees to the user's PDA using the Fourier spectrum-based approach presented earlier. As discussed previously, this Fourier spectrum-based representation is well suited to environments that have limited communication bandwidth.

The Vehicle Data Stream Mining System (VEDAS) [25], is a mobile and distributed data stream mining/monitoring application that taps into the continuous stream of data generated by most modern vehicles. It allows continuous on-board monitoring of the data streams generated by the moving vehicles, identifying the emerging patterns, and reporting them to a remote control center over a low-bandwidth wireless network connection. The system offers many possibilities such as real-time on-board health monitoring, drunk-driving detection, driver characterization, and security related applications for commercial fleet management. While there has been initial work in such constrained environments, we believe that there is still a lot to be done in this area.

8.   Systems Support

A distributed stream mining system can be complex. It typically consists of several sub-components such as the mining algorithms, the communication sub-system, the resource manager, the scheduler, etc. A successful stream mining system must adapt to the dynamics of the data and best use the available set of resources and components. In this section, we will briefly summarize efforts that target the building of system support for resource-aware distributed processing of streams.

When processing continuous data streams, data arrival can be bursty, and the data rate may fluctuate over time. Systems that seek to give rapid or real-time query responses in such an environment must be prepared to deal gracefully with bursts in data arrival without compromising system performance. Babcock *et al* [1] show that the choice of an operator scheduling strategy can have significant impact on the run-time system memory usage. When data streams are bursty, the choice of an operator scheduling strategy can result in significantly high run-time memory usage and poor performance. To minimize memory utilization at peak load, they present Chain scheduling, an adaptive, load-aware scheduling of query operators to minimize resource consumption during times of peak load. This operator scheduling strategy for data stream systems is near-optimal in minimizing run-time memory usage for single-stream queries involving selections, projections, and foreign-key joins with stored relations. At peak load, the scheduling strategy selects an operator path (a set of consecutive operators) that is capable of processing and freeing the maximum amount of memory per unit time. This in effect results in the scheduling of operators that together are both selective and have a high aggregate tuple processing rate.

The aforementioned scheduling strategy is not targeted at the processing of distributed streams. Furthermore, using the Chain operator scheduling strategy has an adverse affect on response time and is not suitable for data mining applications that need to provide interactive performance even under peak load. In order to mine data streams, we need a scheduling strategy that supports both response time and memory-aware scheduling of operators. Furthermore, when scheduling a data stream mining application with dependent operators in a distributed setting, the scheduling scheme should not need to communicate a significant amount of state information. Ghoting and Parthasarathy [16], propose an adaptive operator scheduling technique for mining distributed data streams with response time guarantees and bounded memory utilization. The user can tune the application to the desired level of interactivity, thus facilitating the data mining process. They achieve this through a step-wise degradation in response time beginning from a schedule that is optimal in terms of response time. This sacrifice in response time is used towards optimal memory utilization. After an initial scheduling decision is made, changes in system state may force a reconsideration of operator schedules. The authors show that a decision as to whether a local state change will affect the global operator schedule can be made locally. Consequently, each local site can proceed independently, even under minor state changes, and a global assignment is triggered only when it is actually needed.

Plale considers the problem of efficient temporal-join processing in a distributed setting [41]. In this work, the author's goal is to optimize the join processing of event streams to efficiently determine sets of events that occur together. The size of the join window cannot be determined apriori as this may lead to missed events. The author proposes to vary the size of the join window depending on the rate of the incoming stream. The rate of the incoming stream gives a good indication of how many previous events on the stream can be dropped. Reducing the window size also helps reduce memory utilization. Furthermore, instead of forwarding events into the query processing engine on a first-come first-serve basis, the author proposes to forward the earliest event first to further improve performance, as this facilitates the earlier determination of events that are a part of the join result.

Chen *et al* present GATES [7], a middleware for processing distributed data streams. This middleware targets data stream processing in a grid setting and is built on top of the Open Grid Services Architecture. It provides a high level interface that allows one to specify a stream processing algorithm as a set of pipelined stages. One of the key design goals of GATES is to support self adaption under changing conditions. To support self adaptation, the middleware changes one or more of the sampling rate, the summary structure size, or the algorithm used, based on changing conditions of the data stream. For instance, if the stream rate increases, the system reduces the sampling rate accordingly to maintain a real-time response. If we do not adapt the sampling rate, we could potentially face increasing queue sizes, resulting in poor performance. To support self adaptation, the programmer needs to provide the middleware with parameters that allow it to tune the application at runtime. The middleware builds a simple performance model that allows it to predict how parameter changes help in performance adaptation in a distributed setting.

Chi *et al* [12] present a load shedding scheme for mining multiple data streams, although the computation is not distributed. They assume that the task of reading data from the stream and building feature values is computationally expensive and is the bottleneck. Their strategies decide on how to expend limited computation for building feature values for data on multiple streams. They decide on whether to drop a data item on the stream based on the historic utility of the items produced by the stream. If they choose not to build feature values for a data item, they simply predict feature values based on historical data. They use finite memory Markov chains to make such predictions. While the approach presented by the authors is centralized, load shedding decisions can be trivially distributed.

Conclusions and Future Research Directions

In this chapter, we presented a summary of the current state-of-the-art in distributed data stream mining. Specifically, algorithms for outlier detection, clustering, frequent itemset mining, classification, and summarization were presented. Furthermore, we briefly described related applications and systems support for distributed stream mining.

First, the distributed sources of data that need to be mined are likely to span multiple organizations. Each of these organizations may have heterogeneous computing resources. Furthermore, the distributed data will be accessed by multiple analysts, each potentially desiring the execution of a different mining task. The various distributed stream mining systems that have been proposed to date do not take the variability in the tasks and computing resources into account. To facilitate execution and deployment in such settings, a plug and play system design that is cognizant of each organization's privacy is necessary.

A framework in which services are built on top of each other will facilitate rapid application development for data mining. Furthermore, these systems will need to be integrated with existing data grid and knowledge grid infrastructures [9] and researchers will need to design middleware to support this integration.

Second, next generation computing systems for data mining are likely to be built using off-the-shelf CPUs connected using a high bandwidth interconnect. In order to derive high performance on such systems, stream mining algorithms may need to be redesigned. For instance, next generation processors are likely to have multiple-cores on chip. As has been shown previously [15], data mining algorithms are adversely affected by the memory-wall problem. This problem will likely be exacerbated on future multi-core architectures. Therefore, stream mining algorithms at each local site will need to be redesigned to derive high performance on next generation architectures. Similarly, with innovations in networking technologies, designs that are cognizant of high performance inter-connects (like Infiniband) will need to be investigated.

Third, as noted earlier, in many instances, environments that demand distributed stream mining are resource constrained. This in turn requires the development of data mining technology that is tailored to the specific execution environment. Various tradeoffs, e.g. energy vs. communication, communication vs. redundant computation etc., must be evaluated on a scenario-by-scenario basis. Consequently, in such scenarios, data mining algorithms with tunable computation and communication requirements will need to be devised. While initial forays in this domain have been made, a systematic evaluation of the various design tradeoffs even for a single application domain has not been done. Looking further into the future, it will be interesting to evaluate if based on specific solutions a more abstract set of interfaces can be developed for a host of application domains.

Fourth, new applications for distributed data stream mining are on the horizon. For example, RFID (radio frequency identification) technology is expected to significantly improve the efficiency of business processes by allowing automatic capture and identification. RFID chips are expected to be embedded in a variety of devices, and the captured data will likely be ubiquitous in the near future. New applications for these distributed streaming data sets will arise and application specific data mining technology will need to be designed.

Finally, over the past few years, several stream mining algorithms have been proposed in the literature. While they are capable of operating in a centralized setting, many are not capable of operating in a distributed setting and cannot be trivially extended to do so. In order to obtain exact or approximate (bounded) results in a distributed setting, the amount of state information that needs to be exchanged is usually excessive. To facilitate distributed stream mining algorithm design, instead of starting from a centralized solution, one needs to start with a distributed mind-set right from the beginning. Statistics or summaries that can be efficiently maintained in a distributed and incremental setting should be designed and then specific solutions that use these statistics should be devised. Such a design strategy will facilitate distributed stream mining algorithm design.

References

[1] B. Babcock, S. Babu, M. Datar, and R. Motwani. Chain: Operator scheduling for memory minimization in data stream systems. In *Proceedings of the International Conference on Management of Data (SIGMOD)*, 2003.

[2] B. Babcock, S. Babu, M. Datar, R. Motwani, and J. Widom. Models and issues in data stream systems. In *Proceedings of the Symposium on Principles of Database Systems (PODS)*, 2002.

[3] B. Babcock and C. Olston. Distributed top k monitoring. In *Proceedings of the International Conference on Management of Data (SIGMOD)*, 2003.

[4] V. Barnett and T. Lewis. *Outliers in Statistical Data*. John Wiley and Sons, 1994.

[5] J. Beringer and E. Hullermeier. Online clustering of parallel data streams. *Data and Knowledge Engineering*, 2005.

[6] A. Bulut and A. Singh. SWAT: Hierarchical stream summarization in large networks. In *Proceedings of the International Conference on Data Engineering (ICDE)*, 2003.

[7] L. Chen, K. Reddy, and G. Agrawal. GATES: A grid-based middleware for processing distributed data streams. In *Proceedings of the International Symposium on High Performance Distributed Computing (HPDC)*, 2004.

[8] R. Chen, D. Sivakumar, and H. Kargupta. An approach to online bayesian network learning from multiple data streams. In *Proceedings of the International Conference on Principles of Data Mining and Knowledge Discovery*, 2001.

[9] A. Chervenak, I. Foster, C. Kesselman, C. Salisbury, and S. Tuecke. The data grid: Towards an architecture for the distributed management and analysis of large scientific data sets, 2001.

[10] D. Cheung, J. Han, V. Ng, and C. Y. Wong. Maintenance of discovered association rules in large databases: An incremental updating technique. In *Proceedings of the International Conference on Data Engineering (ICDE)*, 1996.

[11] D. Cheung, S. Lee, and B. Kao. A general incremental technique for maintaining discovered association rules. In *Proceedings of the International Conference on Database Systems for Advanced Applications*, 1997.

[12] Y. Chi, P. Yu, H. Wang, and R. Muntz. Loadstar: A load shedding scheme for classifying data streams. In *Proceedings of the SIAM International Conference on Data Mining (SDM)*, 2005.

[13] M. Ester, H. Kriegel, J. Sander, M. Wimmer, and X. Xu. Incremental clustering for mining in a data warehousing environment. In *Proceedings of the International Conference on Very Large Data Bases (VLDB)*, 1998.

[14] V. Ganti, J. Gehrke, and Raghu Ramakrishnan. Demon–data evolution and monitoring. In *Proceedings of the International Conference on Data Engineering (ICDE)*, 2000.

[15] A. Ghoting, G. Buehrer, S. Parthasarathy, D. Kim, A. Nguyen, Y. Chen, and P. Dubey. A characterization of data mining algorithms on a modern processor. In *Proceedings of the ACM SIGMOD Workshop on Data Management on New Hardware*, 2005.

[16] A. Ghoting and S. Parthasarathy. Facilitating interactive distributed data stream processing and mining. In *Proceedings of the International Parallel and Distributed Processing Symposium (IPDPS)*, 2004.

[17] S. Guha, N. Mishra, R. Motwani, and L. O'Callaghan. Clustering data streams. In *Proceedings of the Symposium on Foundations of Computer Science(FOCS)*, 2000.

[18] Yi-An Huang and Wenke Lee. A cooperative intrusion detection system for ad hoc networks. In *Proceedings of the 1st ACM workshop on Security of ad hoc and sensor networks*, 2003.

[19] G. Hulten and P. Domingos. Mining high speed data streams. In *Proceedings of the International Conference on Knowledge Discovery and Data Mining (SIGKDD)*, 2000.

[20] Sotiris Ioannidis, Angelos D. Keromytis, Steven M. Bellovin, and Jonathan M. Smith. Implementing a distributed firewall. In *ACM Conference on Computer and Communications Security*, 2000.

[21] E. Januzaj, H. Kriegel, and M. Pfeifle. DBDC: Density based distributed clustering. In *Proceedings of the International Conference on Extending Data Base Technology (EDBT)*, 2004.

[22] R. Jin and G. Agrawal. Efficient decision tree construction on streaming data. In *Proceedings of the International Conference on Knowledge Discovery and Data Mining (SIGKDD)*, 2003.

[23] M. Kantarcioglu and C. Clifton. Privacy-preserving distributed mining of association rules on horizontally partitioned data. In *Proceedings of the ACM SIGMOD Workshop on Research Issues in Data Mining and Knowledge Discovery*, 2002.

[24] H. Kargupta. Distributed data mining for sensor networks. In *Tutorial presented at ECML/PKDD*, 2004.

[25] H. Kargupta, R. Bhargava, K. Liu, M. Powers, P. Blair, S. Bushra, J. Dull, K. Sarkar, M. Klein, M. Vasa, and D. Handy. Vedas: A mobile and distributed data stream mining system for real-time vehicle monitoring. In *Proceedings of the SIAM International Conference on Data Mining (SDM)*, 2004.

[26] H. Kargupta and B. Park. A fourier spectrum based approach to represent decision trees for mining data streams in mobile environments. *IEEE Transactions on Knowledge and Data Engineering*, 2004.

[27] H. Kargupta, B. Park, S. Pittie, L. Liu, D. Kushraj, and K. Sarkar. MobiMine: Monitoring the stock market from a PDA. *SIGKDD Explorations*, 2002.

[28] Hillol Kargupta, Souptik Datta, Qi Wang, and Krishnamoorthy Sivakumar. On the privacy preserving properties of random data perturbation techniques. In *Proceedings of the International Conference on Data Mining (ICDM)*, 2003.

[29] E. Knorr and R. T. Ng. Algorithms for mining distance-based outliers in large datasets. In *Proceedings of the International Conference on Very Large Data Bases (VLDB)*, 1998.

[30] S. Lee and D. Cheung. Maintenance of discovered association rules: When to update? In *Proceedings of the Workshop on Research Issues in Data Mining and Knowledge Discovery*, 1997.

[31] Wenke Lee, Rahul A. Nimbalkar, Kam K. Yee, Sunil B. Patil, Pragneshkumar H. Desai, Thuan T. Tran, and Salvatore J. Stolfo. A data mining and CIDF based approach for detecting novel and distributed intrusions. *Lecture Notes in Computer Science*, 2000.

[32] Yehuda Lindell and Benny Pinkas. Privacy preserving data mining. *Lecture Notes in Computer Science*, 1880:36, 2000.

[33] M. Locasto, J. Parekh, S. Stolfo, A. Keromytis, T. Malkin, and V. Misra. Collaborative distributed intrusion detection. Technical report, Columbia University, 2004.

[34] A. Manjhi, V. Shkapenyuk, K. Dhamdhere, and C. Olston. Finding (recently) frequent items in distributed data streams. In *Proceedings of the International Conference on Data Engineering (ICDE)*, 2005.

[35] G. Manku and R. Motwani. Approximate frequency counts over data streams. In *Proceedings of the International Conference on Very Large Data Bases (VLDB)*, 2002.

[36] M. Otey, A. Ghoting, and S. Parthasarathy. Fast distributed outlier detection in mixed attribute data sets. *Data Mining and Knowledge Discovery Journal*, 2006.

[37] M. Otey, S. Parthasarathy, A. Ghoting, G. Li, S. Narravula, and D. Panda. Towards nic-based instrusion detection. In *Proceedings of the International Conference on Knowledge Discovery and Data Mining (SIGKDD)*, 2003.

[38] M. Otey, C. Wang, S. Parthasarathy, A. Veloso, and W. Meira. Mining frequent itemsets in distributed and dynamic databases. In *Proceedings of the International Conference on Data Mining (ICDM)*, 2003.

[39] T. Palpanas, D. Papadopoulos, V. Kalogeraki, and D. Gunopulos. Distributed deviation detection in sensor networks. *SIGMOD Record*, 2003.

[40] S. Papadimitriou, J. Sun, and C. Faloutsos. Streaming pattern discovery in multiple time series. In *Proceedings of the International Conference on Very Large Data Bases (VLDB)*, 2005.

[41] B. Plale. Learning run time knowledge about event rates to improve memory utilization in wide area stream filtering. In *Proceedings of the International Symposium on High Performance Distributed Computing (HPDC)*, 2002.

[42] P. Porras and P. Neumann. EMERALD: Event monitoring enabling responses to anomalous live disturbances. In *Proceedings of the National Information Systems Security Conference*, 1997.

[43] S. Thomas, S. Bodagala, K. Alsabti, and S. Ranka. An efficient algorithm for the incremental updation of association rules in large databases. In *Proceedings of the International Conference on Knowledge Discovery and Data Mining (SIGKDD)*, 1997.

[44] A. Veloso, W. Meira Jr., M. B. De Carvalho, B. Possas, S. Parthasarathy, and M. J. Zaki. Mining frequent itemsets in evolving databases. In *Proceedings of the SIAM International Conference on Data Mining*, 2002.

[45] Wei Yu, Dong Xuan, and Wei Zhao. Middleware-based approach for preventing distributed deny of service attacks. In *Proceedings of the IEEE Military Communications Conference*, 2002.

[46] Yongguang Zhang and Wenke Lee. Intrusion detection in wireless ad-hoc networks. In *Mobile Computing and Networking*, 2000.

[47] Yongguang Zhang, Wenke Lee, and Yi-An Huang. Intrusion detection techniques for mobile wireless networks. *Wireless Networking*, 9(5):545–556, 2003.

Chapter 14

ALGORITHMS FOR DISTRIBUTED DATA STREAM MINING

Kanishka Bhaduri

Dept of CSEE
University of Maryland, Baltimore County
kanishk1@cs.umbc.edu

Kamalika Das

Dept of CSEE
University of Maryland, Baltimore County
kdas1@cs.umbc.edu

Krishnamoorthy Sivakumar

School of EECS
Washington State University
siva@eecs.wsu.edu

Hillol Kargupta

Dept of CSEE
University of Maryland, Baltimore County
hillol@cs.umbc.edu

Ran Wolff

Dept of CSEE
University of Maryland, Baltimore County
ranw@cs.umbc.edu

Rong Chen
School of EECS
Washington State University
rchen@eecs.wsu.edu

Abstract

> The field of Distributed Data Mining (DDM) deals with the problem of analyzing data by paying careful attention to the distributed computing, storage, communication, and human-factor related resources. Unlike the traditional centralized systems, DDM offers a fundamentally distributed solution to analyze data without necessarily demanding collection of the data to a single central site. This chapter presents an introduction to distributed data mining for continuous streams. It focuses on the situations where the data observed at different locations change with time. The chapter provides an exposure to the literature and illustrates the behavior of this class of algorithms by exploring two very different types of techniques—one for the peer-to-peer and another for the hierarchical distributed environment. The chapter also briefly discusses several different applications of these algorithms.

1.   Introduction

A data stream is often viewed as a single source time-varying signal observed at a single receiver [13]. A data stream can be viewed as an unbounded sequence $(x_1, x_2, ..., x_n)$ that is indexed on the basis of the time of its arrival at the receiver. Babcock et al. [4] point out some the fundamental properties of a data stream system such as:

- the data elements arrive continuously,

- there is no limit on the total number of points in the data stream,

- and the system has no control over the order in which the data elements arrive

In some applications, the data also arrive at bursts. In other words, the source occasionally generates the data at a very high rate compared to the rate used for rest of the time. Since data arrives continuously, fast one-pass algorithms are imperative for real-time query processing and data mining on streams. Many data stream algorithms have been developed over the last decade for processing and mining data streams that arrive at a single location or at multiple locations whereby they are sent to one location for processing needs. We refer to this scenario as the centralized data stream mining scenario. Examples of such algorithm include query processing [24], change detection [1][14][6], classification [3][15] and clustering [11][2]. These algorithms, however, are not applicable in settings where the data, computing, and other resources are distributed and cannot or should not be centralized for a variety of reasons *e.g.* low bandwidth, security, privacy issues, and load balancing [5][17][23]. In many cases the cost of centralizing the data can be prohibitive and the owners may have privacy constraints of its data. In order to meet the challenges imposed by these constraints, a new area of data mining has emerged in the last ten years known as the Distributed Data Mining (DDM)[16]. Distributed Data Mining (DDM) deals with the problem of analyzing data by paying careful attention to the distributed computing, storage, communication, and human-factor related resources. Unlike the traditional centralized systems, DDM offers a fundamentally distributed solution to analyze data without necessarily demanding collection of the data to a single central site. In this chapter we will primarily consider loosely-coupled distributed environments where each site has a private memory; the sites can operate independently and communicate by message passing over an asynchronous network. These algorithms focus on distributed computation with due attention to minimizing resource usage (e.g. communication cost) and satisfying application-level constraints (e.g. privacy protection). We focus on a subset of problems of DDM—Distributed Data Stream Mining—where not only the data is distributed, but also the data is non-stationary and arriving in the form of multiple streams. These algorithms pose unique challenges themselves - the algorithms need to be efficient in computing the task, work with a local data, compute the data mining model incrementally, and possibly communicate with a subset of its peer-nodes to compute the result.

This chapter offers two things: (1) provide an overview of the existing techniques for addressing some of the stream mining problems in the distributed scenario, and (2) discuss two very different distributed data stream mining algorithms in greater detail in order to illustrate how these algorithms work.

The chapter is organized as follows. In Section 2 we point out the rationale for the importance of the main topic of our discussion - Distributed Data Stream Mining. Section 3 discusses some of the related papers in the area. In Section 4 we present a *local* algorithm for data mining in a dynamic and peer-to-peer environment. Section 5 discusses a distributed Bayesian network learning algorithm. We conclude the chapter in Section 6.

2. Motivation: Why Distributed Data Stream Mining?

This section presents a few examples for illustrating the need for distributed data stream mining algorithms.

Example 1: Consider a distributed *sensor network* scenario where there are a bunch of sensors deployed in a field. Each of these sensors measures different entities *e.g.* temperature, pressure, vibration etc. Each of these sensors has limited battery power and so developing algorithms with low communication overhead is a must in such a setting. If the task is to monitor a global state of the system, one way would be to centralize all the data at each predetermined time instance and build a model. With this approach there are two major problems:

- due to low battery power of the sensors and low bandwidth, this is prohibitive in a typical sensor network setting, and

- the model represented in such a periodic update scheme can be wrong for a large portion of time before a new model is built (for example, this can happen if the data distribution changes immediately after each periodic update)

Distributed algorithms can avoid these problems.

Example 2: Another example comes from the cross domain *network intrusion monitoring* paradigm. In network traffic monitoring or intrusion detection problem, the goal is to identify inappropriate, incorrect, or anomalous activities that can potentially become a threat to the network. Existing methods of network traffic monitoring/intrusion detection often requires centralizing the sample network packets from diverse locations and analyzing them. This is prohibitive in many real-world situations such as in cross-domain intrusion detection schemes where different parties or companies collaborate to find the anomalous patterns. Hence, the system must be equipped with privacy preserving data mining algorithms so that the patterns can be computed and shared across the sites without sharing the privacy-sensitive data. Also, since the data arrives in the form of continuous streams (*e.g.* TCP/IP packets), transferring the data to a single (trusted) location at each time instance and analyzing them is also not feasible. This calls for the development of distributed privacy preserving data mining algorithms capable of being deployed in streaming environments. The Department of Homeland Security (DHS) is currently funding a number of projects in this related area (see "http://www.agnik.com/DHSSBIR.html" for such a project).

In the next section we present a survey of some of the existing distributed data stream mining algorithms.

3.   Existing Distributed Data Stream Mining Algorithms

There exists a plethora of work in the area of distributed data stream mining. The existing literature provides an excellent starting point for our main topic of discussion in this chapter. Not only have the distributed data mining and databases community contributed to the literature, a bulk of the work also comes from the wireless and sensor networks community. In this section we discuss some of the related papers with pointers for further reading.

Computation of complex functions over the union of multiple of streams has been studied widely in the stream mining literature. Gibbons et al. [10] presents the idea of doing coordinated sampling in order to compute simple functions such as the total number of ones in the union of two binary streams. They have developed a new sampling strategy to sample from the two streams and have shown that their sampling strategy can reduce the space requirement for such a computation from $\Omega(n)$ to $log(n)$, where n is the size of the stream. Their technique can easily be extended to the scenario where there are more than two streams. The authors also point out that this method would work even if the stream is non-binary (with no change in space complexity).

Much work has been done in the area of query processing on distributed data streams. Chen et al. [7] has developed a system 'NiagaraCQ' which allows answering continuous queries in large scale systems such as the internet. In such systems many of the queries are similar. So a lot of computation, communication and I/O resources can be saved by properly grouping the similar queries. NiagaraCQ achieves the same goal. Their grouping scheme is incremental and they use an adaptive regrouping scheme in order to find the optimal match between a new query and the group to which the query should be placed. If none of these match, then a new query group is formed with this query. The paper does not talk about reassignment of the existing queries into the newly formed groups, rather leaves it as a future work.

An different approach has been described in [23]. The distributed model described there has nodes sending streaming data to a central node which is responsible for answering the queries. The network links near the central node becomes a bottleneck as soon as the arrival rate of data the becomes too high. In order to avoid that, the authors propose installing filters which restrict the data transfer rate from the individual nodes. Node O installs a filter of width W_O and of range $[L_O, H_O]$. W_O is centered around the most recent value of the object V ($L_O = V - \frac{W_O}{2}$ and $H_O = V + \frac{W_O}{2}$). Now the node does not send updates if V is inside the range $L_O \leq V \leq H_O$; otherwise it sends updates to the central node and recenters the bounds L_O and H_O. This technique provides the answers to queries approximately and works in the circumstances where we do not require the exact answer to the queries. Since in many cases the user can provide the query precision that is necessary, the filters can be made to work after setting the bounds based on this user input.

The sensor network community provides a rich literature on the streaming algorithms. Since the sensors are deployed in hostile terrains, one of the most fundamental task aims at developing a general framework for monitoring the network themselves. A similar idea has been presented in [27]. This paper presents a general framework and shows how decomposable functions like min, max, average, count and sum can be computed over such an architecture. The architecture is highlighted by three tools that the authors call *digests, scans* and *dumps*. *Digests* are the network parameters (*e.g.* count of the number of nodes) that are computed either continuously, periodically or in the event of a trigger.

Scans are invoked when the *digests* report a problem (*e.g.* a sudden drop in the number of nodes) to find out the energy level throughout the network. These two steps can guide a network administrator towards the location of the fault which can be debugged using the *dumps* (dump all the data of a single or few of the sensors). The other thing that this paper talks about is the distributed computing of some aggregate functions (mean, max, count etc.). Since all these functions are decomposable, the advantage is in-network aggregation of partial results up a tree. The leaf does not need to send all its data to the root and in this way vital savings can be done in terms of communication. The major concern though is maintaining this tree structure in such a hostile and dynamic environment. Also this technique would fail for numerous non-decomposable functions *e.g.* median, quantile etc.

The above algorithm describes a way of monitoring the status of the sensor network itself. There are many data mining problems that need to be addressed in the sensor network scenario. Such an algorithm for multi-target classification in the sensor networks has been developed by Kotecha et al. [17] Each node makes local decisions and these decisions are forwarded to a single node which acts as the manager node. The maximum number of targets is known in apriori, although the exact number of targets is not known in advance. Nodes that are sufficiently apart are expected to provide independent feature vectors for the same target which can strengthen the global decision making. Moreover, for an optimal classifier, the number of decisions increases exponentially with the number of targets. Hence the authors propose the use of sub-optimal linear classifiers. Through real life experiments they show that their suboptimal classifiers perform as well as the optimal classifier under mild assumptions. This makes such a scheme attractive for low power, low bandwidth environments.

Frequent items mining in distributed streams is an active area of research [22]. There are many variants of the problem that has been proposed in the literature (refer to [22] for a description). Generally speaking there are m streams $S_1, S_2, .., S_m$. Each stream consists of items with time stamps $< d_{i1}, t_{i1} >, < d_{i2}, t_{i2} >$, etc. Let S be the sequence preserving union of all the streams. If an item $i \in S$ has a count $count(i)$ (the count may be evaluated by an exponential decay weighting scheme). The task is to output an estimate $\widehat{count}(i)$ of $count(i)$ whose frequency exceeds a certain threshold. Each node maintains a precision threshold and outputs only those items exceeding the precision threshold. As two extreme cases, the threshold can be set to very low (≈ 0) or very high (≈ 1). In the first case, all the intermediate nodes will send everything without pruning resulting in a message explosion at the root. In the second case, the intermediate nodes will send a low number of items and hence no more pruning would be possible at the intermediate nodes. So the precision selection problem is crucial for such an algorithm to produce meaningful results with low communication overhead. The paper presents a number of ways to select the precision values (they call it precision gradients) for different scenarios of load minimization.

In the next two sections we focus our attention on two particular distributed data stream mining algorithms. The first algorithm that we present in Section 4 works in large scale distributed (p2p) systems. The Bayesian network learning algorithm (Section 5) can be used to learn a Bayesian network model when the data is distributed and arriving in the form of data streams.

4. A *local* algorithm for distributed data stream mining

Having presented the general paradigm for distributed data stream mining and some existing algorithms, we now shift our attention to two specific algorithms that have been designed for the distributed streaming environment. The algorithm presented in this section works in a large-scale and dynamic environment *e.g.* a peer-to-peer environment, sensor networks and the like. While this algorithm is truly asynchronous and guarantees eventual correct result, it is restricted to the set of problems where we can define a threshold within which we want our result. As an example let us assume that there are N peers in a network and each peer has a bit b_i (0 or 1). The task is to find out if $\sum_{i=1}^{N} b_i > \epsilon$, where ϵ is a global parameter. There exists an algorithm [26] in the literature that can do this in the streaming scenario whereby each peer needs to contact only a subset of the nodes in its neighborhood. The algorithm that we discuss in this section shares the same philosophy – although it can perform a different set of data mining tasks. Going back to our example, if the problem is modified a little bit and we want to find out the value of $\sum_{i=1}^{N} b_i$ exactly, to the best of the authors' knowledge there exists no algorithm that can do this without collecting all the data. Hence the second algorithm that we have selected for our discussion builds a model incrementally. It first fits a model to the local data and determines the fitness of the model. If the model is not good enough, a sample of its data is taken and along with its neighbors' data used to build a more complex model (details in Section 5).

4.1 Local Algorithms : definition

Before we formally start the discussion of our algorithm let us define what we mean by *local* algorithms since this is the term that will be used throughout the rest of this section.

Local algorithms are ones whose resource consumption is sometimes independent of system size. That is, an algorithm for a given problem is local if, assuming a static network and data, there exists a constant c such that for any size of the system N there are instances of the problem such that the time, CPU, memory, and communication requirements per peer are smaller than c. Therefore, the most appealing property of local algorithms is their extreme scalability. Local algorithms have been presented in the context of graph algorithms [21][19], and recently for data mining in peer-to-peer networks [26][18] [25]. Local algorithms guarantee eventual correctness – when the computation terminates each peer computes the same result it would have computed given the entire data. The main advantage of this definition is that there already exists algorithms that follow this definition of locality. On the contrary, this definition is disadvantageous in the sense that we do not specify precisely how many instances of the problem we need (that can be solved locally) in order to deem an algorithm local.

There are certain characteristics that typify *local* algorithms. These include but are not limited to the ability of each node to compute the result using information gathered from just a few nearby neighbors, the ability to calculate the result in-network rather than collect all of the data to a central processor (which would quickly exhaust bandwidth), and the ability to locally prune redundant or duplicate computations. Needless to mention that these characteristics are extremely important in most large-scale distributed applications.

We are now in a position to present a *local* Distributed Data Stream Mining algorithm.

4.2   Algorithm details

The algorithm that we are going to discuss in this section can be used for monitoring a data model in the distributed and streaming environment. As already mentioned, this algorithm requires a threshold predicate. It also assumes that a tree topology has been laid over the network structure. Before we present the details of the algorithm, let us present an overview. Let X be a dataset which is a union of several data streams. Let $f[X]$ be a function that we want to compute on X. Since it is not always feasible to store and process the entire stream, we select a subset of the stream in a given time frame. We denote this by $\hat{X}_t$. Our main task is to monitor whether $f[\hat{X}_t] > \epsilon$, where ϵ is a user defined threshold. For example, if f is the *average* we can bound the average of a set of vectors within this threshold. The details of this algorithm can be found in [25].

In the actual algorithm description, let $P_1, P_2, ..., P_n$ be n peers connected by an arbitrary communication tree such that the set of $P_i's$ neighbors N_i is known to P_i. Each peer is supplied a stream of points from $\mathbb{R}^d$, where d is the dimension of the problem. The local average of the points at time t is denoted by S_{it}. Each peer maintains the following vectors (each of these vectors is weighted, we omit it in our discussion here for clarity and simplicity):

- $\mathbf{X_i}$: the current estimate of the global average or $\mathbf{X_N}$ (known as the *knowledge* of P_i)

- $\mathbf{X_{i,j}}$: the last vector send by peer P_i to P_j

- $\mathbf{X_{j,i}}$: the last vector received by peer P_i from P_j

- $\mathbf{X_{i\cap j}}$: the *agreement* of P_i and P_j (calculated from $\mathbf{X_{i,j}}$ and $\mathbf{X_{j,i}}$)

- $\mathbf{X_{i\backslash j}}$: the *kept knowledge* of P_i and P_j (calculated from $\mathbf{X_i}$ and $\mathbf{X_{i\cap j}}$)

Initially $\mathbf{X_i}$ is initialized to S_{it}. The first time a data is exchanged between P_i and P_j, it sets the vectors $\mathbf{X_{i,j}}$ and $\mathbf{X_{j,i}}$. Whenever new points arrive, a neighbor is removed or a new neighbor comes in, S_{it} changes. This triggers the changes in the local vectors $\mathbf{X_i}$, $\mathbf{X_{i\backslash j}}$ and/or $\mathbf{X_{i\cap j}}$. The global data mining problem is to find out if $||\frac{\sum_{i=1}^n \mathbf{X_i}}{n}(=\mathbf{X_N})|| < \epsilon$. We will now present different conditions on $\mathbf{X_i}$, $\mathbf{X_{i\cap j}}$ and $\mathbf{X_{i\backslash j}}$ that will allow us to decide if $||\mathbf{X_N}|| < \epsilon$ or $||\mathbf{X_N}|| > \epsilon$. There are two main lemmas that can be used in order to determine this. The first lemmas states that if for **every** peer P_i and **each** neighbor P_j of P_i it holds that both $||\mathbf{X_{i\cap j}}|| < \epsilon$ and $||\mathbf{X_{i\backslash j}}|| < \epsilon$, then $||\mathbf{X_N}|| < \epsilon$. The second lemma states that given d unit vectors, $\hat{u}_1 \ldots \hat{u}_d$ if for a specific one of them $\hat{u}$ **every** peer P_i and **each** of its neighbors P_j have $\hat{u}_i \cdot \mathbf{X_{i\cap j}} \geq \epsilon$ and either $\hat{u}_i \cdot \mathbf{X_{i\backslash j}} \geq \epsilon$ or $\hat{u}_i \cdot \mathbf{X_i} = \hat{u}_i \cdot \mathbf{X_{i\cap j}}$ then $||\mathbf{X_N}|| \geq \epsilon$.

Given these two lemmas, we can now describe the algorithm that decides whether the L2 norm of the average vector $\mathbf{X_N}$ is within a threshold or not. For simplicity we will describe the conditions assuming $\mathbb{R}^2$, the extension to the higher dimension is obvious. For a vector $\mathbf{X}_i = (x_a, y_a) \in \mathbb{R}^2$, the L2-norm is given by

$$\sqrt{x_a^2 + y_a^2}$$

Now if we want to test if the L2-norm is greater than ϵ or not, we want to check if the following condition is true:

$$\sqrt{x_a^2 + y_a^2} > \epsilon$$

which is, in essence, checking if $||\mathbf{X}_i||$ lies inside or outside a circle of radius ϵ.

Consider Figure 14.1. The circle shown in the figure is a circle of radius ϵ. The algorithm needs to check if $||\mathbf{X}_i||$ is inside or outside the circle. In order to do this, it approximates the circles with a set of tangent lines as shown in the figure. The problem of checking inside the circle is relatively simpler - if a peer determines that its local $||\mathbf{X}_i|| < \epsilon$ and for every neighbor $||\mathbf{X_{i\backslash j}}|| < \epsilon$ and $||\mathbf{X_{i\cup j}}|| < \epsilon$, then by the first lemma the peer knows that $||\mathbf{X}_N|| < \epsilon$ as well. Hence the local estimate $||\mathbf{X}_i||$ of $||\mathbf{X}_N||$ is indeed the correct one and thus no message needs to be exchanged. If on the other hand a peer violates this condition, it needs to send a message to **each** violating neighbor.

If on the other hand a peer P_i determines that $||\mathbf{X}_i||$ is outside the polygon, it needs to check the conditions of the second lemma. Note that if two peers simply say that $||\mathbf{X}_i|| > \epsilon$ still the average can be lesser than ϵ (when the two

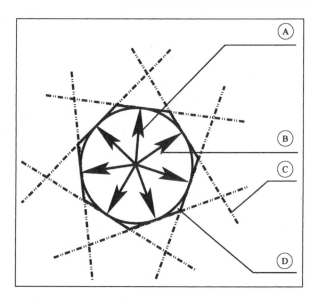

Figure 14.1. (A) the area inside an ϵ circle. (B) Seven evenly spaced vectors - $\mathbf{u_1} \ldots \mathbf{u_7}$.
(C) The borders of the seven halfspaces $\hat{u}_i \cdot \mathbf{x} \geq \epsilon$ define a polygon in which the circle is
circumscribed. (D) The area between the circle and the union of half-spaces.

$X_i's$ are in the opposite directions). So we need to check the conditions based
on the tangent lines. P_i needs to find the first tangent plane such that a neighbor
P_j claims that the point is outside the polygon. More precisely, it needs to
check if there exists some u^* such that $\hat{u}_i \cdot \mathbf{X}_{i \cap j} \geq \epsilon$. Now if P_i enters a
state whereby $||\mathbf{X}_i|| > \epsilon$, it does not need to send a message to its neighbors
if they support its claim. Formally speaking P_i needs to send a message to P_j
whenever $u^* \cdot \mathbf{X}_{i \cap j} < \epsilon$ or $u^* \cdot \mathbf{X}_{i \setminus j} < \epsilon$, else not.

   The last case is when the $\mathbf{X}_i$ lies in the region between the circle and the
polygon (the peer would find u^* to be *nil*. In that case the peer has to resort to
flooding. Also note that, this area can be made arbitrarily smaller using more
number of tangent lines - the trade-off is an increased computation for each
peer.

4.3   Experimental results

   The above algorithm exhibits excellent accuracy and scalability. For each
experiment we introduce new data distribution at every predefined number of
simulator ticks. We start with a particular data distribution and choose ϵ such
that the average vector is within the ϵ-range. As shown in Figure 14.2, the lower
set of bars show the number of peers that report if the average vector is less
than ϵ. After a fixed number of simulator ticks we change the data distribution

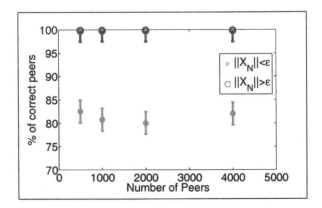

Figure 14.2. Quality of the algorithm with increasing number of nodes

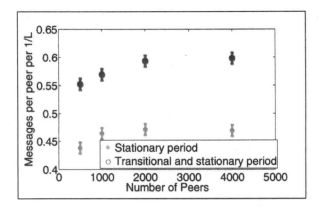

Figure 14.3. Cost of the algorithm with increasing number of nodes

and now, the $||\mathbf{X}_i||$'s are no more inside the circle and so we have the upper set of bars reporting the case when $||\mathbf{X}_i||$ is outside the circle. As we see from the figure, the accuracy remains almost constant. Similarly, Figure 14.3 shows the cost of the simple L2 algorithm. In order to eliminate a message explosion, we have used a leaky bucket mechanism and a peer is not allowed to send more than one message peer leaky bucket duration. Our unit of measurement is messages per peer per unit leaky bucket time (L). It can be easily seen that the number of messages per unit time remains constant - this typifies local algorithms - they are highly scalable.

4.4   Modifications and extensions

We call the above operation mode of the algorithm the 'openloop mode'. It is useful for bounding the average vector within a threshold. If we 'close the loop' we can use the above simple algorithm to monitor data mining results (e.g. eigenstates of the data, k-means of the data etc). Generally, we need to define our $\mathbf{X}_i$ appropriately (we provide examples later) and the open loop algorithm would raise a flag if $||\mathbf{X_i}|| > \epsilon$. We can potentially use this flag to collect statistics of the data and build a new model. This model can then be shipped to the peers and this process can go on. If the data stops changing, the guarantee of the L2 algorithm is that $||\mathbf{X_i}|| > \epsilon$ for every peer iff $||\mathbf{X_N}|| > \epsilon$. Hence for long durations of stationary period we would expect to see local behavior of the algorithm. Only when the data distribution changes, the alert flag would be raised. We can then collect sample of the data from the network and use standard aggregation techniques such as convergecast to propagate the data up the tree to build a new model. As a simple example we show how this algorithm is possible to monitor the eigenstates of the data. We define $\mathbf{X}_i$ as the following:

$$\mathbf{X}_i = A \times S_{it} - \theta \times S_{it}$$

where A and θ are the principal eigenvector and the eigenvalue of the data. A change in the data distribution (S_{it}) would change the $\mathbf{X}_i$ and set up an alert flag. This change might trigger a series of events :

- P_i will communicate with its neighbors and try to solve the problem using the open loop algorithm

- If the alert flag has been there for a long time, P_i sends its data to its parent in the tree. If the alert flag is not there any more, it is considered to be a false alarm

- If a peer receives data from all of its children, it sends the data to its parent. The root after receiving the data from everybody computes the new eigenstates and notifies all the peers about this new model

The above algorithm is what we call the 'closed loop' algorithm. Since the model is built solely on best effort, it may be the case that the model is no longer good enough once it reaches all the peers. All that the algorithm will do is to restart the process once again and build a more up-to-date model. A similar algorithm for monitoring the k-means can be described by simply changing $\mathbf{X}_i$ to $S_{it} - c_i$, where c_i is the current centroid of the global data.

In the next section we present an algorithm that can be used for learning a model from the data in the streaming scenario.

5.   Bayesian Network Learning from Distributed Data Streams

This section discusses an algorithm for Bayesian Model learning. In many applications the goal is to build a model that represents the data. In the previous section we saw how such a model can be build when the system is provided with a threshold predicate. If, however, we want to build an exact global model, development of local algorithms sometimes becomes very difficult, if not impossible. In this section we draw the attention of the reader to a class of problems which needs global information to build a data model (*e.g.* K-means, Bayesian Network,etc). The crux of these types of algorithms lies in building a local model, identifying the goodness of the model and then co-ordinating with a central site to update the model based on global information. We describe here a technique to learn a Bayesian network in a distributed setting.

Bayesian network is an important tool to model probabilistic or imperfect relationship among problem variables. It gives useful information about the mutual dependencies among the features in the application domain. Such information can be used for gaining better understanding about the dynamics of the process under observation. It is thus a promising tool to model customer usage patterns in web data mining applications, where specific user preferences can be modeled as in terms of conditional probabilities associated with the different features. Since we will shortly show how this model can be built on streaming data, it can potentially be applied to learn Bayesian classifiers in distributed settings. But before we delve into the details of the algorithm we present what a Bayesian Network (or Bayes' Net or BN in short) is, and the distributed Bayesian learning algorithm assuming a static data distribution.

A *Bayesian network (BN)* is a probabilistic graph model. It can be defined as a pair $(\mathcal{G}, p)$, where $\mathcal{G} = (\mathcal{V}, \mathcal{E})$ is a directed acyclic graph (DAG). Here, $\mathcal{V}$ is the node set which represents variables in the problem domain and $\mathcal{E}$ is the edge set which denotes probabilistic relationships among the variables. For a variable $X \in \mathcal{V}$, a parent of X is a node from which there exists a directed link to X. Figure 14.4 is a BN called the ASIA model (adapted from [20]). The variables are Dyspnoea, Tuberculosis, Lung cancer, Bronchitis, Asia, X-ray, Either, and Smoking. They are all binary variables. The joint probability distribution of the set of variables in $\mathcal{V}$ can be written as a product of conditional probabilities as follows:

$$P(\mathcal{V}) = \prod_{X \in \mathcal{V}} P(X \mid pa(X)). \qquad (14.1)$$

In Equation (14.1) $pa(X)$ denotes the set of parents of node X. The set of conditional distributions $\{P(X \mid pa(X)), X \in \mathcal{V}\}$ are called the parameters of a Bayesian network. If variable X has no parents, then $P(X \mid pa(X)) = P(X)$ is the marginal distribution of X.

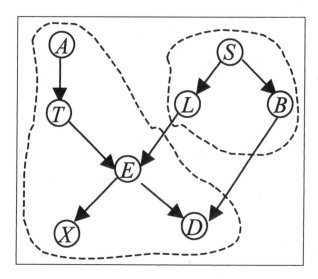

Figure 14.4. ASIA Model

Two important issues in using a Bayesian network are: (a) learning a Bayesian network and (b) probabilistic inference. Learning a BN involves learning the structure of the network (the directed graph), and obtaining the conditional probabilities (parameters) associated with the network. Once a Bayesian network is constructed, we usually need to determine various probabilities of interest from the model. This process is referred to as probabilistic inference.

In the following, we discuss a collective approach to learning a Bayesian network that is specifically designed for a distributed data scenario.

5.1 Distributed Bayesian Network Learning Algorithm

The primary steps in our approach are:

(a) Learn local BNs (local model) involving the variables observed at each site based on local data set.

(b) At each site, based on the local BN, identify the observations that are most likely to be evidence of coupling between local and non-local variables. Transmit a subset of these observations to a central site.

(c) At the central site, a limited number of observations of all the variables are now available. Using this to learn a non-local BN consisting of links between variables across two or more sites.

(d) Combine the local models with the links discovered at the central site to obtain a collective BN.

The non-local BN thus constructed would be effective in identifying associations between variables across sites, whereas the local BNs would detect associations among local variables at each site. The conditional probabilities can also be estimated in a similar manner. Those probabilities that involve only variables from a single site can be estimated locally, whereas the ones that involve variables from different sites can be estimated at the central site. Same methodology could be used to update the network based on new data. First, the new data is tested for how well it fits with the local model. If there is an acceptable statistical fit, the observation is used to update the local conditional probability estimates. Otherwise, it is also transmitted to the central site to update the appropriate conditional probabilities (of cross terms). Finally, a collective BN can be obtained by taking the union of nodes and edges of the local BNs and the nonlocal BN and using the conditional probabilities from the appropriate BNs. Probabilistic inference can now be performed based on this collective BN. Note that transmitting the local BNs to the central site would involve a significantly lower communication as compared to transmitting the local data.

It is quite evident that learning probabilistic relationships between variables that belong to a single local site is straightforward and does not pose any additional difficulty as compared to a centralized approach (This may not be true for arbitrary Bayesian network structure. A detailed discussion of this issue can be found in [9]). The important objective is to correctly identify the coupling between variables that belong to two (or more) sites. These correspond to the edges in the graph that connect variables between two sites and the conditional probability(ies) at the associated node(s). In the following, we describe our approach to selecting observations at the local sites that are most likely to be evidence of strong coupling between variables at two different sites. The key idea of our approach is that the samples that do not fit well with the local models are likely to be evidence of coupling between local and non-local variables. We transmit these samples to a central site and use them to learn a collective Bayesian network.

5.2   Selection of samples for transmission to global site

For simplicity, we will assume that the data is distributed between two sites and will illustrate the approach using the BN in Figure 14.4. The extension of this approach to more than two sites is straightforward. Let us denote by $\mathcal{A}$ and $\mathcal{B}$, the variables in the left and right groups, respectively, in Figure 14.4. We assume that the observations for $\mathcal{A}$ are available at site A, whereas the observations for $\mathcal{B}$ are available at a different site B. Furthermore, we assume that there is a common feature ("key" or index) that can be used to associate a given observation in site A to a corresponding observation in site B. Naturally, $\mathcal{V} = \mathcal{A} \cup \mathcal{B}$.

At each local site, a local Bayesian network can be learned using only samples in this site. This would give a BN structure involving only the local variables at each site and the associated conditional probabilities. Let $p_A(.)$ and $p_B(.)$ denote the estimated probability function involving the local variables. This is the product of the conditional probabilities as indicated by Equation (14.1). Since $p_A(x)$, $p_B(x)$ denote the probability or likelihood of obtaining observation x at sites A and B, we would call these probability functions the likelihood functions $l_A(.)$ and $l_B(.)$, for the local model obtained at sites A and B, respectively. The observations at each site are ranked based on how well it fits the local model, using the local likelihood functions. The observations at site A with large likelihood under $l_A(.)$ are evidence of "local relationships" between site A variables, whereas those with low likelihood under $l_A(.)$ are possible evidence of "cross relationships" between variables across sites. Let $S(A)$ denote the set of keys associated with the latter observations (those with low likelihood under $l_A(.)$). In practice, this step can be implemented in different ways. For example, we can set a threshold ρ_A and if $l_A(x) \leq \rho_A$, then $x \in S_A$. The sites A and B transmit the set of keys S_A, S_B, respectively, to a central site, where the intersection $S = S_A \cap S_B$ is computed. The observations corresponding to the set of keys in S are then obtained from each of the local sites by the central site.

In a sense, our approach to learning the cross terms in the BN involves a selective sampling of the given dataset that is most relevant to the identification of coupling between the sites. This is a type of *importance sampling*, where we select the observations that have high conditional probabilities corresponding to the terms involving variables from both sites. Naturally, when the values of the different variables (features) from the different sites, corresponding to these selected observations are pooled together at the central site, we can learn the coupling links as well as estimate the associated conditional distributions. These selected observations will, by design, not be useful to identify the links in the BN that are local to the individual sites.

Having discussed in detail the distributed Bayesian learning algorithm (assuming a static data), we can now proceed with our discussion on how this algorithm can be modified to work with evolving data.

5.3   Online Distributed Bayesian Network Learning

The proposed collective approach to learning a BN is well suited for a scenario with multiple data streams. Suppose we have an existing BN model, which has to be constantly updated based on new data from multiple streams. For simplicity, we will consider only the problem of updating the BN parameters, assuming that the network structure is known. As in the case of batch mode learning, we shall use techniques for online updating of BN parameters for centralized data. In the centralized case, there exists simple techniques for parameter updating for commonly used models like the unrestricted multinomial model. For example, let us denote by $p_{ijl} = Pr(x_i = l \mid pa_{x_i} = j)$, the conditional probability at node i, given the parents of node i. We can then obtain the estimate $p_{ijl}(k+1)$ of p_{ijl} at step $k+1$ as follows (see [12, Section 5]):

$$p_{ijl}(k+1) = \frac{\alpha_{ijl}(k) + N_{ijl}(k+1)}{\alpha_{ij}(k) + N_{ij}(k+1)}, \qquad (14.2)$$

where $\alpha_{ij}(k) = \sum_l \alpha_{ijl}(k)$ and $N_{ij}(k+1) = \sum_l N_{ijl}(k+1)$. In equation 14.2, $N_{ijl}(k+1)$ denotes the number of observations in the dataset obtained at time $k+1$ for which, $x_i = l$ and $pa_{x_i} = j$, and we can set $\alpha_{ijl}(k+1) = \alpha_{ijl}(k) + N_{ijl}(k+1)$. Note that $N_{ijl}(k)$ are a set of sufficient statistics for the data observed at time k.

For online distributed case, parameters for local terms can be updated using the same technique as in a centralized case. Next, we need to update the parameters for the cross-links, without transmitting all the data to a central site. Again we choose the samples with low likelihood in local sites and transmit them to a central site. This is then used to update the cross-terms at the central site. We can summarize our approach by the following steps:

1 Learn an initial collective Bayesian network from the first dataset observed (unless a prior model is already given). Thus we have a local BN at each site and a set of cross-terms at the central site.

2 At each step k:

- Update the local BN parameters at each site using equation 14.2.

- Update the likelihood threshold at each local site, based on the sample mean value of the observed likelihoods. This is the threshold used to determine if a sample is to be transmitted to a central site (see Section 5.2).

- Transmit the low likelihood samples to a central site.

- Update the parameters of the cross-terms at the central site.

- Combine the updated local terms and cross terms to get an updated collective Bayesian network.

3 Increment k and repeat step (2) for the next set of data.

This section concludes our discussion on the distributed streaming Bayesian learning algorithm. In the following, we point out some of the experimental verifications of the proposed algorithm.

5.4   Experimental Results

We tested our approach on two different datasets. A small real web log dataset was used for batch mode distributed Bayesian learning. This was used to test both structure and parameter learning. We also tested our online distributed learning approach on a simulated web log dataset. Extensive examples for batch mode learning (using both real and simulated web log data), demonstrating scalability with respect to number of distributed sites have been presented elsewhere [8, 9]. In the following, we present our results for BN parameter learning using online data streams.

We illustrate the results of online BN parameter learning assuming the network structure is known. We use the model shown in Figure 14.5. The 32 nodes in the network are distributed among four different sites. Nodes 1, 5, 10, 15, 16, 22, 23, 24, 30, and 31 are in site A. Nodes 2, 6, 7, 11, 17, 18, 25, 26, and 32 are in site B. Nodes 3, 8, 12, 19, 20, and 27 are in site C. Nodes 4, 9, 13, 14, 21, 28, and 29 are in site D. A dataset with 80,000 observations was generated. We assumed that at each step k, 5,000 observations of the data are available (for a total of 16 steps).

We denote by B_{be}, the Bayesian network obtained by using all the 80,000 samples in batch mode (the data is still distributed into four sites). We denote by $B_{ol}(k)$, the Bayesian network obtained at step k using our online learning approach and by $B_{ba}(k)$, the Bayesian network obtained using a regular batch mode learning, but using only data observed upto time k. We choose three typical cross terms (nodes 12, 27, and 28) and compute the KL distance between the conditional probabilities to evaluate the performance of online distributed method. The results are depicted in Figure 14.6.

Figure 14.6 (left) shows the KL distance between the conditional probabilities for the networks $B_{ol}(k)$ and B_{be} for the three nodes. We can see that the performance of online distributed method is good, with the error (in terms of KL distance) dropping rapidly. Figure 14.6 (right) shows the KL distance between the conditional probabilities for the networks $B_{ba}(k)$ and B_{ol} for the three nodes. We can see that the performance of a network learned using our online distributed method is comparable to that learned using a batch mode method, with the same data.

6.   Conclusion

In this chapter we have surveyed the field of distributed data stream mining. We have presented a brief survey of field, discussed some of the distributed data stream algorithms, their strengths and weaknesses. Naturally, we have elucidated one slice through this field - the main topic of our discussion in this

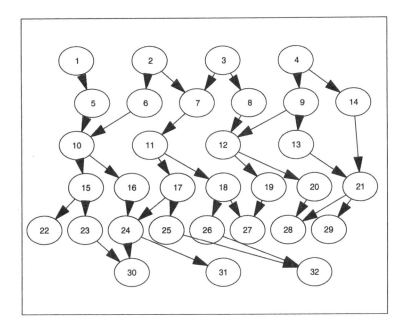

Figure 14.5. Bayesian network for online distributed parameter learning

chapter was algorithms for distributed data stream mining. Many important areas such as system development, human-computer interaction, visualization techniques and the like in the distributed and streaming environment were left untouched due to lack of space and limited literature in the areas.

We have also discussed in greater detail two specific distributed data stream mining algorithms. In the process we wanted to draw the attention of the readers to an emerging area of distributed data stream mining, namely data stream mining in large-scale peer-to-peer networks. We encourage the reader to explore distributed data stream mining in general. All the fields - algorithm development, systems development and developing techniques for human-computer interaction are still at a very early stage of development. On an ending note, the area of distributed data stream mining offers plenty of room for development both for the pragmatically and theoretically inclined.

Acknowledgments

The authors thank the U.S. National Science Foundation for support through grants IIS-0329143, IIS-0350533, CAREER award IIS-0093353, and NASA for support under Cooperative agreement NCC 2-1252. The authors would also like to thank Dr. Chris Gianella for his valuable input.

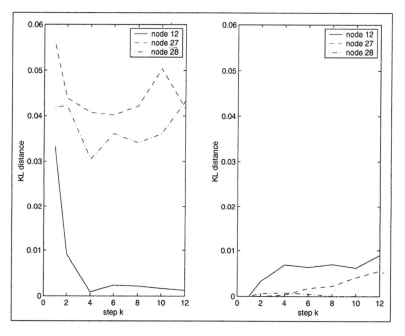

Figure 14.6. Simulation results for online Bayesian learning: (left) KL distance between the conditional probabilities for the networks $B_{ol}(k)$ and B_{be} for three nodes (right) KL distance between the conditional probabilities for the networks $B_{ol}(k)$ and B_{ba} for three nodes

References

[1] C Aggarwal. A framework for diagnosing changes in evolving data streams. In *ACM SIGMOD'03 International Conference on Management of Data*, 2003.

[2] C. Aggarwal, J. Han, J. Wang, and P. Yu. A framework for clustering evolving data streams. In *VLDB conference*, 2003.

[3] C. Aggarwal, J. Han, J. Wang, and P. S. Yu. On demand classification of data streams. In *KDD*, 2004.

[4] B. Babcock, S. Babu, M. Datar, R. Motwani, and J. Widom. Models and issues in data stream systems. In *In Principles of Database Systems (PODS'02)*, 2002.

[5] B. Babcock and C. Olston. Distributed top-k monitoring. In *ACM SIG-MOD'03 International Conference on Management of Data*, 2003.

[6] S. Ben-David, J. Gehrke, and D. Kifer. Detecting change in data streams. In *VLDB Conference*, 2004.

[7] J. Chen, D. DeWitt, F. Tian, and Y. Wang. NiagaraCQ: a scalable continuous query system for Internet databases. In *ACM SIGMOD'00 International Conference on Management of Data*, 2000.

[8] R. Chen, K. Sivakumar, and H. Kargupta. An approach to online bayesian learning from multiple data streams. In *Proceedings of the Workshop on Ubiquitous Data Mining (5th European Conference on Principles and Practice of Knowledge Discovery in Databases)*, Freiburg, Germany, September 2001.

[9] R. Chen, K. Sivakumar, and H. Kargupta. Collective mining of bayesian networks from distributed heterogeneous data. *Knowledge and Information Systems*, 6:164–187, 2004.

[10] P. Gibbons and S. Tirthapura. Estimating simple functions on the union of data streams. In *ACM Symposium on Parallel Algorithms and Architectures*, 2001.

[11] S. Guha, N. Mishra, R. Motwani, and L. O'Callaghan. Clustering data streams. In *IEEE Symposium on FOCS*, 2000.

[12] D. Heckerman. A tutorial on learning with Bayesian networks. Technical Report MSR-TR-95-06, Microsoft Research, 1995.

[13] M. Henzinger, P. Raghavan, and S. Rajagopalan. Computing on data streams. Technical Report TR-1998-011, Compaq System Research Center, 1998.

[14] G. Hulten, L. Spencer, and P. Domingos. Mining time-changing data streams. In *SIGKDD*, 2001.

[15] R. Jin and G. Agrawal. Efficient decision tree construction on streaming data. In *SIGKDD*, 2003.

[16] H. Kargupta and K. Sivakumar. *Existential Pleasures of Distributed Data Mining. Data Mining: Next Generation Challenges and Future Directions.* AAAI/MIT press, 2004.

[17] J. Kotecha, V. Ramachandran, and A. Sayeed. Distributed multi-target classification in wireless sensor networks. *IEEE Journal of Selected Areas in Communications (Special Issue on Self-Organizing Distributed Collaborative Sensor Networks)*, 2003.

[18] D. Krivitski, A. Schuster, and R. Wolff. A local facility location algorithm for sensor networks. In *Proc. of DCOSS'05*, 2005.

[19] S. Kutten and D. Peleg. Fault-local distributed mending. In *Proc. of the ACM Symposium on Principle of Distributed Computing (PODC)*, pages 20–27, Ottawa, Canada, August 1995.

[20] S. L. Lauritzen and D. J. Spiegelhalter. Local computations with probabilities on graphical structures and their application to expert systems (with discussion). *Journal of the Royal Statistical Society, series B*, 50:157–224, 1988.

[21] N. Linial. Locality in distributed graph algorithms. *SIAM Journal of Computing*, 21:193–201, 1992.

[22] A. Manjhi, V. Shkapenyuk, K. Dhamdhere, and C. Olston. Finding (recently) frequent items in distributed data streams. In *International Conference on Data Engineering (ICDE'05)*, 2005.

[23] C. Olston, J. Jiang, and J. Widom. Adaptive filters for continuous queries over distributed data streams. In *ACM SIGMOD '03 International Conference on Management of Data*, 2003.

[24] J. Widom and R. Motwani. Query processing, resource management, and approximation in a data stream management system. In *CIDR*, 2003.

[25] R. Wolff, K. Bhaduri, and H. Kargupta. Local L2 thresholding based data mining in peer-to-peer systems. In *Proceedings of SIAM International Conference in Data Mining (SDM)*, Bethesda, Maryland, 2006.

[26] R. Wolff and A. Schuster. Association rule mining in peer-to-peer systems. In *Proceedings of ICDM '03*, Melbourne, Florida, 2003.

[27] J. Zhao, R. Govindan, and D. Estrin. Computing aggregates for monitoring wireless sensor networks. In *Proceedings of the First IEEE International Workshop on Sensor Network Protocols and Applications*, 2003.

Chapter 15

A SURVEY OF STREAM PROCESSING PROBLEMS AND TECHNIQUES IN SENSOR NETWORKS

Sharmila Subramaniam, Dimitrios Gunopulos
Computer Science and Engineering Dept.
University of California at Riverside
Riverside, CA 92521

sharmi,dg@cs.ucr.edu

Abstract   Sensor networks comprise small, low-powered and low-cost sensing devices that are distributed over a field to monitor a phenomenon of interest. The sensor nodes are capable of communicating their readings, typically through wireless radio. Sensor nodes produce streams of data, that have to be processed in-situ, by the node itself, or to be transmitted through the network, and analyzed offline. In this chapter we describe recently proposed, efficient distributed techniques for processing streams of data collected with a network of sensors.

Keywords: Sensor Systems, Stream Processing, Query Processing, Compression, Tracking

Introduction

Sensor networks are systems of tiny, low-powered and low-cost devices distributed over a field to sense, process and communicate information about their environment. The sensor nodes in the systems are capable of sensing a phenomenon and communicating the readings through wireless radio. The memory and the computational capabilities of the nodes enable in-site processing of the observations. Since the nodes can be deployed at random, and can be used to collect information about inaccessible remote domains, they are considered as very valuable and attractive tools for many research and industrial applications. *Motes* is one example of sensor devices developed by UC Berkeley and manufactured by Crossbow Technology Inc. [13].

Sensor observations form streams of data that are either processed in-situ or communicated across the network and analyzed offline. Examples of systems producing streams of data include environmental and climatological ([39]) monitoring where phenomena such as temperature, pressure and humidity are measured periodically (at various granularities). Sensors deployed in building and bridges relay measurements of vibrations, linear deviation etc. for monitoring structural integrity ([51]). Seismic measurements, habitat monitoring ([7, 38]), data from GPS enabled devices such as cars and phones and surveillance data are further examples. Surveillance systems may include sophisticated sensors equipped with cameras and UAVs but nevertheless they produce streams of videos or streams of events. In some applications, raw data is processed in the nodes to detect *events*, defined as some suitable function on the data, and only the streams of events are communicated across the network.

The focus of this chapter is to describe recently proposed, efficient distributed techniques for processing streams of data that are collected from a sensor network.

1.   Challenges

Typically a large number of sensors nodes are distributed spanning wide areas and each sensor produces large amount of data continuously as observations. For example, about 10, 000 traffic sensors are deployed in California highways to report traffic status continuously. The energy source for the nodes are either AA batteries or solar panels that are typically characterized by limited supply of power. In most applications, communication is considered as the factor requiring the largest amount of energy, compared to sensing ([41]). The longevity of the sensor nodes is therefore drastically reduced when they communicate raw measurements to a centralized server for analysis. Consequently data aggregation, data compression, modeling and online querying techniques need to be applied in-site or in-network to reduce communication across the network. Furthermore, limitations of computational power and inaccuracy and bias in the sensor readings necessitate efficient data processing algorithms for sensor systems.

In addition, sensor nodes are prone to failures and aberrant behaviors which could affect network connectivity and data accuracy severely. Algorithms proposed for data collection, processing and querying for sensor systems are required to be robust and fault-tolerant to failures. Network delays present in sensor systems is yet another problem to cope up with in real-time applications.

The last decade has seen significant advancement in the development of algorithms and systems that are energy aware and scalable with respect to networking, sensing, communication and processing. In the following, we describe some of the interesting problems in data processing in sensor networks and give brief overview of the techniques proposed.

2. The Data Collection Model

We assume a data collection model where the set of sensors deployed over a field communicate over a wireless ad-hoc network. This scenario is typically applicable when the sensors are small and many. The sensors are deployed quickly, leaving no or little time for a wired installation. Nevertheless, there are also many important applications where expensive sensors are manually installed. A typical example is a camera based surveillance system where wired networks can also be used for data collection.

3. Data Communication

The basic issue in handling streams from sensors is to transmit them, either as raw measurements or in a compressed form. For example, the following query necessitates transmission of temperature measurements from a set of sensors to the user, over the wireless network.

"Return the temperature measurements of all the sensors in the subregion R every 10s, for the next 60 minutes"

Typically, the data communication direction is from multiple sensor nodes to a single *sink* node. Moreover, since the stream of measurements observed by sensors are that of a common phenomena, we observe redundancy in the data communicated. For example, consider the following task posed from the sink node to the system:

Due to the above characteristics, along with limited availability of power, the end-to-end communication protocols available for mobile ad-hoc networks are not applicable for sensor systems. The research community has therefore proposed data aggregation as the solution wherein data from multiple sources are combined, processed within the network to eliminate redundancy and routed through the path that reduces the number of transmissions.

Energy-aware sensing and routing has been a topic of interest over the recent years to extend the lifetime of the nodes in the network. Most of the approaches create a hierarchical network organization, which is then used for routing of queries and for communication between the sensors. [27] proposed a cluster based approach known as LEACH for energy-efficient data transmission. Cluster-head nodes collect the streaming data from the other sensors in the cluster and apply signal processing functions to compress the data into a single signal. As illustrated in Figure 15.1, cluster heads are chosen at random and the sensors join the nearest cluster head. Now, a sensor communicates its stream data to the corresponding cluster head, which in turn takes the responsibility of communicating them to the sink (possibly after compressing). A

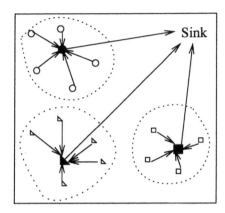

Figure 15.1. An instance of dynamic cluster assignment in sensor system according to LEACH protocol. Sensor nodes of the same clusters are shown with same symbol and the cluster heads are marked with highlighted symbols.

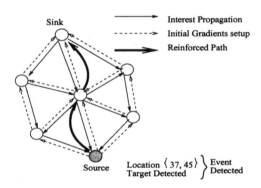

Figure 15.2. Interest Propagation, gradient setup and path reinforcement for data propagation in *directed-diffusion* paradigm. Event is described in terms of attribute value pairs. The figure illustrates an event detected based on the location of the node and target detection.

different approach is the *Directed Diffusion* paradigm proposed by [28] which follows a *data centric* approach for routing data from sources to the sink sensor. Directed diffusion uses a publish-subscribe approach where the inquirer (say, the sink sensor) expresses an *interest* using attribute values and the sources that can serve the interest reply with data (Figure 15.2). As the data is propagated toward the sink, the intermediate sensors cache the data to prevent loops and eliminate duplicate messages.

Among the many research works with the goal of energy-aware routing, *Geographic Adaptive Fidelity* (GAF) approach proposed by [49] conserves energy from the point of view of communication, by turning off the radios of some of the sensor nodes when they are redundant. The system is divided into

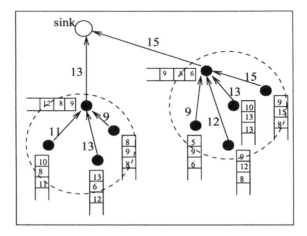

Figure 15.3. Sensors aggregating the result for a MAX query *in-network*

a virtual grid using geographic location information and only one sensor per a cell in the grid is activated to route packets. ASCENT by [8] and STEM by [44] are examples of other available sensor topology control schemes.

Data aggregation is also an integral part of query processing. The results of the posed queries are communicated across the network with in-network aggregation. We discuss more about this in Section 4.

4. Query Processing

In the common query frameworks followed in sensor systems, data collection is driven by declarative queries (examples of such systems include COUGAR ([54]) and TAG ([35])). Users pose declarative queries over the data generated by the sensors. For example, the SQL-like query corresponding to the example task discussed in Section 3 is as follows.

> SELECT S.temperature
> FROM Sensor S
> WHERE S.loc IN R
> DURATION 3600s
> EVERY 10s

The simplest scheme for evaluating such queries on sensor data is to transmit all the data to a centralized database, which can then be used for answering user queries. To improve on this scheme, [6], [34] and [53] suggested the concept of viewing sensor system as a distributed database and proposed incorporating a query layer in the system. The sensors in the system were now query-aware, which paved way for the following. Firstly, the sensors communicated the measurements on-demand i.e., if they satisfied the predicate of the query. Secondly, in-network processing of query results was now possible in query-aware sensors.

A direct implementation of the distributed database techniques are not viable in sensor systems due to their communication and computation constraints. Therefore, [6] and [34] studied the characteristics and challenges of sensor databases and the requirements of a good query plan for such systems. The COUGAR framework proposed by [53] aimed at reducing energy usage by generating efficient query plans for in-network query processing. The authors proposed a general framework for in-network query processing (in [54]) where the query is decomposed into *flow blocks* and a leader node from a set of coordinated nodes collects the query results of a block.

4.1   Aggregate Queries

The class of queries that has received interest in sensor systems is *Aggregate Queries*. Recently various techniques are proposed to efficiently process aggregate queries such as MIN, COUNT and AVG in sensor systems while reducing power consumption. An example of a simple query would be:

"Return the maximum of the temperature measurements obtained from all sensors located within coordinates [0, 100, 100, 0]*."*

The properties of the aggregate functions enables distributed processing of partial data in-network, which can be combined to produce results for the posed queries. Such optimizations reduce energy consumption for query processing by orders of magnitudes. For example, Figure 15.3 shows a routing paradigm where MAX of the observations at the sensors are evaluated efficiently by computing the MAX of different groups of sensors, and communicating only the results to the sink sensor.

One of the first tools for processing aggregate queries in sensor systems, in a distributed and efficient manner, is TAG, presented by [35]. In TAG, queries posed by users are propagated from a base station into the network, piggybacking the existing network protocol. Aggregate results are communicated back to the base station up a spanning tree, with each sensor combining its result with the results obtained from its children. Later, [25] studied the implementation of TAG framework for various sensing applications such as sensor data summarization, vehicle tracking and topographic mapping.

Various improvements and application specific modifications have been proposed recently based on the above query-tree framework suggested for data aggregation and query processing. We give a brief overview of some of the techniques in the following.

Extending the query-tree framework and the work by [40], [14] presented a framework for in-network data aggregation to evaluate aggregate queries in error-tolerant applications. According to this framework, the nodes of the query tree apply their error filters to the partial aggregates of their subtrees, and suppress messages from being communicated to the sink (see Figure 15.4). In ad-

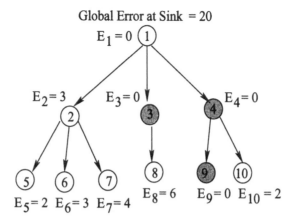

Figure 15.4. Error filter assignments in tree topology. The nodes that are shown shaded are the *passive* nodes that take part only in routing the measurements. A sensor communicates a measurement only if it lies outside the interval of values specified by E_i i.e., maximum permitted error at the node. A sensor that receives partial results from its children aggregates the results and communicates them to its parent after checking against the error interval

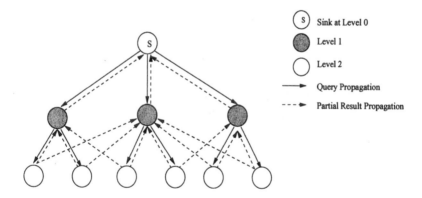

Figure 15.5. Usage of duplicate-sensitive sketches to allow result propagation to multiple parents providing fault tolerance. The system is divided into *levels* during the query propagation phase. Partial results from a higher level (level 2 in the figure) is received at more than one node in the lower level (Level 1 in the figure)

dition, the *potential gain* of increasing the error threshold of nodes is estimated statistically, to guide the allocation of error filters. In another extension, [45] illustrate that the energy consumption of TAG and COUGAR framework can be reduced further through *group-aware* network configuration, where sensors belonging to same group are clustered along the same path, and by suppressing transmissions of measurements with *temporal coherency*.

Focusing on duplicate sensitive queries such as SUM and COUNT, [12] proposed a scalable algorithm that is fault-tolerant to sensor failure and computes approximate answers for the aggregate queries. The duplicate-insensitive sketches used in this work allows communication of results to multiple parents, as illustrated in Figure 15.5, to provide fault tolerance. Recently, [47] extended the class of queries supported to include quantiles, distribution of data in the form of histograms and most frequent items.

4.2   Join Queries

Certain application such as tracking and monitoring a moving object requires execution of *join* queries over data streams produced at different sensors. For example, consider a query of the form

"Return the objects that were detected in both regions R1 and R2".

To evaluate the query, streams of observations from the sensors in regions R1 and R2 are joined to determine if an object was spotted in both the regions.

The applications of join queries in sensor networks were first discussed in the COUGAR framework ([54]) which suggested making an informed decision, based on the selectivity of the join operator, to compute the join results in-network. Studying join queries in detail, [5] proposed a method for effective join operator placement. In this, the authors assume long running queries and propose a technique where the sensors continuously refine the placement of join operator so as to minimize data transmissions over the network. However, the method is restricted to processing queries over pairs of sensors. Recently, [37] proposed REED, an extension of tinyDB for multi-predicate join queries, which can efficiently handle joins queries over multiple sensors and joins of sensor data with external tables.

A non-blocking form of join processing is *sliding time window join* where a time window over the timestamps of the tuples is given as a constraint, in addition to the join conditions. This is studied as a natural way to handle joins on infinite streams such as those from sensors. For example,

"Return the objects that were detected by both sensors S1 and S2 and where $window(S1, S2) = w$"

poses an additional constraint that the timestamps of the values from S1 and S2 should be within window w of time from each other to satisfy the query predicate. [23] studied various forms of such window join queries and proposed *backward and forward evaluation* algorithms BEW-join and FEW-join for executing them. However, the algorithms proposed can potentially produce an unordered stream of tuples as result. The authors address this problem (in [24]) and propose and analyze several methods to provide in-order execution of join queries over sensor streams.

4.3   Top-*k* Monitoring

Another interesting class of problem is to report the k highest ranked answers to a given query. An example to a top-k query in sensor system would be:

"Which sensors have reported the highest average temperature readings over the past month?".

The general problem of monitoring top-k values from streams that are produced at distributed locations is discussed by [3]. The authors propose a technique in which arithmetic constraints are maintained at the stream sources to ensure the validity of the most recently communicated top-k answers. The approach provides answers within a user specified error tolerance and reduces overall communication between the sources. A different approach to providing solution for top-k queries in sensor systems following hierarchical topology is TJA(Threshold Join Algorithm), proposed by [57]. The algorithm consist of the initial phase of setting lower bound for the top-k results in the hierarchies, followed by a join phase that collects the candidate sets in a bottom-up manner. With a fixed number of round trips, in-network processing and fewer readings communicated to the sink, the method conserves energy and reduces delay in answering the query.

Recently, [48] proposed a technique to answer top-k queries approximately by keeping samples of past sensor readings. When querying on a large sample set, the nodes that appear frequently in the answers form a pattern that can assist in the estimation of optimum query plan. Based on this observation, the authors propose a general framework of devising query plans with user defined energy budget, and applies it to answer top-k queries approximately.

4.4   Continuous Queries

Sensors deployed for monitoring interesting changes in the environment are often required to answer queries continuously. For instance, motion or sound sensors might be used to automatically turn lights on by evaluating continuous queries.

When more than one continuous query is evaluated over the readings, we can optimize the storage and computation by taking advantage of the fact that the sources of the query and their partial results could overlap. Continuously Adaptive Continuous Query (CACQ), implemented over Telegraph query processing engine, is an adaptive eddy-based design proposed by [33] which amortized query processing cost by sharing the execution of multiple long running queries. As a related work, we find that the approach proposed by [40] for providing approximate answers for continuous queries is applicable in certain sensor based applications.

In long running queries, streams from different sensors are continuously transmitted to other sensors where the query operator is applied on the data.

Since the rate at which the data is produced by the operator varies over time, a dynamic assignment of operators to nodes reduces communication costs. To achieve this, [5] have worked on optimizing in-network placement of query operators such as aggregate, filtering, duplicate elimination and correlation, where the nodes continuously refine the operator placement.

In many applications in sensor systems, the user is more interested in a macroscopic description of the phenomenon being observed, rather than the individual observations. For example, when sensors are deployed to detect fire hazards, the *state* of the system is either 'Safe' or 'Fire Alarm'. Specific queries regarding the state can be posed to the individual sensors once the state is detected. Recently, [22] have proposed an approach for state monitoring that comprises two processes. The first process is the *learning process* where sensor readings are clustered with user constraints and the clusters are used to define rules describing the state of the system. In the *state monitoring* process the nodes collaborate to update the state of the network by applying the rules to the sensor observations.

5.   Compression and Modeling

In certain applications in sensor systems the type of the query or the characteristics of interesting events is not known apriori. In such scenarios, summaries of the sensor data are stored either in-site or in-network or at the base station, and are used for answering the queries. For example, [18, 19] proposed storage of wavelet based summaries of sensor data, in-network, at various resolutions (spatial) of the system. Progressive aging of summaries and load sharing techniques are used to ensure long term storage and query processing.

A relevant problem is to compress the historical data from multiple streams in order to transmit them to the base station. Recently, [15] proposed the Self Based Regression (SBR) algorithm that provides an efficient *base-signal* based technique to compress historical data in sensors. The base-signals that capture the prominent features of the stream are extracted from the data and are transmitted to the base station, to aid in future reconstructions of the stream. ALVQ (Adaptive Linear Vector Quantization) algorithm proposed by [31] improves on the SBR algorithm by increasing the precision of compression and reducing the bandwidth consumption by compressing the update of the codebook. In a different approach to compressing sensor streams, [43] assume linearity of data over small windows and evaluate a temporal compression scheme for summarizing micro-climactic data stream.

It is clear that all the research contributions discussed here have a common goal: to reduce power consumption of the sensors. Modeling the distribution of the data streams comes in handy when there is a requirement to reduce the power consumption *further*. This approach is highly recommended in *acqui-*

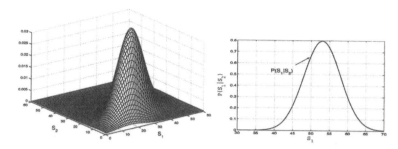

Figure 15.6. (a) Two dimensional Gaussian model of the measurements from sensors S_1 and S_2 (b) The marginal distribution of the values of sensor S_1, given S_2: New observations from one sensor is used to estimate the *posterior density* of the other sensors

sitional systems where considerable energy is consumed even for sensing the phenomenon, apart from the energy consumed in transmitting the values. User queries are answered based on the models, by prediction, and more data is acquired from the system if the prediction is not accurate. The accuracy of the predictions thus serve as a guidance to determine which sensors should be queried to update and refine the models, so that the future queries can be answered more accurately.

5.1   Data Distribution Modeling

Over the recent years, there are many research undertakings in modeling of sensor data. [20] proposed an interesting framework for in-network modeling of sensor data using distributed regression. The authors use linear regression to model the data and the coefficients of kernel-based regression models are computed in-network. This technique exploits the temporal redundancy (the redundancy in readings from a sensor over time) and spatial redundancy (sensors that are close to each other measure similar values) that is common in sensor streams. In [16], a multivariate Gaussian model over the sensors is used for answering queries pertaining to one or more of the sensors. For example, consider a range query that asks:

"*Is the value of a sensor S_1 within the range $[a, b]$?*"

Instead of querying the sensor to obtain its reading for answering the query, it is now possible to compute the probability $P(S_1 \in [a, b])$ by marginalizing the multivariate distribution over the density over only S_1. If this is very high, the predicate is true and the predicate is false if it is very low. Otherwise, the sensor is required to transmit more data and the model is updated. In addition to updating the model with the new observations transmitted, the model is also updated over time with one or more transition models.

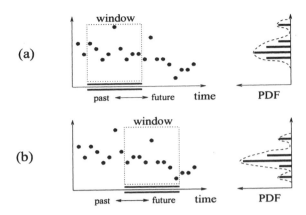

Figure 15.7. Estimation of probability distribution of the measurements over sliding window

Measurements of low-cost attributes that are correlated to an expensive predicate can be used to predict the selectivity (i.e., the set of sensors to query) of the expensive predicate. This observation is utilized in [17] to optimize query plans for expensive predicates.

5.2   Outlier Detection

Sensors might record measurements that appear to deviate significantly from the other measurements observed. When a sensor reports abnormal observations, it might be due to an inherent variation in the phenomenon being observed or due to an erroneous data measurement procedure. In either case, such outliers are interesting and has to be communicated across. [42] proposed an approach for detecting outliers in a distributed manner, through non-parametric modeling of sensor data. Probability distribution models of the data seen over a recent window are computed based on kernel density estimators, as illustrated in Figure 15.7. Since such models obtained at various sensors can be combined efficiently, this approach makes it possible to have models at different hierarchical levels of communication. The models are then used to detect outliers at various levels.

Figure 15.8 graphically depicts the trade-offs between the model size, the desired accuracy of results and the resource consumption common in sensor systems. As seen in the figure, a sensor reporting measurements from dynamic environment such as sounds from outdoor requires large model size and more number of message updates, compared to a sensor reporting indoor temperatures.

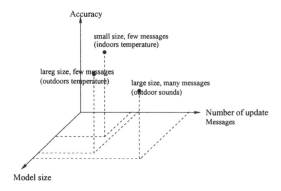

Figure 15.8. Trade-offs in modeling sensor data

6. Application: Tracking of Objects using Sensor Networks

As seen in the above sections, sensor systems are potentially useful in various applications ranging from environmental data collection to defense related monitoring. In this section, we briefly describe some of the recent research works that study surveillance and security management in sensor systems. In particular, we look at *tracking* techniques where data from sensors are processed online, in a real-time fashion, to locate and track moving objects. Tracking of vehicles in battlefield and tracking of spread of wildfire in forests are some of the examples. Typically, sensor nodes that are deployed in a field are equipped with the technology to detect the interesting objects (or in general, *events*). The sensors that detect the event collaborate with each other to determine the event's location and predict its trajectory. Power savings and resilience from failures are important factors to consider while devising an efficient strategy for tracking events.

One of the first works on tracking in sensor systems is by [60] who studied the problem of tracking a mobile target using an information theoretic approach. According to this method, the sensor that detects the target estimates the target state, determines the next best sensor and hands off the state information to it. Thus, only a single node is used to track the target at any time and the routing decision is made based on information gain and resource cost.

Considering the problem in a different setting, [2] proposed a model for tracking a moving object with binary sensors. According to this, each sensor node communicates one bit of information to a base station. The bit denotes whether an object is approaching it or moving away from it. The authors propose a filtering style approach for tracking the object. The method involves a centralized computational structure which is expensive in terms of energy consumption. In the method proposed by [10], the network is divided into clusters and the cluster heads calculates the target location based on the signal readings from the other nodes in the cluster.

The problem of tracking multiple objects has been studied by [26] where the authors propose a method based on stochastic approaches for simultaneously tracking and maintaining identities of multiple targets. Addressing the issue of multiple-target identity management, [46] introduced *identity belief matrix* which is a doubly stochastic matrix forming a description of the identity information of each target. The matrix is computed and updated in a distributed fashion.

Apart from the above works which present tracking techniques *per se*, we also see few methods that employ some communication framework in order to track targets. *Dynamic Convoy Tree-based Collaboration* (DCTC) framework proposed by [59] relies on *convoy tree* which includes the sensors around the moving target. As the target moves, the tree dynamically evolves by adding and pruning some nodes. The node close to the target is the root of the tree where all the sensing data is aggregated.

[32] discuss a group management method for track initiation and management in target tracking application. On detecting the target, sensors send message to each other and a leader is selected among them based on the time stamp of the messages. All sensors that detect the target abandon detection and join the group of the selected leader and the leader gets the responsibility to maintain the collaborative group.

Predictive target tracking based on a cluster based approach is presented by [52] where the target's future location is predicted based on the current location. In order to define the current location, the cluster head aggregates the information from three sensors in its cluster. Then the next location is predicted based on an assumption that it obeys two dimensional Gaussian distribution. [50] proposed a *prediction-based* energy saving scheme for reducing energy consumption for object tracking under acceptable conditions. The prediction models is built on the assumption that the movement of the object usually remains constant for a certain period of time. The heuristics for wake-up mechanism considers only the predicted destination node, or all the nodes on the route from current node to destination node, or all the neighbors of all the nodes along the predicted route. The errors in the estimate of the target's movement are corrected by filtering and probabilistic methods, thus accurately defining the sensors to be notified.

Recently, [21] proposed a two level approach for tracking a target by predicting its trajectory. In this scheme, a low-level loop is executed at the sensors to detect the presence of target and estimate its trajectory using local information, whereas the global high level loop is used to combine the local information and predict the trajectory across the system. The system is divided into cells as

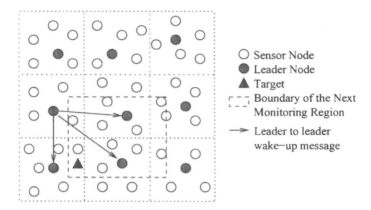

Figure 15.9.   Tracking a target. The leader nodes estimate the probability of the target's direction and determines the next monitoring region that the target is going to traverse. The leaders of the cells within the next monitoring region are alerted

shown in Figure 15.9. Kalman filters are used for predicting the target location locally and the estimations are combined by the leaders of the cells. The probability distribution function of the target's direction and location are determined using Kernel functions and the neighboring cell leaders are alerted based on the probability estimation. [1, 56]

7.   Summary

In this chapter we reviewed recent work on distributed stream processing techniques for data collected by sensor networks. We have focused on the sensor monitoring paradigm, where a large set of inexpensive sensors is deployed for surveillance or monitoring of events of interest.

The large size of data and the distributed nature of the system necessitate the development and use of in-network storage and analysis techniques; here we have focused on the analysis part. However, future systems will operate with larger more expensive and capable sensors (for example video cameras). Consequently, future research work will have to address important and fundamental issues on how to efficiently stoe, index, and analyze large datasets in sensor networks.

The development of efficient techniques for local (in the sensor) storage of the data (perhaps using inexpensive and widely available flash memory), as well as for distributed data storage, and the development and deployment of resource management techniques to manage the resources of the sensor network will be very important in addressing these issues.

References

[1] Ali, M. H., Mokbel M. F., Aref W. G. and Kamel I. (2005) Detectioin and Tracking of Discrete Phenomena in Sensor-Network Databases. *In Proceedings of the 17th International Conference on Scientific and Statistical Database Management.*

[2] Aslam J., Butler Z., Constantin F., Crespi V., Cybenko G. and Rus D. (2003) Tracking a Moving Object with a Binary Sensor Network. *In Proceedings of ACM SenSys.*

[3] Babcock B., and Olston C. (2003) Distributed top-k monitoring. *In Proceedings of the ACM SIGMOD International Conference on Management of Data.*

[4] Biswas R., Thrun S. and Guibas L. J. (2004) A probabilistic approach to inference with limited information in sensor networks. *In Proceedings of the 3rd International Conference on Information Processing in Sensor Networks.*

[5] Bonfils, B. J., and Bonnet P. (2003) Adaptive and Decentralized Operator Placement for In-Network Query Processing. *In Proceedings of 2nd International Conference on Information Processing in Sensor Networks.*

[6] Bonnet P., Gehrke J., and Seshadri P. (2001) Towards Sensor Database Systems. *In Proceedings of the 2nd International Conference on Mobile Data Management,* London, UK.

[7] Cerpa A., Elson J., Estrin D., Girod L., Hamilton M. and Zhao J. (2001) Habitat monitoring: Application driver for wireless communications technology. *In Proceedings of ACM SIGCOMM Workshop on Data Communications in Latin America and the Caribbean.*

[8] Cerpa A. and Estrin D. (2002) ASCENT: Adaptive Self-Configuring sEnsor Networks Topologies. *In Proceedings of IEEE INFOCOM.*

[9] Benjie C., Jamieson K., Balakrishnan H. and Morris R. (2001) Span: An energy-efficient coordination algorithm for topology maintenance in Ad Hoc wireless networks. *In Proceedings of the 7th ACM International Conference on Mobile Computing and Networking.*

[10] Chen W., Hou J. C. and Sha L. (2003) Dynamic Clustering for Acoustic Target Tracking in Wireless Sensor Networks. *In Proceedings of the 11th IEEE International Conference on Network Protocols.*

[11] Chu M., Haussecker H. and Zhao F. (2002) Scalable information-driven sensor querying and routing for ad hoc heterogeneous sensor networks. *International Journal of High Performance Computing Applications.*

[12] Considine J., Li F., Kollios G. and Byers J. (2004) Approximate aggregation techniques for sensor databases. *In Proceedings of the 20th International Conference on Data Engineering.*

[13] *Crossbow Technology Inc.* http://www.xbow.com/

[14] Deligiannakis A., Kotidis Y. and Roussopoulos N. (2004) Hierarchical in-Network Data Aggregation with Quality Guarantees. *Proceedings of the 9th International Conference on Extending DataBase Technology.*

[15] Deligiannakis A., Kotidis Y. and Roussopoulos N. (2004) Compressing Historical Information in Sensor Networks. *In Proceedings of the ACM SIGMOD International Conference on Management of Data.*

[16] Deshpande A., Guestrin C., Madden S. R., Hellerstein J. M. and Hong W. (2004) Model-Driven Data Acquisition in Sensor Networks. *In Proceedings of the 30th International Conference on Very Large Data Bases.*

[17] Deshpande A., Guestrin C., Hong W. and Madden S. (2005) Exploiting Correlated Attributes in Acquisitional Query Processing. *In Proceedings of the 21st International Conference on Data Engineering.*

[18] Ganesan D., Greenstein B., Perelyubskiy D., Estrin D. and Heidemann J. (2003) An Evaluation of Multi-Resolution Storage for Sensor Networks. *In Proceedings of ACM SenSys.*

[19] Ganesan D., Estrin D. and Heidemann J. (2003) Dimensions: Why do we need a new data handling architecture for sensor networks?. *ACM SIGCOMM Computer Communication Review.*

[20] Guestrin C., Bodik P., Thibaux R., Paskin M. and Madden S. (2004) Distributed Regression: an Efficient Framework for Modeling Sensor Network Data. *In Proceedings of the 3rd International Conference on Information Processing in Sensor Networks.*

[21] Halkidi M., Papadopoulos D., Kalogeraki V. and Gunopulos D., (2005) Resilient and Energy Efficient Tracking in Sensor Networks. *International Journal of Wireless and Mobile Computing.*

[22] Halkidi M., Kalogeraki V., Gunopulos D., Papadopoulos D., Zeinalipour-Yazti D. and Vlachos M. (2006) Efficient Online State Tracking Using Sensor Networks. *In Proceedings of the 7th International Conference on Mobile Data Management.*

[23] Hammad M. A., Aref W. G. and Elmagarmid A. K. (2003) Stream Window Join: Tracking Moving Objects in Sensor-Network Databases. *In Proceedings of the 15th International Conference on Scientific and Statistical Database Management.*

[24] Hammad M. A., Aref W. G. and Elmagarmid A. K. (2005) Optimizing In-Order Execution of Continuous Queries over Streamed Sensor Data. *In*

Proceedings of the 17th International Conference on Scientific and Statistical Database Management.

[25] Hellerstein J. M., Hong W., Madden S. and Stanek K. (2003) Beyond Average: Toward Sophisticated Sensing with Queries. *International Workshop on Information Processing in Sensor Networks.*

[26] Hwang I., Balakrishnan H., Roy K., Shin J., Guibas L. and Tomlin C. (2003) Multiple Target Tracking and Identity Management. *In Proceedings of the 2nd IEEE Conference on Sensors.*

[27] Heinzelman, W. R., Chandrakasan A. and Balakrishnan H. (2000) Energy-Efficient Communication Protocol for Wireless Microsensor Networks. *In Proceedings of the 33rd Hawaii Intl. Conf. on System Sciences,* Volume 8.

[28] Intanagonwiwat C., Govindan R. and Estrin D. (2000) Directed diffusion: a scalable and robust communication paradigm for sensor networks. *In Proceedings of the 6th ACM International Conference on Mobile Computing and Networking.*

[29] Krishnamachari B., Estrin D. and Wicker S. (2002) Modelling Data-Centric Routing in Wireless Sensor Networks. *In Proceedings of the 21st Annual Joint Conference of the IEEE Computer and Communications Societies (INFOCOM).*

[30] Lazaridis I., and Mehrotra S. (2003) Capturing sensor-generated time series with quality guarantees. *In Proceedings of the 19th International Conference on Data Engineering.*

[31] Lin S., Kalogeraki V., Gunopulos D. and Lonardi S. (2006) Online Information Compression in Sensor Networks. *In Proceedings of IEEE International Conference on Communications.*

[32] Liu J., Reich J., Cheung P. and Zhao F. (2003) Distributed Group Management for Track Initiation and Maintenance in Target Localization Applications. *IPSN Workshop.*

[33] Madden S, Shah M. A., Hellerstein J. M. and Raman V. (2002) Continuously Adaptive Continuous Queries over Streams. *In Proceedings of the ACM SIGMOD International Conference on Management of Data.*

[34] Madden S. R. and Franklin M. J., (2002) Fjording the Stream: An Architecture for Queries Over Streaming Sensor Data. *In Proceedings of the 18th International Conference on Data Engineering.*

[35] Madden S. R., Franklin M. J. and Hellerstein J. M. (2002) TAG: A Tiny Aggregation Service for Ad-Hoc Sensor Networks. *In Proceedings of the 5th Symposium on Operating System Design and Implementation.*

[36] Madden S. R., Franklin M. J., Hellerstein J. M. and Hong W. (2003) The Design of an Acquisitional Query Processor for Sensor Networks. *In Pro-*

ceedings of the ACM SIGMOD International Conference on Management of Data.

[37] Madden S. R., Lindner W. and Abadi D. (2005) REED: Robust, Efficient Filtering and Event Detection in Sensor Networks. *In Proceedings of the 31st International Conference on Very Large Data Bases.*

[38] Mainwaring A., Polastre J., Szewczyk R., Culler D. and Anderson J. (2002) Wireless Sensor Networks for Habitat Monitoring. *In Proceedings of ACM International Workshop on Wireless Sensor Networks and Applications.*

[39] Omar S. A. Assessment of Oil Contamination in Oil Trenches Located in Two Contrasting Soil Types, (2001) *Conference on Soil, Sediments and Water.* Amherst, MA, USA.

[40] Olston C., Jiang J. and Widom J., (2003) Adaptive filters for continuous queries over distributed data streams. *In Proceedings of the ACM SIGMOD International Conference on Management of Data.*

[41] Pottie G. J., and Kaiser W. J. (2000) Wireless integrated network sensors. *In Communications of the ACM,* Volume 43, Issue 5.

[42] Palpanas T., Papadopoulos D., Kalogeraki V. and Gunopulos D. (2003) Distributed deviation detection in sensor networks. *ACM SIGMOD Records,* Volume 32, Issue 4.

[43] Schoellhammer T., Osterweil E., Greenstein B., Wimbrow M. and Estrin D. (2004) Lightweight Temporal Compression of Microclimate Datasets. *In Proceedings of the 29th Annual IEEE Conference on Local Computer Networks.*

[44] Schurgers C., Tsiatsis V., Ganeriwal S. and Srivastava M. (2002) Topology management for sensor networks: exploiting latency and density. *In Proceedings of the 3rd ACM international symposium on Mobile ad hoc networking & computing*

[45] Sharaf M. A., Beaver J., Labrinidis A. and Chrysanthis P. K. (2004) Balancing energy efficiency and quality of aggregate data in sensor networks. *The VLDB Journal.*

[46] Shin J., Guibas L. and Zhao F. (2003) Distributed Identity Management Algorithm in Wireless Ad-hoc Sensor Network. *2nd Int'l Workshop on Information Processing in Sensor Networks*

[47] Shrivastava N., Buragohain C., Agrawal D. and Suri S. (2004) Medians and Beyond: New Aggregation Techniques for Sensor Networks. *In Proceedings of ACM SenSys.*

[48] Silberstein A., Braynard R., Ellis C., Munagala K. and Yang J. (2006) A Sampling-Based Approach to Optimizing Top-k Queries in Sensor Networks. *In Proceedings of the 22nd International Conference on Data Engineering.*

[49] Xu Y., HeidemannJ. and Estrin D. (2001) Geography-informed energy conservation for Ad Hoc routing. *In Proceedings of the 7th ACM International Conference on Mobile Computing and Networking.*

[50] Xu Y., Winter J. and Lee W. (2004) Prediction-Based Strategies for Energy Saving in Object Tracking Sensor Networks. *IEEE International Conference on Mobile Data Management.*

[51] Xu N., Rangwala S., Chintalapudi K. K., Ganesan D., Broad A., Govindan R. and Estrin D. (2004) A wireless sensor network For structural monitoring. *In Proceedings of the 2nd international conference on Embedded networked sensor systems.*

[52] Yand H. and Sikdar B. (2003) A Protocol for Tracking Mobile Targets using Sensor Networks. *IEEE International Workshop on Sensor Networks Protocols and Applications.*

[53] Yao Y. and Gehrke J. The cougar approach to in-network query processing in sensor networks. *ACM SIGMOD Records.*

[54] Yao Y. and Gehrke J. (2003) Query Processing for Sensor Networks. *Conference on Innovative Data Systems Research.*

[55] Ye F., Luo H., Cheng J., Lu S. and Zhang L. (2002) A Two-Tier Data Dissemination Model for Large-Scale Wireless Sensor Networks. *In Proceedings of the 8th ACM International Conference on Mobile Computing and Networking.*

[56] Yu X., Niyogi K., Mehrotra S. and Venkatasubramanian N. (2004) Adaptive Target Tracking in Sensor Networks. *The Communication Networks and Distributed Systems Modeling and Simulation Conference.*

[57] Zeinalipour-Yazti D., Vagena Z., Gunopulos D., Kalogeraki V., Tsotras V., Vlachos M., Koudas N. and Srivastava D. (2005) The threshold join algorithm for top-k queries in distributed sensor networks. *In Proceedings of the 2nd international workshop on Data management for sensor networks.*

[58] Zeinalipour-Yazti D., Kalogeraki V., Gunopulos D., Mitra A., Banerjee A. and Najjar W. A. (2005) Towards In-Situ Data Storage in Sensor Databases. *Panhellenic Conference on Informatics.*

[59] Zhang W. and Cao G. Optimizing Tree Reconfiguration for Mobile Target tracking in Sensor Networks. *In Proceeding of IEEE INFOCOM.*

[60] Zhano F., Shin J. and Reich J. (2002) Information-Driven Dynamic Sensor Collaboration for Tracking Applications. *IEEE Signal Processing Magazine,* Vol. 19.

[61] Zhano J., Govindan R. and Estrin D. (2003) Computing Aggregates for Monitoring Wireless Sensor Networks. *IEEE International Workshop on Sensor Network Protocols Applications.*

Index

Printed in the United States